ENVIRONMENT, KNOWLEDGE AND GENDER

ENVIRONMENTAL KNOWLEDGE AND GENDER

This book is dedicated to my mother and late father for all their help and support over the years.

Environment, Knowledge and Gender

Local development in India's Jharkhand

SARAH JEWITT
University of Nottingham

Routledge
Taylor & Francis Group

LONDON AND NEW YORK

Contents

List of Figures

List of Tables

List of Plates

Preface

The ideas contained within this book have evolved over a period of almost ten years and represent a personal as much as an intellectual journey. During this period, community-based forest management has attracted a wide range of research and has found for itself a niche in the agendas of many development donors and environmentally-oriented NGOs. Likewise, issues of gender and indigenous knowledge have come to the forefront of environmental policy-making in developing countries. Focusing, as it does, on two tribal (*adivasi*) dominated villages in Jharkhand, this study can hope to add little to the broader theoretical literature on common property resource management, political ecology and development thinking more generally (Agrawal, 2001). Rooted less in theoretical conundrums than in 'practical realities and dilemmas' (Scheper-Hughes, 1992, p. 4). it takes a more pragmatic stance using detailed field-based research to examine the relevance of populist ideas on community, gender, and indigenous agro-ecological knowledges to local development priorities. Reflecting my own training as a geographer, the importance of place is strongly emphasised throughout the book.

Given recent crises of confidence within the social sciences over issues of truth, (mis-) representation and the hidden political agendas informing research in developing countries (Clifford, 1986), it is important to address issues of my own 'positionality' with relation to my research subjects. As a 'white Western woman' attempting to study tribal societies in a developing country, it is imperative to acknowledge the attention that relativist and post-modern critiques have drawn both to the impossibility of speaking the 'truth' about other cultures and the power relations inherent in 'speaking for' and translating the reality of others. In Salman Rushdie's words 'every story one chooses to tell is a form of censorship, it prevents the telling of other tales' (Rushdie, 1983). Clearly, the story that I tell reflects my own cultural baggage and intellectual bias. I share Meera Nanda's views about the emancipatory role of science and the dangers of romantic populist emphases on traditional lifestyles and knowledge systems. At the same time, I believe that unless and until more 'modern ways of knowing' (Nanda, 2001, p. 2561) are adopted by India's resource poor villagers, it is important to show sensitivity to and work around their cultural norms and belief systems whenever possible. In more practical terms, my study was clearly subject to a number of constraints resulting from my own background and culture: constraints that could arguably have been overcome better by a researcher from the same linguistic, cultural and ethnic background as the research subjects. Although my spoken Hindi is competent, it is far from fluent. This, coupled with the fact that many of my respondents spoke either a dialect of Hindi or a tribal language (Kurukh or Mundari) with a completely different structure and vocabulary left me unable to understand the nuances of discussions. Although my research assistants (Harish, Pyari and Subani) did an excellent job of linguistic and cultural translation, my role as an outsider made my understanding of (and attempted integration into) village culture a much slower and more cumbersome process than it would have been for a local researcher.

Whilst acknowledging and accepting these limitations, I also feel that it is important not to lose sight of the idea that 'the critique of a privileged vantage point need not imply a critique of the possibility of knowledge' (Corbridge, 1993, p. 456). Although relativist and post-modern deconstruction has been extremely valuable in promoting a recognition that 'cultural analysis is always enmeshed in global movements of difference and power' (Clifford and Marcus, 1986, p. 22), it is, when pressed to its logical conclusion, nihilistic in its suggestion that 'no systematic knowledge of human action or trends of social development is possible' (Giddens, 1990, p. 47). Also, if deconstruction 'prevents us from asserting or stating or identifying anything, then surely one ends up not with 'difference', but with indifference' (Kearney, 1984, p. 4).

Like many social scientists who have conducted research in developing countries, therefore, I hold on to the realist belief that although all our representations of the world are imperfect, some accounts are better (and closer to the 'truth') than others. Following Scheper-Hughes, I recognise that although ethnographic research is 'necessarily flawed and biased', for many resource-poor and powerless people in developing countries it represents not 'a hostile gaze but rather an opportunity to tell a part of their life story' (Scheper-Hughes, 1992, p. 28). To justify my own presence in Jharkhand, I can do little more than put forward the argument that outsiders often see and question things that insiders do not. Where possible, I have tried to 'give voice' to respondents (particularly women) who ordinarily face constraints in making themselves heard (Agarwal, 2001) and in an effort triangulate the many conflicting and different 'truths' surrounding the micro-political ecology of resource management in Jharkhand, I have used a variety of different research methodologies. Despite my attempts to 'listen and observe carefully, empathetically, and compassionately' (Scheper-Hughes, 1992, p. 28), however, what emerges is inevitably partial and subjective.

Acknowledgements

There are many people who have contributed towards the realisation of this book. I am deeply grateful to my parents, Mary and Mike Jewitt, my sister, Judy, and her partner, Bob, for their help and assistance. I also wish to thank Michael Maurice for being so wonderfully supportive during the final stages of this book.

I am deeply indebted and extremely grateful to Stuart Corbridge, who introduced me to Jharkhand, stimulated my interest in the region and encouraged me to do fieldwork there. I also wish to thank Tim Bayliss-Smith for supervising my Ph.D. I am very grateful to my SOAS colleagues, notably Bob Bradnock for his encouragement, guidance and apparently boundless knowledge of South Asia, and Kathy Baker for her great wisdom and also for being such a brilliant friend. Many thanks to other friends and colleagues who contributed in diverse ways: Tony Allan, Margaret Baluch, Huw and Donna Evans, Carolyn Jones, Steve Jones, Emma Mawdsley, Debbie Potts, Viv Slinger, Philip Stott, Alison Warner and Glyn Williams. Catherine Lawrence and Ines Baker deserve special mention for doing such a splendid job with the maps and index. I also gratefully acknowledge the financial assistance that I received for my doctoral fieldwork from the Economic and Social Research Council.

In India, I am deeply grateful to Saraswati Raju, 'O.P.' and Manju Bohra for their support and help while I was in New Delhi. In Ranchi, Sanjay and Nishi Kumar and their family were a great source of information, encouragement and hospitality and Professor Ram Dayal Munda was an important (if elusive) source of support and information on the Jharkhand movement. I would also like to thank Anup Sarkar and Father Francken from the Xavier Institute of Social Services in Ranchi who helped me to set up my fieldwork and find research assistants. Also Mukund Nayak who introduced me to Jharkhandi music and dance. Special thanks are due to Pyari Lakra and Subani Kujur who were excellent research assistants, invaluable guides to village life and who became two of my best friends. My biggest debt, however, is to the villagers of Ambatoli and Jamtoli: especially Gandri ('ma') and Pothwa Bhagat and Simon Oraon who invited me into their homes, looked after me, put up with my ignorance and constant questions and generally helped to make my stay one of the most enlightening, stimulating and rewarding experiences of my life.

List of Abbreviations

BAU	Birsa Agricultural University
BC	Backward Caste
BDO	Block Development Officer
BPFA	Bihar Private Forests Act
DGSS	Dashauli Gram Swaraj Sangh
EOI	Export Oriented Industrialisation
FAO	Food and Agriculture Organisation
GNP	Gross National Produce
HFC	Hindustan Fertilizer Corporation
HYV	High Yielding Variety
IDS	Institute of Development Studies
ISI	Import Substitution Industrialisation
JFM	Joint Forest Management
JMM	Jharkhand Mukti Morcha
NCA	National Commission on Agriculture
NGO	Non Governmental Organisation
NTFPs	Non Timber Forest Products
SC	Scheduled Caste
ST	Scheduled Tribe
VFPMC	Village Forest Protection and Management Committee
XISS	Xavier Institute of Social Services

Glossary

adivasi	indigenous people of India, also known as tribals
andolan	people's movement
Badhya	professional herbalist
bari zamin	homestead land
bhagat	ghost finder
Bhagat	an honorific term, usually used by Oraon men who belong to the Tana Bhagat faith
Bhaktain	an honorific term, usually used by Oraon women who belong to the Tana Bhagat faith
bhuinhar	Oraon pioneer settler
bhut	ghost or spirit
Chaala Pachcho	the 'old lady of the grove': the principal village deity in the *Sarna* pantheon, also known as *Jakra Burhia* or *Sarna Burhia*
churil	the spirit or ghost of a woman who died during pregnancy or childbirth
coupe	a system of forest management which usually involves the annual felling of a certain amount of the forest area so that by the time the 'coupe rotation' has been completed, the first coupe to have been cut is ready for felling again
dain	witch or wizard
dal	lentil
Danda Katta	'cutting of the evil teeth' ceremony
dawai	medicine or (usually when the term is used ironically) liquor
dhan	paddy
Dharmes	the supreme deity in the *Sarna* pantheon
dhuwan	incense
diku	an abusive Jharkhandi term for (exploitative) 'outsider'
don	relatively wet and fertile lowlands
dowli	a sickle-shaped tool for cutting wood
garmajurwar	common land
garmi dhan	summer paddy
gora dhan	upland paddy
gotra	clan
gram	village
hanria	rice beer
hat	village market

hisingar	jealousy of other people's good fortune
hul	rebellion or uprising
jagirdar	traditional rent collector
Jakra	sacred grove
Jakra Burhia	the 'old lady of the grove': the principal village deity in the *Sarna* pantheon, also known as *Sarna Burhia* or *Chaala Pachcho*
jaree bootee	herbs or other small plants
jarru	sweeping brush
jatra	fair
katat	a system of landuse whereby villagers had unrestricted use of local forests upon payment of a fixed 'tribute' to the zamindar
Kattiyan	a village-level record of rights that contains details of local land holdings (Kattiyan Part I) and fuelwood, timber and minor forest produce allowances (Kattiyan Part II)
kharif	the monsoon season (June-November)
khuntkattidar	Munda pioneer settler
Kurukh	the name by which Oraons call themselves and their language
lac	shellac
latha kunta	a pivot device used for drawing water from wells
madad	help or assistance
Mahto	the secular headman of an Oraon village
maidan	an open stretch of grassland
mandar	a special type of clay drum which is particularly popular with the Oraons
manjhihas	land owned by the village community
Mukhia	elected Panchayat leader
Nagpuri	a common tribal *lingua franca*, also known as *Sadri*
otanga	someone who hunts people for human sacrifice
Pahan	the main religious leader in Oraon villages
Panch	village Panchayat committee
Panchayat	village-level local government.
Parha	a traditional Oraon form of village administration run by village elders and consisting of up to 21 villages
Parha Raja	a *Parha* chief
puja	worship
Pujar	*Pahan*'s assistant
purdah	seclusion (refers to the veil worn by women in seclusion)
rabi	the winter season (November to March)
rajhas	land reclaimed from the forest

Rakhat	a system of landuse whereby villagers' rights to forest produce were subject to certain restrictions imposed by the zamindar
sadan	'artisan' or 'service' castes which have developed reciprocal relationships with the *adivasis* and share a common cultural (and often religious) outlook
Sadri	a common tribal *lingua franca*, also known as *Nagpuri*
Samiti	committee
Sarna	a semi-Animist faith to which most Oraons and Mundas in the research area subscribe (the term also means 'sacred grove')
Sarna Burhia	the 'old lady of the grove': the principal village deity in the *Sarna* pantheon, also known as *Jakra Burhia* or *Chaala Pachcho*
Sarpanch	head of the *Panch* (village Panchayat committee) in the research area, but in other areas, this title may be given to the elected Panchayat leader
sikar	hunt
soop	winnowing basket
Soussar	*Pahan*'s assistant
tangiya	axe
tanr	relatively dry and poor quality uplands
tola	hamlet
zamin	land
zamindar	traditional rent collectors made into landlords by the British Permanent Settlement Act

Note: a glossary of plant names can be found in Appendix I.

1 Introduction

Since the early 1980s, dissatisfaction with deterministic models and 'grand theories' of development (Booth, 1985; 1994; Corbridge, 1990; 1993) has spread from predominantly academic circles into mainstream policy-making. At the same time, there has been increasing recognition of the detrimental socio-economic, political and environmental impacts associated with many externally imposed, 'top-down' and locally insensitive development programmes (Escobar, 1995) Political ecology approaches that analyse 'the biogeographical outcomes of social relations in the context of particular spatial and political configurations' (Rangan, 2000 p. 63) have been particularly important in revealing the different agendas that inform development and environmental policy- (and myth-) making (Blaikie and Brookfield, 1987; Bryant, 1992; 1997; Peet and Watts, 1996a; Rocheleau et al, 1996; Bryant and Bailey, 1997; Singh, 1999; Sivaramakrishnan, 1999; Agrawal and Sivaramakrishnan, 2000a; Bradnock and Saunders, 2000; Stott and Sullivan, 2000a; Rangan 2000a; Turton, 2000). Recognition of the difficulties inherent in transferring capital intensive technology from 'developed' to 'developing' countries has also helped to undermine the idea that the West is the model for the rest of the world to follow (Escobar, 1995; Pretty and Shah, 1999).

Much emphasis has been placed upon the wider socio-environmental problems associated with industrial development (Meadows, 1972; Goldsmith et al, 1972; Shiva, 1988; Apffel Marglin and Marglin, 1990). Commercial forestry for the production of raw materials for industry, for example, has often threatened the subsistence needs of forest-dependent populations (Guha, 1989; Hecht and Cockburn, 1989). Also, the 'green revolution' agricultural technology which, despite substantially increasing agricultural productivity,[1] creating extra employment and restraining food prices, has been criticised for its geopolitical origins, ecological impacts, potential for increasing economic differentiation and promotion of mechanisation-induced landlessness (Johnson and Kilby, 1975; Byres, 1981; 1983; Bates, 1988; Shiva, 1991a; Alvarez, 1994; Yapa, 1996).[2] More recent attempts to maintain increases in food production using biotechnology and genetic modification, meanwhile, have generated widespread fears about potential environmental and public health disasters (Krishnakumar, 1999; Lipton, 1999).

For many radical populist, eco-feminist and 'anti-development' theorists, such problems are seen as symptomatic of fundamental flaws with the concept of development (Esteva, 1987; Shiva, 1988; 1991a; Banuri, 1990; Marglin, 1990a; Parajuli, 1991; Escobar, 1992; 1995;

Sachs, 1992; Alvarez, 1994; Yapa, 1996). Eco-feminist writers have been especially critical of development's 'patriarchal' and dominating world view which exploits both nature and women; emphasising instead women's supposedly innate link with nature and their role in environmental stewardship (Ortner, 1974; Daly, 1978; Merchant, 1989; Mies and Shiva, 1993). The Indian populist eco-feminist writer, Vandana Shiva, goes further, drawing attention to the dangers of reductionist science, environmentally-destructive development and the 'patriarchal model of progress which pushes inexorably towards monocultures, uniformity and homogeneity' (Shiva, 1992, p. 205).[3]

Highlighting the damage caused to Third World eco- and social systems by transfers of Western technology and ideas, some anti-developmentist writers have likened development to a Western disease that destroys its Third World hosts (Rahnema, 1988) and compared development assistance to 'the AIDS virus; a pathogen that destroys the ability of the host country to resist the invasion of a foreign socio-economic system' (Shucking and Anderson, 1991, p. 21). Others have become concerned about the displacement of 'traditional' indigenous knowledge and culture which could represent the key to a more environmentally-sensitive and locally-appropriate form of development (Appadurai, 1990; Gupta, 1996; Yapa, 1996) Instead, they propose a radical and often highly romanticised populist alternative for the Third World based on small scale enterprise, labour intensive technology, environmental knowledge and community organisation (Esteva, 1987; Marglin, 1990a; Alvarez, 1994; Escobar, 1992; 1995).

In more mainstream development circles, concern about development failures has helped to stimulate a shift in emphasis away from industrialisation towards rural development and the desirability of using grassroots or 'bottom-up' and participatory approaches to bring this about. Particularly since the late 1970s, wider discourses of peasant moral economies have placed increasing emphasis on the benefits of investigating and attempting to build upon indigenous technologies, local environmental knowledge and village-based (often common property) resource management systems (Chambers, 1979; 1981; Brokensha, 1980; Korton, 1980; Gujit and Kaul Shah, 1998a; Devavaram et al, 1999).

A major leader in this field has been Robert Chambers who has long argued that local people should co-operate and participate in development projects and has been instrumental in developing fairly rapid yet participatory research methodologies (Rapid and Participatory Appraisal (PA)) for use by NGOs and project planners (Chambers, 1978; 1992; 1995). In collaboration with others, he has also called for a new 'learning paradigm' which views investigations into local knowledge systems as an important basis for promoting more sensitive, appropriate and sustainable development (Pretty and Chambers, 1993; 1994). Indeed, the adoption of these ideas into mainstream development thinking represents an important sea-change away from traditional top-down approaches. Since the early 1990s, most major donor agencies,

NGOs and Third World governments have taken on board the rhetoric of participatory development and PA techniques are now widely used (or at least paid lipservice to) by development practitioners (World Bank, 1996; 2001; Narayan and Rietbergen-McCracken, 1998; FAO, 1999; DfID, 2001; Vedeld, 2001). Increasing emphasis has also been placed on the need to make development research and planning more geographically specific and sensitive to the socio-economic, political and environmental complexities of different local settings (Escobar, 1995; Sivaramakrishnan, 1999; Neefjes, 2000). In part, this reflects a recognition that 'it is at least partially the continuing ignorance and occlusion of alternative knowledges - of local environmental and other narratives - which allow the globalising institutions of the north to uphold the hegemonic and normalising discourses they do' (Sullivan, 2000, p. 34).

Although participatory approaches can give development planners important insights into local development constraints and priorities (as well as allowing local people greater control over the process), they have been somewhat plagued by their own success. For Third World government departments with well-established hierarchies, participatory development is often seen as little more than a fad that they must seen to be promoting to get external funding (Mosse, 1994; Gujit and Kaul Shah, 1998b). According to Crawley, many institutions 'have viewed PRA and related approaches as a mechanism for cost recovery in projects initiated externally to local communities, for reducing initial outlay, and for improving the accuracy and acceptability of research' (Crawley, 1998, p. 31).

In practice, many of the lowest-level field staff who are responsible for conducting PA have little training in either its concepts or techniques and without the correct approach (or an awareness of the wider socio-political context into which projects are to be implemented) are likely to achieve mixed results.[4] Another problem is that PA can easily become a standardised technique, incapable of investigating the complexity of social change and intra-community power structures (Blackburn with Holland, 1998a; Cornwall, 1998; Gujit and Kaul Shah, 1998a; Hinchcliffe et al, 1999). The time constraints placed on PA tend to exacerbate this problem by on the one hand forcing it to be rather superficial and on the other hand discouraging more traditional (and expensive) in-depth surveys that could fill in missing details. As Gujit and Kaul Shah point out, 'It takes time for people and groups to decide what they want to see changed, and why, and then to act ... Real consensus is not reached in a two-week planning exercise ... Given the entrenched nature of hierarchical structures and relationships, it is not possible to expect far-reaching changes with every intervention. This begs the question of what can be expected from externally-determined participatory research and planning processes when they are driven by speedy disbursement of funds' (Gujit and Kaul Shah, 1998b, p. 12). As part of an argument for integrating PA with more long-term fieldwork methodologies, the empirical chapters that follow highlight the value of theoretically grounded qualitative geographical and ethnographic research in examining the 'embeddedness'[5] (Granovetter, 1985) of locally-specific, socio-political relations and development priorities.

Tracing global shifts in development thinking through to national level policy-making in India and its local scale implications, the main aim of this book is to investigate the practical value of radical populist and eco-feminist alternatives to more mainstream forms of development. Using detailed empirical data on forest (and to a lesser degree, agricultural) development issues from two *adivasi* (tribal) villages[6] in Jharkhand (until recently, part of Bihar State) India, an investigation is made of the extent to which populist and eco-feminist discourses reflect local realities and offer practical solutions to real development problems. Responding to previous failures to identify and clarify the 'blurred boundaries' and interdependence between agrarian and so-called natural environments (Agrawal and Sivaramakrishnan, 2000a) efforts are made to frame forest and agricultural management issues within the context of wider village development priorities.

Particular attention is given to the displacement (or otherwise) of traditional agricultural and forest management practices by modern techniques and the impacts of this on local livelihoods. An attempt is also made to address traditional biases towards the technical aspects of indigenous knowledge by emphasising their interconnections with socio-cultural knowledges and illustrating the importance of understanding both when seeking to promote more sensitive and appropriate development. Using the spatially sensitive approaches central to political ecology and its recent offshoot, institutional ecology, the book aims to enable 'excluded voices' to be heard (Foucault, 1990) as well as to provide an in-depth analysis of differences within communities (Stott and Sullivan, 2000b; Agrawal and Sivaramakrishnan, 2000a).[7] Another major theme is the importance of empirical research for revealing the contextually embedded but temporally fluid nature of environmental decision-making and knowledge distribution by gender, class and community.

Chapter Two is devoted to a global overview of changing ideologies within development thinking since the 1950s. Particular attention is placed upon the political ecology behind the shift away from the 'transfer of technology' model towards more bottom-up approaches. The Chapter also examines the importance and potential of recent emphases upon local environmental knowledges[8] and resource management systems and the problems of putting into practice the more radical and romantic views of some of their populist, eco-feminist and anti-developmentist supporters.

Chapter Three maintains this emphasis with respect to the adoption of eco-feminist and associated 'women, environment and development' (WED) discourse into development planning. In particular, it considers the dangers of overestimating women's agro-ecological knowledges and assuming that they can easily participate in (or be empowered by) environmentally-oriented development projects.

It also challenges the perception of women as environmental guardians, highlighting the socio-cultural constraints that women face in both accumulating environmental knowledge and vocalising it in public fora.

Chapter Four investigates the ecological, economic and socio-cultural implications of forest decline in India and how recent shifts in development thinking have influenced agricultural and forest policy-making. It also examines changes in forest policy since the British colonial period, the causes (and implications) of forest decline and the state's attempts to address these problems through a transition from top-down to more participatory forest management approaches. Particular emphasis is placed on the difficulties that have arisen from the Indian state's conflicting desires to exploit forests commercially, conserve them as a source of biodiversity or timber for future industrial use and provide forest-dependent populations with their subsistence requirements for forest produce.

Chapter Five links back to Chapter Three with an examination of how gender issues have been integrated into India's environmentally-oriented (especially forest) development initiatives. Particular attention is focused on the problems of adopting romanticised eco-feminist discourse into policy-making without a thorough understanding of local gender-environment relations and their wider political ecologies.

In Chapter Six, the various different methodologies used during the course of this research are discussed and the selection of the two fieldwork villages (Ambatoli and Jamtoli) is explained. In addition, the geographical, socio-cultural, political and ecological characteristics of Jharkhand are described in detail. In preparation for the empirical Chapters on agriculture (Eight and Nine) and on forestry (Ten, Eleven and Twelve), Chapter Seven provides more specific background information on the fieldwork villages with particular reference to inter- and intra-village wealth differentials and the influences that these have on household subsistence and survival strategies.

Chapter Eight examines agricultural practices in Ranchi District, Jharkhand, in light of populist emphases on indigenous agricultural knowledge and concerns about the displacement of traditional technology. Using village-level data plus agricultural statistics dating from 1901, an attempt is made to examine the extent to which traditional farming methods have been displaced by modern (Western) agricultural practices and the implications of this for local livelihoods and environments. Attention is also focused on the linkages between the technical and socio-cultural aspects of local agricultural knowledges.

Chapter Nine continues this agro-ecological focus with an investigation of gender variations in agricultural participation, decision-making and knowledge transfers between villagers' natal and marital places. In an attempt to highlight the limitations of eco-feminist/WED discourse about a special women-environment link, emphasis is placed on the micro-political ecologies (notably economic, socio-cultural and 'actor' related factors) that supplement gender as an influence on agricultural task allocation, decision-making and knowledge articulation. Particular attention is drawn to the effects of patrilineal inheritance systems and patrilocal residence patterns on women's familiarity with and control over local agricultural environments.

Chapter Ten echoes Chapter Seven in its use of village-level and historical data on forest use and management in the research area to examine the impact of 'scientific' forestry practices on local (socio-cultural as well as technical) silvicultural knowledge systems and forest quality. It then goes on to examine the cultural, subsistence-related and ecological importance of forests to local livelihoods and cultures, paying particular attention to the reasons behind forest decline in the research area.

Chapter Eleven examines different theoretical perspectives on community-based forest management in light of empirical data on autonomous and joint forest management from Jharkhand. Firstly the Chapter contrasts Jamtoli's success in sustaining autonomous forest management with Ambatoli's difficulties in establishing it on a village-wide basis. It then examines the impact of the Bihar joint forest management initiative in the research area. Jamtoli's re-alignment of its existing forest protection system along JFM lines is discussed first, followed by a description of Ambatoli's more prolonged efforts to establish JFM which finally bore fruit in 1998. Particular attention is placed on local attitudes towards the programme and its achievements in uniting different stakeholder groups, adapting to pre-existing institutional ecologies, empowering local people and promoting forest regeneration in villages where autonomous initiatives have been unsuccessful.

Chapter Twelve picks up on the themes raised in Chapter Eight with an examination of gender-based variations in forest use, silvicultural knowledges and forest management coupled with an investigation into the wider socio-economics and micro-political ecologies of gender-environment relations. Emphasising the shortfalls of eco-feminist/WED emphases on women as knowledgeable and active environmental custodians, it highlights the temporal and spatial fluidity of forest-based task allocation and identifies significant gender- and community-based variations in decision-making and control over environmental resources. Chapter Thirteen presents a general overview of the issues raised in Chapters One to Twelve and a conclusion to the book.

Notes:

[1] The green revolution raised Asian per capita food production by 27 per cent and enabled India to become food self-sufficient (Lipton and Longhurst, 1989).

[2] The green revolution's geopolitical origins lie primarily in its promotion as a tool to prevent a communist 'red revolution' in Asia (by keeping the peasants well fed). For the USA, it was an attractive commercial proposition as it provided a market for US agro-chemicals and agricultural machinery. For domestic governments, meanwhile, the prospect of intensifying production was politically much less fraught than the alternative of substantial land reform (Yapa, 1996).

3 Although Shiva has been heavily criticised in the academic literature, I feel that it is important to engage with her work as she has been so influential, both within and outside India, in getting the idea of a special women-environment link into mainstream public consciousness and development policy-making.

4 In 1997 I attended a PA training session at Edinburgh University along with a number of quite senior Indian Forest Service officers who were attending the session as part of their MA programme. None of them could see the point of PA and regarded the whole exercise as a waste of time. They had no intention of passing on what they had learned to their subordinates back in India. Instead, they planned to 'hold more regular village meetings and call it PA'.

5 According to Granovetter (1985), 'embeddedness' refers to the contextualisation of regimes in ways that reflect the socio-political role of local actors working within larger social structures and cultural practices.

6 To protect the identities of local people, I have given these villages the pseudonyms of 'Ambatoli' and 'Jamtoli'.

7 Mindful of the criticism that has been aimed at political ecology for its lack of theoretical cohesion (Peet and Watts, 1996a), emphasis here is placed on the use of political ecology as an analytical approach for examining human-environment interactions.

8 The words 'knowledge', 'knowledges' and 'knowledge systems' are used interchangeably throughout the book.

2 Changing Ideologies in Development Thinking: the Shift to a more Appropriate Approach

Introduction

This Chapter seeks to examine some of the many changes that have taken place within development thinking over the last half century; most notably the shift away from 'grand theories' towards more locally-oriented initiatives that highlight the importance of indigenous knowledge systems and local development priorities. More specifically, the first two sections point to some of the political ecologies behind traditional emphases on industrialisation and transfers of technology while the rest of the Chapter traces the shift towards a more grassroots approach. In particular, an investigation is made into the linkages between a growing interest in indigenous knowledge and common property resource management and the move towards more bottom-up and locally appropriate forms of development. Particular attention is drawn to the role of populist discourse in stimulating the adoption of more participatory approaches as well as in drawing attention to the loss of traditional technology and the potential for uniting local and scientific knowledge systems. Emphasis is also placed upon the inequalities that exist in different stakeholders' environmental knowledge possession and the problems of neglecting the socio-political and ritual-symbolic embeddedness of local knowledges in favour of their technical elements. Towards the end of the Chapter, an attempt is made to evaluate radical populist and anti-developmentist critiques of development as well as the role of peasant resistance and 'new social movements' in formulating more demand-led types of development.

The Drive for Development through Industrialisation

The popularity of industrial development, according to Gavin Kitching, has its roots in nationalism as much as in a desire to alleviate poverty and promote greater economic equality (Kitching, 1989). The reasoning behind this is that without industrialisation, nation-states would be unable to produce the armaments that would allow them to defend themselves against more powerful industrial nations. During the

1950s and 1960s, the case for a transfer of technology from the West was strengthened by the technical optimism of that period coupled with a growing geopolitically-rooted development discourse that linked poverty with political instability and generated grand theories for achieving development within a decade (Escobar, 1995). Industrialisation was favoured as the primary development mechanism by the US and major European powers as unlike populist alternatives based on agriculture, industrialisation was deemed capable of promoting modernisation and 'destroying archaic superstitions and relations' (Escobar, 1995, p. 40). An additional argument furthering the case for industrialisation at the expense of populist alternatives based primarily on agricultural development was that there are limits to the amount of growth that can be obtained from agriculture alone.

For the governments of many developing countries, too, industrialisation provided a positive vision for the future by comparison with agriculture which was seen by many as being 'backward'. It also provided a means by which newly independent nations could, in theory, assert their independence from the ex-colonial powers. In order to achieve this, many countries adopted import substitution industrialisation (ISI) policies which protected local industry against foreign competition, but this often resulted in inefficiency, limited employment generation and the production of low-quality, expensive goods (Booth, 1985; Elson, 1988; Corbridge, 1995b). A further problem related to the inability of some states to resist capture by dominant proprietary elites (Bardhan, 1984; Wade, 1990). Where this occurred, it hindered the removal of protection measures, exacerbated the problem of industrial inefficiency and reduced the possibility of a successful shift to export-oriented industrialisation (EOI) with the greater independence and freedom from foreign exchange constraints that this could bring.

The seeming failures of ISI led to different changes in development policies in different countries. In the East Asian 'newly industrialised countries', there was a successful transition from ISI to EOI from about the mid-1960s. In many poorer countries, the capacity to re-shape ISI internally was more limited (Wade, 1990), but planners and activists did begin to rethink the relationships between poverty, agriculture and industry in the process of development (Corbridge, 1995b). In South and Southeast Asia, the beginnings of a more agriculture-oriented strategy of rural development (the green revolution in the mid-1960s) reflected US efforts to control the spread of communism and coincided with the emergence of populist ideas on alternative forms of development in the West. Early examples of this genre include the works of Goldsmith et al (1972) and Schumacher (1973).

The Transfer of Technology

For many planners in developing countries during the 1950s and 1960s, however, the rural sector was seen more often from an orientalist rather than a populist perspective. By conceptualising the West as the role model for the rest of the world, development discourse 'became a powerful instrument for normalizing the world' (Escobar, 1995, p. 26). Supposedly 'childlike' indigenous people 'had to be "modernized", where modernization meant the adoption of the "right" values, namely those held by the white minority ... and, in general, those embodied in the ideal of the cultivated European' (Escobar, 1995, p. 43). Farmers in developing countries were therefore represented as 'backward', 'irrational' or 'stupid' for their failure or unwillingness to respond to cash crop and export markets and, later, to adopt green revolution agricultural technologies (Blaikie, 1985; Blaikie and Brookfield, 1987c; Hecht, 1987; Marglin, 1990b; Banuri and Apffel Marglin, 1993; Escobar, 1995; 1996; Yapa, 1996; Zimmerer, 1996). There were also doubts about their efficiency, economic rationality, environmental stewardship and willingness to work: doubts that early (mostly unsuccessful) community development initiatives failed to assuage (Blaikie, 1985; Korton, 1986; Pretty and Shah, 1999). Given the conceptualisation of development as the 'task of the white fathers to introduce the good but backward Third World people into the temple of progress' (Escobar, 1995 p. 158), the top-down development model was rarely questioned. Instead, the opinion that Third World farmers were 'ignorant and needed to be taught how to farm' (Hecht, 1987, pp. 18-19) helped to justify the continuation of technology transfers and to discourage other explanations for the slow growth of agricultural productivity. As Escobar points out, it was not realised that a transfer of technology 'would depend not merely on technical elements but on social and cultural factors as well. Technology was seen as neutral and inevitably beneficial, not as an instrument for the creation of social and cultural orders' (Escobar, 1995, p. 36).

By the mid-1960s, attitudes towards peasant farmers became a little less derogatory as research into peasant decision-making strategies suggested that farmers would act in an economically rational manner given favourable market conditions (Schultz, 1964; Cohen, 1967; Escobar, 1995). This revelation stimulated research into the reasons why many farmers were initially unable or unwilling to adopt modern varieties (MVs) of seed. Much of the literature produced focused upon issues such as the wider social environments of farmer decision-making (Barlett, 1980; Bennett, 1980; Berry, 1980; Cancian, 1980); the varying levels of farmer efficiency in different areas (Nair, 1979); the inefficiency of agricultural extension (particularly its ability to improve the flow of information to farmers without increasing their capacity to make use of it - Berry, 1980); and concern about the dangers of an emphasis on production without a simultaneous consideration for ecological and socio-cultural conditions. Other areas of interest have included the 'yield gap' between research stations and farmers fields, the constraints responsible for this and the identification of infrastructural difficulties as possible culprits for the lack of success.

Much of the Marxist literature on the green revolution has highlighted its lack of scale neutrality, inaccessibility to poor farmers and potential for agrarian class differentiation (Johnston and Kilby, 1975; Byres, 1981; 1983; Brass, 1990). The green revolution technology itself was one of the last factors to be questioned for its appropriateness as a means of rural development, yet later research brought to light a number of problems with the new technology; especially in capital intensive wheat-based systems. Particularly significant were the expense of the green revolution 'package' (seeds, irrigation, chemical fertilisers, herbicides, pesticides) for resource poor farmers and the pest susceptible nature of early MVs (Berry, 1980; Lipton and Longhurst, 1989).

Although plant scientists were able to breed much greater pest and disease resistance into later MVs and thereby reduce the need for large applications of agro-chemicals (Greenland, 1997), it nevertheless became apparent that intensive green revolution agriculture had significant environmental costs. Irrigation systems were often wasteful and poorly drained causing waterlogging, salinisation and soil toxicity. Mono-cropped cereals such as rice and wheat often displaced less valuable crops and brought a reduction in varietal diversity as well as (in the case of double and triple cropping) causing soil exhaustion. Water resources, meanwhile, became polluted as the availability of heavily subsidised agro-chemicals encouraged farmers to 'insurance spray' their crops against increased weed growth, disease and pest persistence. In addition to finding their way into water sources, these chemicals often killed natural predators and stimulated the development of resistance amongst surviving pests (International Rice Research Institute, 1994).

Even more worrying from the point of view of future food security has been the general decline in MV cereal output growth (Khush, 1995; Pingali et al, 1997). Aggregate rice output growth has declined to one point five per cent per annum and there has been a steady decline in yields even at major research centres. The main cause of this is thought to be the intensive and continuous nature of wet rice cropping which reduces nitrogen availability by slowing organic matter mineralisation (Greenland, 1997). Salinity build-up and increases in soil pests and micro-nutrient deficiencies are also important in some areas (Swaminathan, 1996).

As a means of alleviating some of these problems, much recent emphasis has been placed on biotechnology and genetic modification (GM) which, in theory at least, could improve yield sustainability and labour-intensive employment in less favourable rainfed areas not reached by the green revolution. GM staples enriched with Vitamin A, iron, zinc or iodine could also have important nutritional and health benefits for the poor and could be grown in currently unusable (often rainfed) environments if weed inhibiting characteristics, salt tolerance and moisture stress resistance were introduced. Indeed, they have been promoted as the key to a 'new doubly green revolution' (Conway, 1998) that could increase yields in an ecologically sustainable manner. According to Lipton, they might, in tandem with current declines in total fertility rates, also represent the means to eliminating world hunger if '"one more heave" from agricultural research ... can raise yields over a wide area ... providing income from labour-intensive farm work to just two more generations of the poor' (Lipton, 1999, p. 12).

Given the anti-GM lobby's concern about health risks and existing evidence of environmental problems associated with the green revolution, however, biotechnology has attracted much scepticism. Moreover, GM research has attracted a lot of bad press for its focus on the demands of multinational agribusiness and rich farmers rather than the needs of poor farmers and consumers (Krishnakumar, 1999). Particularly contentious has been the insistence by GM suppliers on germplasm patenting and intellectual property rights: rights that favour the final developers and ignore both farmers who have been selecting seed for generations and agricultural research institutions which have traditionally improved and distributed it for free (Lipton, 1999).

Interest in Indigenous Knowledge as a Route to more Appropriate Development

From the early 1980s when the socio-economic and environmental impacts of green revolution agriculture started to be taken seriously, the tone of much development studies and agronomic literature shifted away from a desire to 'educate' the so called 'tradition-bound' Third World farmers to act in an economically 'rational' manner and adopt the new technology, towards a sense of disillusionment with the transfer of technology model. As a result, more reflective questions about the suitability of green revolution packages to Third World farmers' ecological and socio-economic circumstances started to be asked.

At around the same time, ecocentric environmental discourse coupled with romantic populist emphases on self-reliance, co-operation and small scale technology helped to stimulate interest in 'intermediate' technology as a more appropriate way forward (Schumacher, 1973). These ideas were echoed within broader development emphases on peasant moral economies and rights to subsistence which in turn helped to highlight the richness of many indigenous knowledges and environmental management systems (Chambers, 1983; Chambers and Ghildyal, 1985; Hecht, 1987; Richards, 1985; 1986; Farmer and Bates, 1996).

As a result, research efforts shifted to the investigation of factors such as the complexity and heterogeneity of farming communities and the non-economic aspects of farmer decision-making (Barlett, 1980; Bennett, 1980; Berry, 1980; Cancian, 1980; Chibnik, 1980; Nair, 1979; Watts, 1982). Historical factors, too (particularly colonialism, and its emphasis on cash crop production at the expense of traditional food plants) were examined for their influence on agricultural productivity (Bernstein, 1978; Shenton and Freund, 1978; Watts, 1982). Increasing emphasis was also placed on the need to consider regional specificity, local farmer input into agricultural development programmes, and biases favouring resource rich farmers (Chambers, 1983; Chambers and Ghildyal, 1984; Rhoades and Booth, 1982).

Bottom-up Development

Disillusionment with the ideas that the benefits of industrial development would 'trickle down' to the rural poor and that capital intensive technology could be transferred successfully from developed to developing countries paved the way for interest in more bottom-up and rural-oriented 'alternatives to outsider-driven development approaches' (Gujit and Kaul Shah, 1998b, p. 4).[1] These have aimed, primarily, to increase understandings of the ecological, social, cultural, political and economic milieux of the areas which development projects are attempting to assist, and to promote there more appropriate and sustainable forms of development (Chambers, 1979; 1983; Brokensha et al, 1980; Blaikie, 1985; Chambers, Pacey and Thrupp, 1989). The need for the co-operation and participation of local people in development projects has also received increased attention from development donors and practitioners (Chambers, 1978; 1995; World Bank, 1996; 2001; FAO, 1999; 2001; DfID, 2001; Gujit and Kaul Shah, 1998a; Blackburn with Holland, 1998a; Pretty and Ward, 2001). Two early models that were introduced to enhance understanding of local ecological and socio-economic constraints were farming systems research and rapid rural appraisal (Chambers, 1981). These approaches were later criticised for emphasising questioning (or informing) rather than listening to local people (Chambers, 1987; 1992; 1994; 1995; Fairhead, n.d.). This, it was argued, tended to dissuade local communities from participating fully in development projects and from voicing their own concerns and requirements.

To overcome these problems, a number of 'farmer first' (Rhoades and Booth, 1982; Chambers and Ghildyal, 1985; Chambers, Pacey and Thrupp, 1989), 'farmer participatory research' and 'participatory rural appraisal' models were developed (Farrington, 1988; Chambers, 1992; 1994). These emphasised the need to enable and encourage local people to help define development projects' research agendas, evaluate their results and disseminate their findings. They also aimed to increase the technological options available to local people and to improve researcher awareness of individual farming and environmental management systems. Chambers and Ghildyal, for example, stressed that research priorities should be 'identified by the needs and opportunities of the farm family rather than by the professional preferences of the scientist' (Chambers and Ghildyal, 1985, p. 13).

Although approaches like farmer first have been tremendously valuable for helping to set in motion a 'paradigm shift' from top-down to bottom-up forms of rural development (Scoones and Thompson, 1994a), they have sometimes been found, in practice, to be rather politically naive and wanting in 'analytical depth' (Long and Long, 1992; Scoones and Thompson, 1993). In their place, 'beyond farmer first' approaches have been proposed (Scoones and Thompson, 1994a). Whilst maintaining the farmer first emphasis on active participation, empowerment and poverty alleviation, these approaches aim to 'challenge the simple notion that social processes follow straightforward and systemic patterns and can thus be manipulated with a transfer of power from 'outsiders' (e.g. researchers, extension workers, development agents, etc.). to 'insiders' (i.e. local people)' (Scoones and Thompson, 1993, p. 1). In particular, beyond farmer first emphasises the need to adopt a more politically and sociologically differentiated view of development which explores unequal distributions of power, knowledge and control over resources. (Scoones and Thompson, 1993; 1994a). It also stresses the need for a more actor oriented approach (Long and Long, 1992; Long and Villareal, 1994) which takes into account the role of 'situated agents' as well as wider-scale structural factors (Bebbington, 1994a; 1996).

Community-based Resource Management

In tandem with the growth of interest in bottom-up and participatory development, there has been an increase in emphasis upon common property resource management. This stemmed in part from the renewed popularity of environmentalism after the mid-1960s which helped to romanticise community-based resource management as part of a reaction against 'conventional exploitative development' (Regier et al, 1989). Much of the environmental movement's concern then was with pollution, population growth and resource depletion - themes emphasised by the popular literature of the time such as Rachel Carson's 'Silent Spring' (1964).

The Ehrlichs' 'The Population Bomb' (1966), Hardin's 'The Tragedy of the Commons' (1968) and 'Living on a Lifeboat' (1974) and the Club of Rome's 'Limits to Growth' (Meadows, 1972) represented some of the most right-wing reactions to this concern. A more left-wing response to perceived population-resource imbalances and the evils of industrialisation was the development of populist, utopian visions of small scale, community-oriented and decentralised societies such as Goldsmith et al's 'Blueprint for Survival' (1972) and Schumacher's 'Small is Beautiful' (1973). An important political ecology factor underlying their continued popularity was the 1970s oil price hike, after which the concept of sustainability within both resource management and development thinking received a great deal of attention.

At the same time, opposition to the 'eco-fascist' and 'eco-doomster' turn that some of the radical right-wing environmental literature had taken encouraged more critical research into the relationships between population increase and environmental problems and (later) between poverty and environmental degradation (Eckholm, 1984; Blaikie, 1985; Blaikie and Brookfield, 1987b; World Commission for Environment and Development, 1987; World Bank, 1992a; 2001; Holmgren et al, 1994; Tiffen et al, 1994; Reardon and Vosti, 1995; Peet and Watts, 1996a; DfID, 2001). Reardon and Vosti (1995), for example, draw a distinction between 'welfare poverty' (based on income or calorific intake) and 'investment poverty', arguing that the alleviation of the former will not necessarily help households to address environmental problems as they may still be too 'investment poor' to afford environmental protection measures. Presumably by the same token, such households may also be too investment (and welfare) poor to refrain from activities that cause environmental degradation.

Particularly influential in the growth of interest in community-based resource management has been the criticism directed at Hardin's 'tragedy of the commons' model (Hardin, 1968) which showed that his so-called commons were more akin to open access resources (Berkcs, 1989b; Berkes and Taghi Farvar, 1989; Singh, 1991a; Stevenson, 1991; Dhar, 1999). The distinction between common property and open access is crucial as community access to true commons is restricted, whereas that to open access resources is not. According to Stevenson 'common property is a form of resource management in which a well-delineated group of competing users participates in extraction or use of a jointly held, fugitive resource according to explicitly or implicitly understood rules about who may take how much of the resource' (Stevenson, 1991, p. 46). Open access resources, by contrast, are characterised by an 'absence of rights and duties' which 'means that the institution of property does not exist' (ibid., p. 49).

This vital distinction makes the degradation of open access resources far more likely than that of common property. According to Stevenson, true open access resources are quite rare. More often, resource degradation results from the fact that state-owned property (often land that was formerly managed as part of a common property system) is exploited by local people who resent its privatisation and act in accordance with the Turkish saying: 'one would have to be a despicable fool not to help oneself to state property' (quoted in Berkes and Taghi Farvar, 1989, p. 10).

This realisation helped to stimulate further research into the political ecology of community resource management institutions operating under different property ownership regimes and conditions of resource scarcity. Shepherd, for example, has emphasised the failure of many colonial and post-colonial governments to recognise group-based property systems and argues, like Hardin, that population increase can be a major factor in the decline of common property resource management in the absence of state intervention (Shepherd, 1992c; 1993a). In a model illustrating indigenous and participatory forest management (see figure 2.1), Shepherd shows that common property resource management develops in 'stage two' where 'some resource scarcity is experienced', but starts to decline in 'stage three' when 'there are too many people or too many interests for the size of a resource and the management costs begin to outweigh the benefits' (Shepherd, 1993a, p. 1).

This is where the similarity between Shepherd's and Hardin's model ends, however, as Shepherd is very critical both of Hardin's concept of 'individual rationality leading to mass irrationality' (Shepherd, 1992c, p. 73) and the extent to which Hardin's assumptions about commons have been retained by many development practitioners. Instead, she argues that if a suitable approach can be found, it may be possible to maintain certain 'stage three' systems as functioning common property resource management systems or even to 'roll them back towards stage two' (Shepherd, 1992a).

Shepherd also argues that given helpful state intervention (Shepherd, 1993a; 1993b) in the form of participatory or joint resource (particularly forest) management, it may be possible to re-create common property resource management in certain situations. To assist development practitioners and state officials who are thinking of investigating participatory approaches, Shepherd has developed a second model which indicates where and in which situations common (autonomous or joint) resource management is most likely to be successful (see figure 2.2).

As in the first model, increasing population densities and intensity of land use are accompanied by a decline in functioning common property resource management systems (Shepherd, 1993a; 1993b). The communities with the best potential for participatory management are therefore those in area types one and two that have relatively low population densities, working systems of common property resource management and fairly extensive agriculture. The problem, of course, is that this model is only an indicator which, as Chapters Ten and Eleven will illustrate, cannot substitute for detailed empirical research into the specific institutional ecologies and environmental realities of different resource management strategies and 'area types'. It is also very difficult, without theoretically grounded local level research into the socio-political embeddedness of resource management institutions, to assess whether a community has progressed too far towards intensive agriculture, individual land ownership and open access systems to make a return to some form of common property resource management. By combining resource management theory with locally grounded investigations into the socio-political embeddedness and temporal fluidity of resource management regimes, political ecology approaches are in an ideal position to identify spatially specific recipes for sustainable resource management and account for

Figure 2.1   Shepherd's 'first model' - 'Indigenous and participatory forest management'

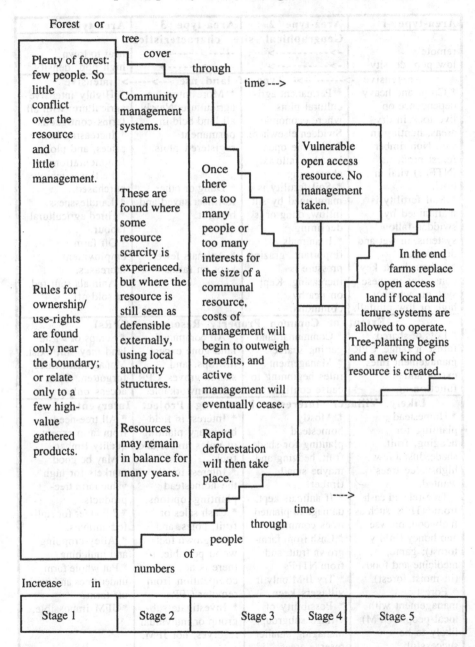

Source: Shepherd, 1993a.

Figure 2.2   Shepherd's 'second model' - 'Participatory tree projects - when to try what: a range of indicators'

Area-type 1	Area-type 2	Area-type 3	Area-type 4
Geographical site characteristics			
remote low pop. density	->----------> ->---------->	-----> -----><------>	near to town high pop. density
extensive ->---------->- Type of land use --->-----> intensive			
* Crops and heavy dependence on livestock in dry areas, hunting in wet. Non timber forest products (NTFPs) vital in both. * Soil fertility is maintained by swidden-fallow systems, in wet and dry areas. * Labour is the key constraint in these systems whether high or low rainfall.	* Permanent agricultural plots where economic. Swidden elsewhere. Still some open land but fallows shortening. * Soil fertility is maintained by silt inflow, dung or is declining. * If animals important, grazing pressure is increasing. Kept on nearby commons.	* More intensive agriculture. Nearly all land held in permanent registered plots. * Dung or other fertiliser saved and bought. * Animals fewer, kept on farms.	* Highly intensive agriculture; with all farms contiguous. * Increasing land prices, and plot fragmentation. * Fertiliser mainly purchased. * Landlessness. * Hired agricultural labour. * Off-farm employment increases. * Animals stall fed or sold off.
Extent of Common Property Resources (CPRs)			
* Lots of shared forest or range. CPR management rules extant and functioning.	* Communal land getting scarcer. * Management rules beginning to cause conflict.	* All communal land gone except hilltops, and sacred groves. CPR rules unworkable.	* Scraps of waste land may still exist. Management rules forgotten. Open access only.
Likely Villager Interests: Promising Project Interventions			
* Homestead planting for hedging, fruit, shade. Just a few high-value trees wanted. * Tree-related cash from NTFPs: such as fuelwood, browse and honey (in dry forest); game, medicine and foods (in moist forest). * Forest management with local people (JFM) likely to be successful.	* Mostly homestead planting for shade, fruit, hedging and maybe small timber. * If animals kept, damage to planted trees common. * Cash from farm-grown fruit, and from NTFPs. * Try JFM only if villagers keen. * Possibility of village subgroups managing smaller reserve areas.	* Interest in field-boundary planting of poles timber and fuel. * Interest in all the homestead planting options. * Cash sales of fruit, poles, and farm-grown fuel-wood possible, if there is no competition from remoter CPRs. * Investigate sub-group or individual reserves, not JFM.	* All tree-needs from farm except quality timber. * May be good markets for high value farm tree-products. * ? Fodder for stall-fed animals. * Alley-cropping and mulching? * Put whole farm under trees and work off-farm? * JFM impossible.

Source: Shepherd, 1993b.

variations in the performance of different environmental management systems.

Situations that resource management theorists most commonly identify with successful common property management include a recognition by local communities that a resource must be protected from further exploitation and a perception that the benefits of collusion are likely to exceed the institutional and socio-political costs of establishing such a system (Ostrom, 1990; 1999; Stevenson, 1991; Ostrom et al, 1993; Baland and Platteau, 1996; Kaufman, 1997a; Agrawal, 2001; Pretty and Ward, 2001). There should also be a belief that the benefits to be realised from community management would exceed those that could be achieved by individual-level exploitation. A good example of this is where common property management makes financial sense in that infrastructural costs such as fencing, labour, transport and buildings can be shared by the community.

Other factors that encourage the establishment of successful common property resource management systems include a scarcity of resources, an easily defined user group and a recent history of community-based resource management. A powerful or well respected local leader who is able to mobilise villagers' support for common property resource management and provide a means for ensuring that the system of rules and duties for resource exploitation are enforced can also be very important (Wade, 1986; 1988). Management is made easier if both the resource and its users are well defined and the system of limiting resource use (such as when access to the resource is permitted, how much of it can be taken, what maintenance duties are expected in return and what punishments will accrue to offenders) is well understood by the community (Blaikie et al, 1986; Singh, 1999). A degree of local autonomy in defining rules and undertaking monitoring activities (and the recognition of this by external governmental authorities) has also been acknowledged as an important factor in long-term common property resource management systems (Ostrom, 1990; 1999). Relatively small, socio-culturally or economically homogenous groups that have a mutual interest in the future management of a resource can encourage resource users to treat it as a type of extended private property from which outsiders are excluded (Gadgil and Iyer, 1989; Singh, 1991a; Fernandez, 1999). This is particularly the case where levels of environmental uncertainty are high and the dependability of yield (and potential for intensification) is low as this will reduce the likelihood of conflict between members (Netting, 1976; Shepherd, 1993a; 1993b).[2]

Criteria that assist in the initiation (as opposed to the maintenance) of common resource management institutions have been neatly summarised by Elinor Ostrom (1999) in her paper on self-governance and forest resources. Building on the eight 'design principles' (Ostrom, 1990) illustrated by long-standing resource management institutions, she identifies four resource attributes and seven user attributes that can encourage the emergence of resource management systems (see table 2.1). Like Shepherd, she considers the quality of the resource (notably its perceived scarcity) and a high but uniform degree of user dependence on it to be important criteria influencing its likely protection. She also regards relatively small (manageable) resources about which users share a common image as being important along with user characteristics such as prior organisational experience, trust in one another and autonomy from external authorities.

Common Property Resources and Development

As Shepherd's second model illustrates, interest in and enthusiasm for common property resource management has not been confined to academic circles (Shepherd, 1992a; 1993a; 1993b). Development planners within government bodies, donor agencies and non-governmental organisations (NGOs) have all come to see the potential of common property resource management for future participatory development initiatives. Of particular interest are the reciprocal obligations inherent within common property resource management systems that have potential to ensure greater livelihood security, equitable resource distribution, participation in the production system, resource conservation and ecological sustainability: in short, the very things that much rural development is trying to achieve (see, for example, Berkes and Taghi Farvar, 1989; Warren et al, 1995a; Kaufman 1997a; Netting, 1997; Chambers, 1998; Pretty and Ward, 2001). The favourable financial implications associated with the strengthening of existing forms of community-based resource management or the re-establishment of traditional systems through participatory or joint management have also been attractive.

Common property resources have also received attention as part of recent emphases within rural development on the value of learning from local environmental knowledges and management strategies, encouraging the participation of local people in development projects and ensuring that such projects are acceptable to them. Some of the findings from this work have shown how commons in many developing countries are an important resource for poorer and landless community members who rely heavily upon them for fuel wood, fodder and, to some extent, food (Korton, 1986; Shiva, 1988; Bahaguna, 1991; Kumar, 1991; Singh, 1991a; 1991b; Singh, 1999; Agarwal, 2001).[3]

Other work has emphasised how community-based systems of common property (particularly forest) management can ensure the sound management and equal access of community members to vital subsistence resources such as fuel wood, construction timber, animal fodder and irrigation water (Arnold and Campbell, 1986; Blaikie et al, 1986; Singh, 1986; Shiva, 1986; Wade, 1988; Gadgil and Iyer, 1989; Sengupta, 1991; Ghate 1998). In response to this interest, a number of countries have tried to re-establish systems of common property

Table 2.1   Attributes important to the emergence of common resource management institutions

Attributes of the Resource

Attributes of the Users

Feasible improvement: The resource is not at a point of deterioration such that it is useless to organise or so underutilised that little advantage results from organising.

Indicators: Reliable and valid information about the general condition of the resource is available at reasonable cost.

Predictability: The availability of resource units is relatively predictable.

Spatial Extent: The resource is sufficiently small, given the transportation and communication technology in use, that users can develop accurate knowledge of external boundaries and internal microenvironments.

Salience: Users are dependent on the resource for a major proportion of their livelihood or other variables of importance to them.

Common understanding: Users have a shared image of the resource and how their actions affect each other and the resource.

Discount Rate: Users have a sufficiently low discount rate in relation to future benefits to be achieved from the resource.

Distribution of interests: Users with higher economic and political assets are similarly affected by a current pattern of use.

Trust: Users trust each other to keep promises and relate to one another with reciprocity.

Autonomy: Users are able to determine access and harvesting rules without external authorities countermanding them.

Prior organisational experience: Users have learned at least minimal skills of organisation through participation in other local associations or learning about ways that neighbouring groups have organised.

Source: Ostrom, 1999.

resource (particularly forest) management as a means of arresting deforestation and meeting local fuel wood and fodder needs.

Local Agro-ecological and Silvicultural Knowledge Systems

In tandem with and partly responsible for the recent interest in common property resource management, there has been a more general acceptance within development studies that many Third World communities possess long histories of agricultural experimentation and innovation and complex systems of agro-ecological knowledge (Brokensha et al, 1980; Chambers, 1983; 1998; Richards, 1985; 1986; Rhoades and Bebbington, 1988; 1995; Chambers, Pacey and Thrupp, 1989; Verhelst, 1990; Chambers and Conway, 1992; Scoones and Thompson, 1994a; Warren et al, 1995b). By the same token, it has become widely understood that people in developing countries are neither stupid nor irrational for not adopting wholesale the agricultural technologies of the West (Blaikie, 1985; Altieri et al, 1987; Hecht, 1987). Also apparent has been a growing recognition from government and non-government based development agencies of the need to both 'mainstream' environmental considerations and learn from (rather than ridicule) Third World agriculture and environmental management strategies (World Bank, 1996; Narayan and Rietbergen-McCracken, 1998; DfID, 1999; 2001; FAO 1999; 2001). Britain's Department for International Development (DfID), for example, has recently adopted a strong pro-poor strategy emphasising the management of natural resources by local people and highlighting 'the need for poverty focused development activity to be sustainable, people centred, responsive and participatory, multi-level, dynamic, and conducted in partnership' (DfID, 2001, p. 57).

This is not to say that interest in the potential of indigenous agro-ecological knowledge as an aid for implementing development projects in the Third World is particularly new, however. Anthropology has a long history of studying 'ethnoscience' (Sturtevant, 1964; Spradley, 1972; Warren, 1975; Werner and Begiste, 1980; Werner and Manning, 1979) and the geographer, Carl Sauer recommended in the early 1940s that local knowledge should be built into any programme of technology generation (Bebbington, 1992).

By the early 1970s, radical populist and ecocentric strands of the environmental movement had started to highlight the ecological impacts of industrialised agriculture (Carson, 1964). At the same time, empirical research on tropical agriculture brought recognition of the fact that tropical ecosystems are quite different to temperate ones and may be damaged seriously by transfers of Western agricultural methods and environmental management techniques (Janzen, 1973; Hecht, 1987; Escobar, 1995; Pretty and Shah, 1998; Sullivan, 2000). This work stimulated, in turn, agro-ecological research on the nutrient cycling, soil conservation and pest regulating capabilities of modern agroecosystems - typically mono-crop systems - compared to those of traditional (usually Third World) polycropping systems. On all counts except their short term productivity levels, the polycrops were found to be more effective. They were also shown to be more energy efficient, ecologically stable, disease resistant and better at reproducing themselves without human assistance than were the mono-crops (Bayliss-Smith, 1982; Altieri et al, 1987; Hecht, 1987).[4]

Interest in indigenous environmental knowledges more specifically was stimulated by key works such as Brokensha et al's (1980) book entitled 'Indigenous Knowledge Systems and Development'. This details work done, *inter alia*, on indigenous folk taxonomies (Brush, 1980; Richards, 1980), farmers' selection and improvement of plant varieties (Brush, 1980), indigenous systems of time, soil and topography classification (Howes, 1980) and the skills of different cultural groups in plant identification, breeding, propagation and utilisation (Brokensha and Riley, 1980; Howes, 1980).

In association with populist disillusionment over the 'technical fix' approach came more direct concern about the danger of a transfer of Western technology to Third World eco- and social systems. An important response to this was the concept of sustainable development which has succeeded in providing links between technocentric environmentalist discourse and more traditional economic-growth led development ideology (Lele, 1991). These discourses have in turn encouraged mainstream development agencies to take on board local agro-ecological knowledge systems and environmental management strategies as part of an emphasis on participatory and 'sustainable livelihoods' approaches (World Bank, 1996; 2001; DfID, 1999; 2001; FAO, 1999; 2001; Neefjes, 2000; Rahman, 2001).[5] Of particular interest has been the possibility of achieving a more holistic approach to resource management by uniting traditional environmental knowledge systems with Western science (Berkes, 1989a; Berkes and Taghi Farvar, 1989; Chambers, Pacey and Thrupp, 1989; Scoones and Thompson, 1993; 1994a; Agrawal, 1995). As Chapter Four will show, this approach is now being taken very seriously in India by local and external development agencies, particularly with regard to local resource management systems and forest-related issues.

Local Knowledges and Appropriate Development

Encouraged by advocates of both bottom-up and sustainable forms of development, much of the early research on indigenous knowledges emphasised the cultural, ecological and financial appropriateness of local agricultural methods and the equitable resource distribution and conservation associated with many community-based environmental management systems (Brokensha, 1980; Altieri et al, 1987; Berkes, 1989; Chambers, Pacey and Thrupp, 1989). Indeed, their potential seemed so great that by the early 1990s, 'mainstream' development organisations such as UNESCO (Hadley and Schreckenberg, 1995) and the World Bank (World Bank, 1992a) had started to fund research on indigenous agriculture and forestry management systems.

Many NGOs and development donors have also come to accept the belief that projects stand a much better chance of success if development practitioners are willing to listen to, work with and learn from recipient communities (World Bank 1996; FAO 1999; DfID 2001). At the same time, there has been greater interest in the linkages between poverty and environmental degradation and the possibility that attempts to build upon indigenous knowledges could help to empower local people (World Commission on Environment and Development, 1987; Chambers, 1992; Vivian, 1992; Agrawal, 1995; Reardon and Vosti, 1995; DfID, 2001; Neefjes, 2000).

The Dangers of Technology Displacement

Another reason behind recent emphases on indigenous knowledge systems stems from populist concerns about the need to obtain as much information as possible on traditional agro-ecological practices and crop varieties before the 'dominating knowledges' and practices of the West displace them completely (Apffel Marglin and Marglin, 1990). Gupta (1992), for example, has pointed out the tendency for biodiversity to be associated with the more disadvantaged areas of the world, which, if lost through the impact of modernisation, could seriously reduce the capacity of these areas to secure future autochthonous development. There has also been concern about the psychological effects of modern technological developments on indigenous people who sometimes come to feel that their own knowledges are worthless (Chapman, 1983; IDS Workshop, 1989a; Scoones and Thompson, 1993; Salas, 1994; Agrawal, 1995; Mundy and Compton, 1995; Gupta, 1996; Yapa, 1996).

Western-style education, in particular, can cause local knowledge to be rejected for not being 'scientific' with the result that local people become embarrassed about using indigenous agricultural and resource management practices (Chapman, 1983; Gupta, 1996). Similarly, Westernised or 'city based' food preferences can have an important impact on cropping patterns by encouraging demand for rice and wheat at the expense of more traditional (yet often nutritious and hardy) pulses that come to be perceived as 'poor peoples' crops' (Shiva, 1988; 1991a; 1991b; Dixon, 1990). The rapid rate at which both new plant varieties become available and food preferences change has generated fears that valuable indigenous genetic material could be displaced permanently (IDS Workshop, 1989b).

Many populist writers also see potentially quite serious dangers associated with the loss of so called 'traditional technology'. Appadurai (1990), for example, has warned that skills such as the construction and maintenance of non-motorised wells may soon be lost, yet wells powered by electric motors will remain unreliable in many developing countries until continuous access to electricity is guaranteed. A reliance upon these motorised wells also helps to make traditional bullock-powered wells obsolete and with them traditional forms of bullock sharing and farmer co-operation. Another example given by Appadurai is the replacement of traditional leather water vessels with the metal water containers which are now used widely by farmers of all economic groups. He argues that should the price of metal water containers (or fertilisers, or most other farm inputs for that matter) rise suddenly, then resource poor farmers would suffer the most from the loss of traditional technology.

To prevent the danger of the current situation from becoming any worse, there have been calls for the establishment of 'national archives' to collect information about traditional knowledge systems (Brokensha, 1980; Altieri et al, 1987; Hecht, 1987; Agrawal, 1995; Warren et al, 1995a). Others have been concerned with the need to encourage farmers to be proud of their local knowledge systems (rather than allowing themselves to be stigmatised by outsiders) and to feel able to speak out for the inclusion of their agro-ecological and silvicultural practices in development strategies without fear of appearing 'backward' (Howes and Chambers, 1980; Richards, 1980; Blauert, 1988; Thrupp, 1989; Salas, 1994; Gupta, 1996).

The reality of these fears is illustrated by Chapman (1983), who notes how indigenous farmers often feel very embarrassed about their farming practices in front of outsiders and strive hard to reduce 'dissonance' in conversations with extension workers or agricultural scientists. Examples of how they do this include typical 'peasant resistance strategies' (Scott, 1985) such as hiding traditional crop varieties from visiting extension workers, giving stock answers that they do not understand but which they have learned will make extension workers go away, and rejecting indigenous farming practices altogether (pretending that they never practised them) in favour of 'all modern' methods (Chapman, 1983; IDS Workshop, 1989b).

To alleviate such problems, many writers have advocated learning from farmers rather than telling them how they should be farming (Brokensha et al, 1980; Howes and Chambers, 1980; Box, 1989; Chambers, Pacey and Thrupp, 1989; Verhelst, 1990; Fernandez, 1999). Box recommends the testing and refining (if possible) of farmers' practices and experiments at research stations. The reasoning behind these recommendations is that a more sensitive approach might enable researchers to gain a clearer picture of farmer constraints and how these could be alleviated in ways that the technical and socio-cultural aspects of indigenous knowledges could cope with (Rhoades and Bebbington, 1988; 1995; Millar, 1994; Stolzenbach, 1994). They may also help to forge stronger links between scientific and local knowledges and maximise the appropriateness of development strategies for local men and women.

The Problems and Possibilities of Integrating Local and Scientific Knowledge Systems

Since the early 1980s, many bottom-up development writers have sought to emphasise the similarities between indigenous and scientific knowledges because of their interest in combining scientific knowledge (from agricultural research stations) with 'the ethnoscientific knowledge of farmers to effect widespread distribution of agricultural innovation' (Meehan, 1980, p. 387). Local people's knowledge about soil types, plant varieties, silviculture, ecology and botany plus their skill in experimentation and risk minimisation have received the most emphasis; the idea being to utilise these knowledges for use in environmental management. Particular attention has therefore been given to the ability of local people to theorise about their agro-ecological practices, empirically test hypotheses and conduct experiments to help decide whether or not to adopt new seeds or technologies (Brokensha et al, 1980; Chambers, 1983; Richards, 1985; 1986; Rhoades and Bebbington, 1988; 1995; Chambers, Pacey and Thrupp, 1989; Okali et al, 1994). Relatively little attention, by comparison, has been paid to the different socio-political and cultural contexts in which these knowledges are embedded and how their 'de-contextualisation' might affect them or their possessors (Bebbington, 1990a; Cornwall et al, 1994; Scoones and Thompson, 1993; 1994b; Escobar, 1996).

In recent years, however, more sophisticated analyses of indigenous knowledge systems have questioned the assumption that they are easily definable stocks of knowledge that can readily be extracted and combined with scientific knowledges (Scoones and Thompson, 1993; 1994b; Agrawal, 1995; Springer, 2000). They have also highlighted the different epistemological frameworks that surround scientific and local knowledges (Banuri and Apffel Marglin 1993b; Alvarez, 1994; Scoones and Thompson, 1994a; Agrawal, 1995). Rather than sharing the characteristics of openness and objectiveness which is supposed to enable scientific knowledge to question its underlying foundations, indigenous knowledges are seen as more intuitive and less prone to analysis. They are also seen as being substantively different from scientific knowledge because they relate to everyday activities rather than to abstract theories and ideas. A third major difference is that indigenous knowledges are thought of as being much more deeply rooted in local environments and the socio-cultural moral milieux of local communities whereas scientific knowledges are supposed to be universal and able to distinguish between technical and cultural factors (Agrawal, 1995; Escobar, 1995).

According to Agrawal (1995), these divisions between indigenous and scientific knowledge are often more apparent than real. Nevertheless, they have been helpful in stimulating attempts to recognise and understand the linkages between the socio-cultural contextualisation of local knowledge and its technical, hypothesis testing and experimental aspects. After all, it should not be forgotten that so-called 'universal' modern science is not uninfluenced by social relations of power nor are its proponents totally divorced from more embedded beliefs and superstitions.

Emphasis on the possibilities for linking indigenous and scientific knowledges in ways that are culturally appropriate have also been taken on board as part of a wider 'learning approach' (Pretty and Chambers, 1993; 1994) which emphasises the need to understand the complexities of local knowledge systems and encourage local participation in development projects. Before a really constructive dialogue can be achieved with respect to local and scientific knowledge systems, however, further empirical investigations are necessary into the temporal, spatial and socio-political embeddedness of agro-ecological knowledge, the inequalities that exist in its possession and the micro-political ecologies that influence environmental resource management more generally.

Community-based Variations in the Possession and Control over Agro-ecological Knowledge

Anticipating more recent concern over simplistic conceptualisations of participation and empowerment (Blackburn with Holland, 1998a; Crawley, 1998; Gujit and Kaul Shah, 1998a; Agrawal and Sivaramakrishnan, 2000a), two significant forces advocating awareness of inter- and intra-community variations in knowledge possession and control are proponents of the beyond farmer first (Long and Long, 1992; Scoones and Thompson, 1993; 1994a) and gender analysis (GA) approaches (Agarwal, 1992; Jackson, 1993a; 1993b; 1994; Braidotti et al, 1994). In particular, these writers have sought to challenge simple unified representations of indigenous knowledge systems, pointing out that far from existing in a political vacuum, local knowledge is politically and socially constructed and is therefore constantly undergoing change (Bebbington, 1990a; 1992; 1994a; Scoones and Thompson, 1993; 1994b; Long and Villareal, 1994; Marsden; 1994; Salas, 1994).[6] Indeed, the efforts of GA writers to update gender-environment theorisation and challenge prevailing views of women and men as undifferentiated categories are sufficiently important to the content of this book that they will be examined separately in Chapter Three.

Some of the most important influences on the extent and types of knowledge possessed by particular socio-cultural groups are historical factors that determine people's exposure to different types of environmental management (for example, through experience in *hacienda* or plantation systems - Bebbington, 1990a; 1996). Another significant cause of knowledge variation is the existence of inequalities in land ownership or resource management systems which affect the extent to which local people are able to utilise or expand their knowledges (Norem et al, 1989; Jeffery, 1998b; Agrawal and Sivaramakrishnan, 2000b; Gururani, 2000; Jackson and Chattopadhyay,

2000). Likewise, the communication and transmission of local knowledge, (like that of scientific knowledge) is politically-determined as it occurs within 'socially and politically constituted networks of different actors, organisations and institutions' (Long and Villareal, 1992 quoted in Scoones and Thompson, 1993, p. 16). It can be very revealing, therefore, to view local people as 'situated agents' who acquire knowledge in 'cultural, economic, agroecological and socio-political contexts that are products of local and non local processes' (Bebbington, 1994, p. 89).

Any attempt to encourage a meaningful dialogue between local and scientific knowledges must therefore steer away from romanticised narratives about indigenous societies informed by ideas of moral economy. Instead, it is important to recognise the 'internal stratifications and oppression within communities, and the politically asymmetrical position of all communities attempting to behave as autonomous actors ranged against powerful political or economic interests' (Agrawal and Sivaramakrishnan, 2000b, p. 9). This is particularly so for participatory development approaches which frequently gloss over intra-community (especially gender) variations in the degree and level of participation (Agarwal, 2001). When attempting to build upon local knowledges or resource management strategies, therefore, participatory development approaches must recognise the spatial specificity and temporal fluidity of indigenous knowledge systems. They must also be sensitive to the micro-political ecologies of indigenous knowledge distribution as well as inequalities in the power of different stakeholders to utilise or communicate that knowledge. After all, participation has little meaning 'unless it simultaneously addresses the power structures that appear to perpetuate poverty' (Blackburn and Holland, 1998b, p. 2).

Cultural Aspects of Indigenous Knowledge Systems

As research into local environmental knowledges has become more widespread, greater efforts have been made to bring about a replacement of 'etic' (from outside a culture) with 'emic' (from inside) explanations for local agricultural and resource management systems (Bebbington, 1990a; 1990b; 1994a; Banuri and Apffel Marglin, 1993a; Alvarez, 1994; Scoones and Thompson, 1993; 1994a). Rather less emphasis, however, has been placed on how farmers might use their indigenous knowledge politically as, for example, part of a demand for more autonomy or resources from the state. Less scarce, but still not abundant in the current literature, is work on the symbolic importance of indigenous knowledges (Blauert, 1988; Bebbington, 1992).

In practical terms, increased levels of interest in the social and cultural systems within which indigenous knowledges are embedded have enabled significant insights to be gained into the rationalities (socio-cultural, political, technical and environmental) behind many local agro-ecological and silvicultural practices. They have also helped to generate interest in the technical, historical and socio-cultural aspects of local knowledge systems. A significant related benefit has been the promotion of interest in more sensitive forms of development which recognise the close association that indigenous knowledges have, through symbolism, with the socio-cultural values of their possessors (IDS Workshop, 1989a). Such approaches have also stressed the need to prevent local knowledge from being 'de-legitimised' in the eyes of its possessors, or worse, trivialised if 'isolated from its cultural context and forced into the framework of western epistemology' (IDS Workshop, 1989a, p. 38).

The main reason for the relative dearth of literature on the socio-cultural *vis-à-vis* the technical aspects of indigenous knowledges has been that many authors have had to emphasise the latter in order to guarantee them a hearing on the agendas of food policy-makers (Bebbington, 1992; Scoones and Thompson, 1993; 1994b; Cornwall et al, 1994; Agrawal, 1995). Nevertheless, some lateral interest in the political and socio-cultural aspects of local knowledges (and the resulting respect for local cultures and concern for their preservation that this has brought) has been stimulated by research into their technical aspects; a consequence which may not have been unintended by the writers of this literature (Bebbington, 1992). As a result, a few researchers have drawn attention to how local people think about their own agro-ecological practices and knowledges and how or why they might change (or differently represent) their views in the face of changing social, ecological or political circumstances (Blauert, 1988; Baines, 1989; Blauert and Guidi, 1992; Bebbington, 1990a; 1992; 1996; Millar, 1994; Matose and Mukamuri, 1994; Scoones and Thompson, 1994b; Jeffery 1998a; Karlsson, 2000).

An example of this shift is the examination of local knowledges as hermeneutic science (Bebbington, 1990a). This has been expressed particularly in terms of the symbolic and cultural importance of local knowledges to their possessors and the ways in which this 'informs their interpretation of their past, present and future, and of their relationships to each other, to the environment and, above all, to the rest of society' (Bebbington, 1990a, p. 5). Related to this is an interest in how concepts grounded in Western epistemologies (such as plant nutrients and diseases) compare with indigenous epistemologies which 'make sense within their particular socio-cultural, economic and ecological milieu' (Fairhead, n.d., p. 2).

Also of growing interest to students of indigenous knowledges is their historical context and the ways in which this reflects the socio-cultural and environmental relations which have influenced farming, land use, resource management and the symbolic interpretation of the environment over time. Indigenous agro-ecological knowledges and management systems represent, therefore, not only important sources of technological knowledge and agro-ecological practices (which may be appropriate for diffusion to other environments and cultures), but also spatially rooted systems of symbolic features and meanings which form the basis of mutual understanding and communication between local communities (Bebbington, 1990a; Escobar, 1995; Bhattacharaya, 1998; Freeman, 1998; Kalam, 1998; Linkenbach, 1998).

Such understandings are particularly significant as local knowledge systems can be defined only in terms of the knowledge that groups have at a particular time in a particular place. In other words, because systems of indigenous agro-ecological and silvicultural knowledge reflect the socio-cultural and environmental relations of their possessors, they are constantly being re-defined to accommodate new ideas, technologies and situations (Bebbington, 1990a; Horta, 2000).

To many Western systems of thought, however, these linkages can appear illogical and disconcerting with the result that they have often been ignored for the sake of 'legitimising' the more 'technical' aspects of the knowledge system (Bebbington, 1992; Scoones and Thompson, 1993; 1994b; Cornwall et al, 1994). One example of such behaviour is cited by the IDS Workshop (1989a, p. 38) and concerns the emphasis of Western medical science on alternative medical practices. Significantly, the interest lies almost entirely with the technicalities of such systems (such as herbal practices) and not in the rituals and symbolism associated with their socio-cultural aspects. Yet to the people practising these types of medicine, their ritual-symbolic aspects are as important as (and inseparable from) their herbal ones (Fairhead, n.d.).

Despite increasing recognition of this fact and the now quite long-standing interest in indigenous knowledge systems, there has been little research done on the linkages between their ritual-symbolic and technical aspects.[7] Nevertheless, quite a lot of interest has been shown in the cultural importance of land, the iconography of landscapes and environmental symbolism. A significant amount of this work has been concentrated upon aboriginal 'dreamings' in Australia (Chatwin, 1988; Tilley, 1994) although some interesting research has also been done on the symbolism of forest environments to *adivasi* populations in India (Fernandes and Kulkarni, 1983; Chandrakanth et al, 1990; Corbridge, 1991a; Kelkar and Nathan, 1991; Gadgil and Subash Chandran, 1992; Mahapatra, 1992; Vannucci, 1992; Vatsyayan, 1992; Xaxa, 1992; Jeffery, 1998a).

According to Fairhead, these socio-cultural knowledges are important for linking 'the ecological processes which enjoin agro-ecological, social and human productivity relations' (Fairhead, n.d., p. 19). If, however, disturbance occurs due to an area being 'developed' by (say) a logging company, the result can be socially very traumatic for local people as important landmarks and sacred sites can be lost along with a local community's 'historical links with the past' (Baines, 1989, p. 282).

In addition to this work in Asia, some very interesting research has been done on the spiritual and religious importance of Trique and Mixtec socio-cultural environments in Mexico (Blauert, 1988; Blauert and Guidi, 1992) and the relationships between indigenous Andean agriculture and symbols of wider spiritual and cultural importance such as the Earth Mother (see, for example, Isbell, 1978). Another interesting focus of investigation has been the ways in which particular indigenous agricultural systems can be understood as 'products of a local cultural system which simultaneously plays a part in its ongoing reproduction' (Bebbington, 1990a, p. 22). A good example of this can be found in Blauert's (1988) description of Trique hunting practices in which local people's respect for forest habitats and the spirits of game animals acts as a kind of inbuilt conservation strategy. The cultural significance of these local knowledges and practices therefore influences the ways in which they are used and the manner in which they are likely to respond to change; factors that both act independently of the technical content of the Trique knowledge system.

Radical Populism and Local Knowledges

Recent emphases on the value of local knowledges and resource management strategies have, not surprisingly, found much favour with certain radical populist and eco-feminist writers who seek alternatives to reductionist, patriarchal science and its conquest of nature (Shiva, 1988; Merchant, 1989; Mies and Shiva, 1993; Alvarez, 1994; Escobar, 1995). Romantic populist visions of rural 'wholesomeness' (Byres, 1979) where peasants and artisans live self-sufficient lives and control their own work schedules are greatly embellished by research showing that local technology in developing countries is often far more appropriate than capital intensive technology imported from the West. As a result, it has much appeal amongst Western and Westernised populist intellectuals who see indigenous technologies and community-based resource management strategies as important building blocks in an alternative, sustainable and more equitable development strategy (Schumacher, 1973; Shiva, 1988; Escobar, 1992; 1995; Esteva, 1987; Parajuli, 1991; Alvarez, 1994; Banuri and Apffel Marglin, 1993b; Gupta, 1996).

Gandhi's vision of a 'village centred economic order' (Guha, 1992, p. 58) characterised by full employment, self-sufficiency, labour intensive technology, decentralisation, collective land ownership, resource conservation and non-violence (Oommen, 1981) has been particularly central to radical populist thinking on development. Also influential have been romantic accounts of traditional forms of community-based property management that 'apparently not only regulated the harvesting practices of their own members but also excluded others from access to resources controlled by them' (Gadgil and Iyer, 1989, p. 246). Of particular appeal to many populists, however, are accounts that describe the destruction, by colonialism, of an idyllic rural golden past characterised by egalitarian and sustainable common property management systems and local technologies (Guha, 1983; 1989; Arnold and Campbell, 1986; Gupta, 1986; Shiva, 1986;

Singh, 1986; Berkes and Taghi Farvar, 1989a; Gadgil and Iyer, 1989; Kumar, 1991; Gadgil and Guha, 1995).

The Development of Anti-development

Since the late 1970s, in particular, these types of account have received much emphasis as part of a more general reaction against the dominance of Western ideas in development thinking. In particular, the idea that the West has set the standard by which the rest of the world should be measured has received harsh criticism for its ethnocentrism (Fanon, 1967; Marglin, 1990a; Banuri and Apffel Marglin, 1993a; Escobar, 1992; 1995; Alvarez, 1994). Many of these ideas have been articulated as part of an emerging 'tradition' of anti-development which sometimes equates development with a Western disease (Rahnema, 1988; Shucking and Anderson, 1991) that disturbs the authenticity of non-Western, 'non-modern' societies. The recent emergence of 'new social movements' (Scott, 1990) whose leaders specifically demand 'greater autonomy over the decisions that affect their lives' (Escobar, 1992, p. 421), and aim 'to counteract state power by applying their own indicators to assess the desirability of development' (Parajuli, 1991, p. 177) has been seen by many anti-developmentists as providing support for their ideas (Slater, 1985; Esteva, 1987; Parajuli, 1991; Alvarez, 1994; Escobar, 1992; 1995; Sittirak, 1998). According to Sittirak, anti-developmentist writers like Vandana Shiva have 'by eloquently turning the table and showing us a powerfully different perspective, participated in the historical revolutionary process of 'returning the gaze' by reconstructing and thereby demystifying the centre's or development experts' own assumed privilege, which cannot be completed without also retrieving the capacity for generating local knowledge' (Sittirak, 1998, p. 32).

There has also been some cross-fertilisation between older 'new' social movements, particularly the environmental movement, and peasant resistance movements which seek alternatives to the 'ethos of expansion' (Goldsmith et al, 1972) and scientific dominance associated with the Western model of development (Ghai and Vivian, 1992b; Gadgil and Guha, 1992; 1994; Vivian, 1992; Alvarez, 1994; Colchester, 1994; Omvedt, 1994; Peet and Watts, 1996a; Yapa, 1996; Arora, 1997; Karlsson, 2000; Rangan 2000). According to Esteva 'the experiment is over ... development is dead' (Esteva, 1987, p. 137). Anti-development 'not so much a development alternative, as an alternative to development' (Watts, 1993, p. 258) is proposed in its place.

Influential in the evolution of these ideas have been books such as Edward Said's 'Orientalism' (1978) which discusses and critiques Western views and stereotypes of the 'orient', and work by Baudet (1965), Inden (1986), Kiernan (1969), Pagden (1982), Barker et al (1986), Pinney (1988) and others which focuses upon European views about colonial (and neo-colonial) 'others'. Central to this work is the argument that 'without examining Orientalism as a discourse we cannot possibly understand the enormously systematic discipline by which European culture was able to manage - and even produce - the Orient politically, sociologically, ideologically, scientifically, and imaginatively during the post-Enlightenment period' (Said, 1978, p. 3). Particularly revealing in many colonial and early development narratives are orientalist representations of a silent and helpless 'other' requiring translation, vocalisation and presentation by a more objective, scientific and logical Europe. Said's work has also been effective in drawing attention to past and present orientalists who attempt to 'speak out' for their 'others' whilst retaining the 'cultural baggage' of their Western education and social upbringings.

Implicit in much anti-development discourse is the idea that Western representations of non-Western 'others' are dangerous as they provide justification for a post-colonial exploitation of the Third World in the form of development. Equally damaging in the eyes of some are the governments of developing countries themselves, which, taken in by the so called 'dream' of development, subjugate their own 'others' to development's hegemonic discourses (Parajuli, 1991). According to Escobar, for example, development discourse 'has created an extremely efficient apparatus for producing knowledge about, and the exercise of power over, the Third World' (Escobar, 1995, p. 9). Responding to a more general anti-scientific perspective amongst the humanities which portrays an 'archaic picture of science as an unreflexive endeavour in which scientists are ignorant of the contexts in which they work' (Sullivan, 2000 p. 16), many anti-developmentist writers have taken a strongly relativist viewpoint on the linkages between power and knowledge (Foucault, 1990); especially on the validity of scientific analyses of environmental problems.[8]

Populist-cum-anti-developmentist writers including Apffel Marglin and Marglin (1990) and Banuri and Apffel Marglin (1993a) have enlarged upon some of these ideas, pointing to the way in which development is likened to the 'transformation of a child into an adult precisely because adult behaviour is an agreed standard against which to measure the progress of the child' (Marglin, 1990b, p. 2). A central theme in Apffel Marglin and Marglin's book 'Dominating Knowledge. Development, Culture and Resistance' is a contrast between the 'dominant knowledge system of the West, *episteme*, with the traditional knowledge systems or *technai*' of developing countries (Marglin, 1990b, p. 25). The chapter by Nandy and Visvanathan, for example, emphasises the 'imperialistic pretension to universality made on behalf of Western *episteme* and the total inability of its adherents to regard competing systems with anything but contempt, the inability indeed even to contemplate the existence of competing systems' (Marglin, 1990b, p. 25).

An Evaluation of Populism and Anti-developmentism

Many of these radical populist and anti-developmentist critiques have been both beneficial to and influential in changing recent development thinking and practice. Indeed, it is undoubtedly true that they have helped to bring about a 'paradigm shift' away from traditional derogatory perceptions of local knowledges towards more bottom-up, participatory and 'empowering' approaches to development (Ghai and Vivian, 1992b; Vivian, 1992; Scoones and Thompson, 1994a; Agrawal, 1995).

Of particular importance have been the ways in which these discourses have questioned the socio-economic and environmental appropriateness of Western development and highlighted the psychological and material impacts that it can have (through the association of Western technology with social power) upon its recipients (Shiva, 1988; 1991a; Apffel Marglin and Marglin, 1990; Banuri and Apffel Marglin, 1993a). An additional benefit has been the attention drawn to the danger of local knowledges being displaced or de-legitimised by Western 'dominating knowledges' and modern farming practices (Apffel Marglin and Marglin, 1990; Marglin, 1990a; Banuri and Apffel Marglin, 1993a; Scoones and Thompson, 1993; 1994a; Agrawal, 1995; Gupta, 1996).

While this emphasis on local knowledge systems has helped to shift attention towards indigenous communities and marginalised groups, however, it has also 'congealed a series of dichotomies that threaten to become naturalised. "Woman", "indigenous", "community", and "local" have become central as building blocks for specific streams of environmental writings. The populist potential of these constructs is enhanced by using them in opposition to others such as "man", "Western/scientific", "state", and "global"' (Agrawal and Sivaramakrishnan, 2000b, p. 8). Unfortunately, the resulting romanticised focus on the 'cooperative and harmonious ideal promised by the imagery of "community"' (Gujit and Kaul Shah, 1998b, pp. 7-8) can mean that socio-economic inequalities and oppressive social hierarchies are overlooked.

There is also a possibility that this combination of anti-developmentist and romantic populist ideas could both promote the conviction that 'every item of local knowledge about the environment conceals grains of scientific truth' (IDS Workshop, 1989a, p. 36) and greatly overrate the potential of indigenous knowledge for 'solving' the Third World's development and environmental problems (Ostrom et al, 1993).[9] It could also encourage the use of inappropriate practices: a classic example being the common Latin American and African custom of manually extracting and throwing away antibody-rich maternal colostrum because it is regarded as 'dirty' (Scheper-Hughes, 1992). Also, as Lipton (1977) points out, a continued emphasis on indigenous medicines could encourage policy-makers to condemn rural people to ineffective traditional medical care and reserve more costly scientific medicine for urban elites.

Another concern is that populist romanticisations of rural life, indigenous knowledge and local cultures exhibit a kind of 'green orientalism' (Lohmann, 1993) in their assumptions that rural people in developing countries desire a return to traditional agricultural and resource management practices. Indeed, there are numerous discourses on how 'prior to colonialism and industrialisation, India's culture was imbued with a set of beliefs and practices that naturally held human demands on the environment in check, so that populations and their forest environment existed in a kind of ecologically sustainable homeostasis' (Freeman, 1998, p. 56). Yet bearing in mind the repressive nature of many traditional societies (Vivian, 1992) plus their inability to satisfy local requirements for carbohydrates, water and basic health care (Kelkar and Nathan, 1991), this assumption seems rather dubious. During their research in Jharkhand, Jackson and Chattopadhyay found little support for eco-populist discourse about *adivasi* sacralization of nature, as tribals are concerned far less with environmental harmony than they are with land and forest entitlements. Indeed, these authors critique such narratives as 'caricatures of adivasi lives, distortions of actors' own perceptions of their struggles, and as appropriations that make use of adivasis in the project of urban academics and activists seeking an authentic and indigenous critique of development' (Jackson and Chattopadhyay, 2000, p. 151).

In addition, populist romanticisations of indigenous knowledge often ignore the fact that many community-based resource management systems and traditional farming practices have already been undermined as a result of economic development, histories of colonialism, cultural hybridisation or even population increase (Blaikie, 1985; Peet and Watts, 1996a; Sivaramakrishnan, 1999; Karlsson, 2000; Rangan, 2000a; Agrawal and Sivaramakrishnan, 2000a; Nanda, 2001b). When analysed from a political ecology perspective, romanticised narratives of tribals living in harmony with nature can often be seen to 'represent "inventions of tradition", selective presentations of historical and social data in order to advance particular political projects' (Jeffery, 1998b, p. 7). An excellent example of this is Rangan's illustration of how narratives of pre-colonial ecological harmony and moral economy in Uttarakhand have been employed by populist eco-feminists and environmental historians (most notably Vandana Shiva and Ramachandra Guha) 'to eulogize the passing of simple and self-sufficient peasant lifeways' (Rangan, 2000, p. 23) upon contact with the British colonial state's imposition of 'western patriarchy', 'scientific forestry' and 'capitalist maldevelopment'.

From a more theoretical perspective, Shepherd's first model indicates that when population increase or resource degradation occurs beyond a certain point, the institutional costs of maintaining common property resource management cause many systems to break down (Shepherd, 1993a; 1993b). Bebbington (1992), meanwhile, shows that for many indigenous groups in the central Andes of Ecuador, a 'return to the past' may be neither possible nor desirable. This is because land subdivision and the sale of cattle has caused a shift away from traditional food crop varieties grown with organic manure towards more market-oriented intensive farming and a dependence upon chemical fertilisers.

Bebbington also points out that many local people have become very attracted to modernity and affected by its 'devaluation of the past' (Bebbington, 1990b, pp. 9-10): a situation that has created tensions for some of the populist NGOs working with these indigenous groups as it conflicts with their emphasis upon the need to revive more traditional forms of indigenous agriculture as part of a 'cultural determination' strategy. Instead of an emphasis on traditional farming practices, many local people would prefer greater access to modern agricultural technology which would enable them to maintain family incomes through cash cropping, reduce their dependence upon migration and thereby strengthen indigenous culture as a result of their ability to participate more in community life (Bebbington, 1996).[10]

Linkenbach (1998) and Rangan (2000a) write of a similar situation in India's Uttarakhand where many local people feel marginalised by the region's famous Chipko *andolan* (movement) and oppose its representation by writers such as Guha as a 'decentralised, environmentally sensitive alternative to the present patterns of "destructive development"' (Rangan, 2000, p. 42). According to Rangan, local people 'were voicing the need *for* development, they were not *against* it. They did not want to celebrate their marginal status, nor did they want alternatives *to* development. They demanded to be included, made part of the developmental process that they felt had been denied them so far' (Rangan, 2000a, p. 139, emphasis in the original). In the words of one villager, Chipko brought: 'No benefit for the people, nothing for the Forest Department, no good for the forests, no development in the area. Just a bundle of laws around our necks' (Kalbali Khan, quoted in Rangan, 2000a, p. 155). Examples like these obviously raise the question of whether populist representations of what people in poorer countries might want are any more accurate than those found in more traditional development discourses. The resentment caused by Nyrere's attempt to bring about populist development through Tanzania's *Ujamaa* project is another case in point (Kitching, 1989).

One of the main criticisms that can be aimed at the anti-developmentist school, meanwhile, is that its rejection of the West is rather myopic insofar as it refuses to recognise that it has brought any benefits to the Third World. Anti-development also fails to raise counterfactual questions with regard to what the situation in the Third World would be like in the absence of development assistance from the West or Western style development from within. According to Kitching, development is and must be 'an awful process' (Kitching, 1989, p. 195) which requires us to take a responsible course of action by accepting rather than avoiding its implications. In contrast to this, anti-developmentists justify their rejection of development with the argument that it has 'in many ways created the hungry, the illiterate, the marginals, the migrants, those belonging to the informal economy, the excluded women and indigenous peoples' (Escobar, 1992, p. 431). Yet there is plenty of evidence to suggest that this view is both blinkered and unrepresentative of a majority of men and women in developing countries (Singh, 1990; World Bank, 1991).

Much of the popularity of anti-development discourses stems perhaps from their ability to play upon guilt complexes about the privilege and wealth enjoyed by many Western and Westernised Third World elites compared to the poverty of the rural masses (Kitching, 1989). In addition, the perceived humiliation associated with a history of colonial domination (and, many would argue, neo-colonialism and 'dependent development') has also made many Third World intellectuals quite anti-Western and nationalistic in their outlook (see, for example, Nandy, 1983; Shiva, 1988; 1991a; 1991b; Esteva, 1992; Alvarez, 1994; Escobar, 1995). This in turn has evoked feelings of guilt about imperialism in many radical academic and intellectual circles within developed countries. Indeed, many of the most vocal proponents of anti-development are based in the West, including Apffel Marglin and Marglin, Parajuli, Pieterse, Escobar, Norgaard, Sachs and Slater.

Amongst more mainstream development thinkers, there is much scepticism about these discourses. Kitching, for example, provides a compelling argument for the non-viability of populism as an 'alternative development strategy to large-scale industrialism and urbanisation' (ibid., p. 180). Instead, he argues that some form of industrialisation, preferably starting with the expansion of primary products exports, is necessary for 'a really effective "agricultural" or "rural development" effort which will improve the human well being of peasants and other rural dwellers' (ibid., p. 178). He also makes the point that if the populist 'syndrome' is emphasised further, any benefits that come from the rural sector are likely to be squandered on the increase in the number of state departments deemed necessary to direct research into rural development and local knowledge systems.

Given the extreme romanticisation inherent in much radical populist discourse on Third World livelihoods plus the tendency of many anti-developmentists to 'speak for' their 'others' without consulting them first, the practical value of these approaches must surely be in doubt. This is particularly the case for what might be called 'supply side' populism (Richards, 1990b), promoted and imposed by radical academics, development workers and (increasingly) NGOs. More promising in many respects are the 'demand side' (Richards, 1990b) forms of populism (or development) that have been articulated as part of broader new social movements (Scott, 1990). In many cases, these have consisted of attempts by local people to organise themselves politically (sometimes with the help of NGOs) to achieve greater local control over socio-cultural, economic and environmental resources as well as their future development (Ghai and Vivian, 1992b; Vivian, 1992; Gadgil and Guha, 1992; 1994; Peet and Watts, 1996b).[11] As examples from Chapter Four and empirical data from Jharkhand will show, such initiatives have often shown themselves quite capable of determining their own development priorities and the extent to which (if at all) they want help in achieving them.

Peasant Resistance, Indigenous Knowledge and Politics

In some cases, local socio-cultural and technical knowledges have been mobilised as part of a wider strategy of peasant resistance (Scott, 1985) against the state and/or state-imposed development programmes. This type of resistance has often been accompanied by demands for autochthonous development, increased autonomy from the state, greater power over the resources that local people need for subsistence and increased access to employment (Blauert and Guidi, 1992; Gadgil and Guha, 1992; 1994; Ghai and Vivian, 1992b; Schuurman, 1993; Escobar, 1995; Kaufman, 1997a; Rangan, 2000a).

Bebbington's (1990a; 1990b; 1992) work on the demands for more locally-defined forms of development by indigenous people in Ecuador provides a good illustration of this type of peasant resistance. There, increased levels of access by indigenous groups to modern agricultural technology have been associated by them with a rise in socio-cultural standing which has allowed them to start to live 'like white people' (Bebbington, 1990b, p. 11) and has encouraged them to 'claim the full rights of citizenship from which they have so long been excluded' (Bebbington, 1990a, p. 256).[12] Demanding from the state equal access to modern technology (which is associated with social prestige and power) is one way of asserting this desire and it is one which may have radical political implications in future years (Bebbington, 1990b).

A famous Indian case of peasant resistance to restrictions on local forest management systems is Chipko which is often represented as a struggle between India's (virtuous) forest citizens and the (repressive) state. According to Guha, the movement can be seen as the culmination of a long history of opposition to commercial forestry practices which, by limiting local people's access to forests and forest produce, deprived them of their right to subsist (Guha, 1989; 1993). The movement itself began in the early 1970s as a protest against the state's refusal to recognise local people's claims to forest produce and its preference for auctioning timber from local forests to outside contractors.

The most widely accepted accounts of the movement state that it commenced in early 1973, when the Dashauli Gram Swaraj Sangh (DGSS), a voluntary organisation based near Gopeswar, asked the Forest Department for an allotment of ash trees to make agricultural instruments (Guha, 1993; Pathak, 1994). Its request was turned down, but soon afterwards, the DGSS found that the same ash trees had been allocated to a sports equipment company called Symonds. Its members therefore decided to hug the trees to prevent contractors from felling them.[13] In spite of the DGSS's continued protests, however, the auction system continued. In November 1973, the Forest Department made plans to fell part of Reni forest, despite the fact that villagers were convinced that forest felling had been responsible for a devastating flood in Reni village in 1970 (Bandyopadhyay, 1992; Guha, 1989; 1993; Pathak, 1994). A demonstration was held to oppose the felling but the Forest Department tried to break up local opposition by persuading Chandhi Prasad Bhatt, the leader of the DGSS, to stay in Gopeswar to meet the Conservator of Forests. At the same time, the men of Reni village were called to Chamoli to collect long overdue compensation for land acquired by the state (Pathak, 1994). While the coast was clear, the contractors moved into Reni forest, but they were seen by the head of the village women's committee who urged other village women to go into the forest to hug the trees and plead with the contractors not to fell them. Eventually the contractors were forced to withdraw (Guha, 1989; 1993; Bandyopadhyay, 1992; Pathak, 1994).

After the Reni incident, Chipko spread to Tehri and Kumaun and after the 1974 Nainadevi fair at Nainital, Chipko developed into a more radical populist movement under the leadership of Sundarlal Bahuguna (Guha, 1993; Pathak, 1994). Subsequently, forest auctions were opposed in several places and in 1977, Chipko won its first victory when the committee investigating the Reni flooding concluded that it had been caused by deforestation in the neighbouring catchment. In 1980, the Forest Conservation Act further restricted the conversion of forests to non-forest uses and a 15 year ban was imposed on the felling of green timber above 1000 metres in the Himalayas. The 1988 amendment to the Act made it illegal to undertake afforestation on state-owned land without permission from the Ministry of Environment and Forests (Pathak, 1994; Rangan, 2000a).

Chipko's success captured the imagination of both the national and international media and attracted the interest of established environmental groups. As these linkages increased and the movement matured, it developed different internal emphases which, according to Guha, 'reflect in a microcosm, different strands in the modern environmental debate' (Guha, 1993, p. 103). Firstly, there is Chandi Prasad Bhatt's broadly populist emphasis on a socially just and environmentally sensitive route to development through the use of appropriate technology. Secondly, there is Sunderlal Bahuguna's proposal for a more fundamental populist alternative to development in the form of a 'return to pre-industrial modes of living' (Guha, 1993, p. 103). Thirdly, there is the Marxist ideology of the Uttarakhand Sangharsh Vahini which has shifted away from a predominantly environmental focus towards an emphasis on greater equality and economic redistribution through grassroots challenges to capitalist penetration (Guha, 1993).

Rangan develops these ideas, using a political ecology approach to investigate Chipko's transformation into myth and simultaneously highlighting the need for caution in the (mis)representation of local development priorities. Moving away from the simplistic state-citizen dichotomies that dominated many early accounts of Chipko, she shows sensitivity to the many different voices within the movement and highlights how the views of elite (often strongly populist and/or post-materialist) leaders came to dominate those of the main (or original) participants.[14] In exposing Chipko's different environmental narratives as *fictions*, 'produced for social and political purposes' (p. 41), Rangan makes a striking contrast between Chipko's success in the eyes of Gandhian activists, environmentalists, eco-feminists and populist intellectuals and its failures from the viewpoints of Garhwali villagers. Taking a critical look at anti-development discourse and its romanticisation of new social movements, she documents the employment and infrastructural development opportunities denied to Garhwali villagers by Chipko's tighter environmental legislation. She also highlights local demands for 'regional economic development, rather than environmental protection' (p. 42), arguing that Chipko's failure to achieve local development goals has been a major factor in the demands of the Uttarakhand Kranti Dal (Uttarakhand Revolutionary Party) for a separate hill state.

To most outsiders, however, such internal conflicts have been concealed by Chipko's international success in appealing to discourses of environmental catastrophe and populist emphases on the potential of grassroots activists (especially women) in promoting an alternative environmentally-sensitive development pathway. Indeed, to many old and new social movements, Chipko has come to be seen as a successful challenge to the dominant ideologies of state-imposed development and as an alternative (demand side populist) model of socially just and environmentally sustainable development.

Another Indian example of the politicisation of indigenous knowledge systems as a form of peasant resistance can be found in Jharkhand where my own fieldwork has been carried out. There, as in a number of other tribal areas of India (Singh, 1982; Hardiman, 1987), a local sense of identity and place has been mobilised politically by the region's tribal and, more recently, non-tribal inhabitants. Proving themselves capable 'not only of resisting the dominant knowledge but also of articulating their own world views' (Parajuli, 1991, p. 185) they demanded the separation of the Jharkhand region from the State of Bihar. Initially, their arguments for this have rested upon political, economic and cultural grounds although more recently, environmental concerns (particularly those relating to the area's forest lands) have found an important place in both the region's consciousness and politics.

In the same way that the Chipko movement can be seen as a form of 'opposition to outside interests trying to appropriate local common property resources' (Berkes, 1989a, p. 237), the Jharkhandis' refusal to respect the privatisation of forest resources (and the resulting conflict with the Forest Department) represents, in part, an assertion of traditional woodland use rights. Several studies in Bihar and Orissa have shown that many local (particularly tribal) communities have used their indigenous silvicultural knowledges and forest management strategies as a means of challenging both the Forest Department's existing ideology of scientific forest management and the illegal (but unstopped) felling carried out by corrupt forest officials and commercial contractors (Fernandes, 1983; Mehrotra and Kishore, 1990; Kelkar and Nathan, 1991; Kant et al, 1991).

One of the main ways in which they have done this is to control their own and restrict outsiders' use of forests through the establishment of locally-based forest protection and management systems. Most of these schemes are organised by village level institutions and are often based upon a strong sense of allegiance to traditional forms of tribal leadership. A number of them, however, have been established with the support and encouragement of various NGOs and Jharkhand political parties; groups which were also influential in mobilising a Jharkhandi sense of place and identity which stands in opposition to the existing ideology of state-sponsored development through industrialisation.

Conclusion

These examples show just some of the ways in which local people have succeeded in finding a way out of their own development 'impasse' by determining locally-based development strategies that reflect their main priorities. They also show how skilful local activists can be in playing the 'environmental' or 'indigenous' card (often forging links with NGOs and 'old social movements' in the process) to ensure that their concerns are heard on wider political, economic and development agendas (Ghai and Vivian, 1992b; Gadgil and Guha, 1992; 1994; Vivian, 1992; Colchester, 1994; Omvedt, 1994; Karlsson, 2000; Rangan, 2000a). In particular, they draw attention to the ways in which local people have used not only their indigenous technology but also the socio-cultural aspects of their knowledge systems as means of asserting their identity and right to self-determination.

In addition to these grassroots challenges to traditional development models, criticism from radical academics, development practitioners and NGOs has had a significant influence on global development thinking and policy-making over the last fifty years. Most donor agencies have now moved away, in theory at least, from traditional top-down and transfer of technology approaches, adopting instead more bottom-up and participatory approaches which emphasise the need to understand and build upon local agro-ecological knowledge and resource management systems. More recently, this shift has been taken a step further as policy-makers have stressed the need for local empowerment as well as participation.

As Chapters Four and Five will illustrate, these shifts in development thinking have also been taken on board at the national level by government policy-makers. In India, populist and eco-feminist emphases on indigenous agro-ecological knowledges and resource management strategies have stimulated interest in traditional agricultural practices and community resource management as means of achieving more sustainable and appropriate environmental development. Efforts by intellectuals and action groups to highlight the shortfalls of India's agricultural and forest development policies, meanwhile, have been influential in encouraging the adoption of more participatory and gender-sensitive approaches (Corbridge and Jewitt, 1997). Before going on to examine this cross-fertilisation of development theory and practice in India along with its practical implications for environmentally-oriented policy-making, Chapter Three will take a closer look at the theoretical evolution of gender-environment debates and their influence on mainstream development thinking.

Notes:

[1] Although many bottom up approaches are broadly populist in orientation, they still belong mainstream development thinking and practice and can therefore be differentiated from more romantic or radical populist approaches that often critique and influence mainstream development thinking but are not part of it. The term populism here is used in the context of a broadly defined development strategy based on small-scale, decentralised and environmentally sensitive programmes designed in opposition to industrial capitalism (Kitching, 1989).

[2] A good synopsis of favourable conditions for sustainable common resource management is provided by Arun Agrawal who draws together the work of Ostrom, Wade and Baland and Platteau but argues for 'a shift towards comparative and statistical rather than single-case analyses' (Agrawal, 2001, p. 1649).

[3] During the 1980s in India's semi-arid regions, landless and resource poor families obtained over 90% of their fuel wood and between 69-89% of their grazing requirements from commons (Jodha, 1986). Firewood provided between 65-67% of the total domestic energy in hill and desert areas of India (Agarwal, 2001).

[4] Even shifting cultivation which has long been criticised for its environmental damage (see, for example, Stebbing, 1922) has been found by more recent research to be ecologically sustainable if suitable rotation cycles (of around ten years) are maintained. Indeed, the abundance of weeds in shifting cultivation plots that used to be seen as evidence for it being a 'lazy form of cultivation' (Stebbing, 1939, p. 111) are now recognised for their role in encouraging nutrient cycling and reducing soil erosion (Argawal and Narain, 1992; Ramakrishnan and Patnaik, 1992).

[5] The FAO, for example, defines participation as 'a process of equitable and active involvement of all stakeholders in the formulation of development policies and strategies and in the analysis, planning, implementation, monitoring and evaluation of development activities' (FAO, 2001, p. 1). It goes on to state that it is necessary for disadvantaged stakeholders to be 'empowered to increase their level of knowledge, influence and control over their own livelihoods, including development initiatives affecting them' (FAO, 2001, p. 1).

6 Other criticisms of agrarian populism include its neglect of political economy perspectives; particularly its failure to recognise that problems such as exploitative land tenure relations, market dependencies and state-minority group tensions play a significant role in Third World poverty and low agricultural productivity (Bebbington, 1992; Redclift, 1987; Watts, 1983; 1989; Gubbels, 1994; Matose and Mukamuri, 1994). It frequently also takes a simplistic view of 'empowering' participatory approaches which fail to acknowledge the intra- and inter-community inequalities that exist in local people's possession of and control over agro-ecological knowledge (Scoones and Thompson, 1993; 1994b; Bebbington, 1994a; Blackburn, 1998; Blackburn with Holland, 1998b; Crawley, 1998; Cornwall, 1998; Gujit and Kaul Shah, 1998; Sarin, 1998b).

7 Indeed, Warren et al's (1995) follow-up volume to Brokensha et al's (1980) book is entitled 'The Cultural Dimension of Development: Indigenous Knowledge Systems', yet pays very little attention to these issues.

8 As Nanda points out, the 'danger of social constructivism and the postmodern theories of science is that they deny that we can ever break out of the prison of our myths, biases and cultural assumptions: for them, even the best attested knowledge at any time is "local knowledge" of a particular place, time and people. This idea of all science as local science or "ethnoscience" opens the way to indigenist defence of traditions as legitimate sciences' (Nanda, 2001a, p. 2557). She goes on to argue that Indians 'cannot afford to forget or trivialise the liberatory potential of knowledge. This connection has been lost in the unilateral collapse of knowledge into power and ideology of the west' (ibid., p. 2559).

9 Meera Nanda is more outspoken in her critique of the populist romanticisation of local knowledges, stating that such knowledges are 'more often than not, patently irrational, obscurantist and downright oppressive to the same subaltern on whose behalf these intellectuals claim to speak' (Nanda, 2001a, p. 2554).

10 Bebbington argues that '*runas* in Chimborazo do not have time to wait for the dawn of new utopias. They demand liberation from where they are now' (Bebbington, 1996, p. 105).

11 According to Linkenbach, 'ecological movements are not creating a new economics for a new civilization, they are not presenting a solution for the crisis of the modern world, and they do not have the capacity ... for ending development. But they can show the difficulties, shortcomings and limited scopes of the dominant as well as the alternative models for development at the level of action' (Linkenbach, 1995, pp. 81-2).

12 Presumably, however, such actions would be regarded by poststructuralist and anti-development writers as examples of these peasants having been 'discursively regulated' into 'replacing their aspirations with Western mimicry' (Peet and Watts, 1996, p. 17).

13 The movement got its name from the imperative (*Chipko!*) of the Hindi verb to hug, stick or adhere (*chipakna*).

14 Of particular interest is the shift by both Bhatt and Bahuguna from their initial protests against the Forest Department's use of outside contractors (depriving local people of income and employment), to later emphases on the spiritual and ecological role of forests and the subsistence needs of local people (Rangan, 2000).

3 The Adoption of Gender Issues into Development and Environmental Policy-making

Introduction

Until the establishment of 'women in development' (WID) approaches in the early 1970s, much of the emphasis on grassroots development and indigenous agro-ecological knowledge was strongly male-biased; primarily because of the relative invisibility of women's work to outsiders. Thereafter, increasing attention was drawn to the unequal impacts of development on women and the equality- (and later efficiency-) related consequences of excluding women from the development process (Moser, 1993; Braidotti et al, 1994).[1] In particular, WID researchers highlighted women's role as major subsistence producers with well developed systems of agro-ecological knowledge. They also increased awareness of prevailing gender divisions of labour in the South that often gave women a double or even triple work burden consisting of reproductive (childbearing and rearing) work, subsistence production (often including unpaid community management duties) and, for many women, paid employment (Boserup, 1970). Further research during the UN Decade for Women (1976-85) highlighted persistent gender inequalities in sex ratios, divisions of labour, participation and remuneration in the labour force, inheritance systems, residence patterns and control over important subsistence resources (including land), plus major intra-household inequalities in the allocation of income, food, health care, education and other basic resources (Boserup, 1970; Harriss and Watson, 1987; Sen and Grown, 1987; Moser, 1993; Braidotti et al, 1994; Wieringa, 1994; Agarwal, 1992; 1997; Razavi, 1997).

Echoing the wider environmental concerns of the time, WID stimulated investigations into the differential impacts of resource shortages on women and men which showed that women often bear the brunt of environmental degradation: primarily as a result of prevailing gender divisions of labour that identify women as family subsistence providers and make them responsible for collecting environmental resources such as fuel wood, fodder and water (Agarwal, 1992; 1997; Jackson, 1993a; 1993b; Mies and Shiva, 1993; Moser, 1993; Braidotti et al, 1994). WID research on environmental crises in the developing world, meanwhile, documented how environmental degradation contributes to women's drudgery by increasing the time taken to collect vital subsistence resources such as fuel wood and water (Eckholm, 1984; Agarwal, 1986a; Dankelman and Davidson, 1988; Braidotti et al, 1994; Moser, 1993). Increasingly, this emphasis on women as victims of environmental problems coupled with media images of 'poor women from the South, burdened by heavy loads of fuel, fodder and water, against a backdrop of barren landscapes' (Braidotti et al, 1994, p. 87) helped to spawn the view that women could make an important contribution to environmental protection.

The Influence of Eco-feminism

To some extent, WID helped to stimulate further interest in eco-feminist ideology which posits a close bond between women and nature, and, in the more radical 'spiritual' form of eco-feminism, links women's supposedly innate connection with nature to their cycles of menstruation and birth (Ortner, 1974; Daly, 1978). Also, as mothers, and especially if they are victims of environmental degradation themselves, women are thought to be more concerned about the impact of environmental problems on the next generation and are therefore more inclined to mobilise in support of environmental protection and regeneration. Indeed, much eco-feminist ideology is premised on the belief that women's environmental knowledges are better developed than those of men.

Third World women are seen to be particularly vulnerable to the destruction of the environmental resources that they need for 'staying alive' (Shiva, 1988). A major protagonist of this view is Vandana Shiva who suggests that women are able to offer an alternative, more co-operative, nurturing and caring relationship with nature based on what she calls the 'feminine principle' or *prakriti*. Central to her argument is the idea that women 'are an intimate part of nature, both in imagination and practice. At one level, nature is symbolised as the embodiment of the feminine principle, and at another, she is nurtured by the feminine to produce life and provide sustenance' (Shiva, 1988, p. 38). She goes on to suggest that in addition to having better-developed knowledges than men, the stability, diversity and sustainability inherent in the feminine principle enables women to live in harmony with the natural environment 'as sylviculturalists, agriculturalists and water managers, the traditional natural scientists' (ibid., p. 41).[2] Support for these ideas comes from her claim that many agricultural innovations from all over the world were made by women. Some of the main examples that she cites include mulching, irrigation, the propagation of seedlings from cuttings, the construction of seedling beds, the creation of the hoe, simple plough and spade and the domestication of many food crops (including wheat, rice, barley, maize, millet and sorghum).

'Women, Environment and Development' (WED) Approaches

Influenced by such discourses and responding to concern by donor agencies and NGOs about women's role in development, these ideas helped to stimulate the development of a 'women, environment and development' (WED) perspective which posited a 'special link' between women and the environment (Braidotti et al, 1994; Joekes et al, 1994; Leach, Joekes and Green, 1995). By the late 1980s, these accounts had merged into a broader emphasis on the role of women as the primary natural resource users, environmental custodians and possessors of highly-developed systems of environmental knowledge. They emphasised the 'closeness to nature' and environmental knowledges that these responsibilities bring, arguing that 'women's subsistence concerns make them the agents for conservation and this will be good both for them and for the environment' (Leach, Joekes and Green, 1995, p. 2). WED discourse also argued that women are more 'community spirited' than men and their concern for environmental protection forms an extension of their wider role as community carers (Women's Environment Network, 1989). According to Braidotti et al, WED accounts 'depict women as privileged environmental managers because of their intimate knowledge of natural processes due to their closer relationship with nature: therefore women are seen as the answer to the crisis. Women have the solutions; they are privileged knowers of natural processes' (Braidotti et al, 1994, p. 97).

Reflecting this view, significant recent attention has been placed on the need to 'empower' women by challenging unequal social relations to enable their greater participation in development (Wieringa, 1994; Braidotti et al, 1994; Mayoux, 1995).[3] This shift has developed in tandem with 'gender and development' (GAD) initiatives which focus primarily on unequal social relations and the need to increase women's participation in development (Braidotti et al, 1994). In addition to placing women at the forefront of environmentally-oriented development projects, empowerment initiatives aim to give them greater opportunities to determine their own futures (Moser, 1993; Mayoux, 1995; DfID, 1997; 2001). DfID's new strategy for 'Poverty Elimination and the Empowerment of Women', for example, 'underlines clear and direct links between the empowerment of women and the reduction of poverty. Upholding the rights of women as full and equal members of society not only brings benefits to them, but to the whole development process' (DfID, 2001, p. 44).

Echoing this theme, WED's emphasis on the benefits of involving women (as both the main victims of environmental degradation and custodians of environmental resources) has highlighted an appealing complimentarity of interests that has found much popularity within both sustainable and participatory development discourses. Indeed, the adoption of certain parts of eco-feminist ideology into WED discourse and thereafter into development policy-making is clearly visible. It is also possible to trace how eco-feminist assumptions that women have well developed environmental knowledge systems and 'relate to the environment positively except when forced by poverty to do otherwise' (Jackson 1993a, p. 398) have been taken on board by development policy makers and donors 'which see the interests of women and the environment as coterminous' (Jackson 1993a, p. 398). Cecile Jackson, for example, documents how the World Bank has identified 'win-win' gender-environment policies (World Bank 1992b; 1994; 1995a) which highlight 'synergistic interventions' that 'can effectively tackle poverty alleviation, women and development, and environmental conservation goals simultaneously' (Jackson, 1993a, p. 398).

This trend is also apparent in those sustainable development discourses which emphasise the key role that women's environmental stewardship - alongside their more traditional 'social reproduction' roles - can play in achieving more equitable and environmentally sensitive development (UNCED, 1992; Braidotti et al, 1994). As a result, women have been portrayed as *the most valuable resource* in the process towards achieving sustainable development' (Braidotti et al, 1994, p. 97 emphasis in the original) and increasing attention has been paid to the need for their participation in the development process (Braidotti et al, 1994). One of Agenda 21's objectives regarding 'global action for women towards sustainable and equitable development', for example, was to 'establish mechanisms to assess the implementation and impact of development and environment policies and programmes on women and to ensure their contributions and benefits' (UNCED, 1992, p. 191). In response to this, the Food and Agriculture Organisation (FAO) has promoted the mainstreaming of gender in its programmes through the implementation of the FAO/Women-in-Development (WID) Plan of Action (FAO, 1999). Likewise, DfID has stated that women 'have a vital role in environmental management and development. Their full participation is therefore essential to achieve sustainable development' (DfID, 1997).[4] The World Resources Institute, meanwhile, stressed that 'for reasons of both equity and sustainable development, [women] must be given the opportunity to become full and equal participants in their societies and economies and in all development efforts' (ibid., pp. 43-44). In a similar vein, DfID and major NGOs like Oxfam and the International Institute of Environment and Development (IIED) all emphasise women's provision of 'primary environmental care' as an important route to sustainable development (Jackson, 1994; Leach, Joekes and Green, 1995). DfID's White Paper on International Development also states that the goal of 'achieving equality between men and women is based on principles of human rights and social justice. Empowerment of women is moreover a prerequisite for achieving effective and people-centred development' (DfID, 1997).

An Evaluation of the 'Special' Women-environment Link

In support of the eco-feminist/WED proposition of a special relationship between women and the environment, there are numerous Asian and African examples of subsistence agriculture and the collection of water, fuel wood, fodder and 'minor' or non-timber forest products (NTFPs) being culturally defined as 'women's work' (Dankelman and Davidson, 1988; Shiva, 1988; Centre for Women's Development Studies, 1991; Mies and Shiva, 1993; Sarin, 1995). There is also empirical support for the related argument that women have quite specialised and well developed environmental knowledge systems. Indeed, there is a tendency in many Third World countries for gender divisions of labour to make women responsible for the lion's share of subsistence production: a situation that places them almost unavoidably in contact with the natural environment (Centre for Women's Development Studies, 1991; Braidotti et al, 1994; Dankelman and Davidson, 1988; Jackson, 1993a; Sen and Grown, 1987; Shiva, 1988; 1993b).[5] A good example is Boster's (1984) description of how Aguaruna Jivaro women in Peru often cultivate over 100 varieties of manioc purely for the sake of maintaining diversity.

In many respects, eco-feminist/WED emphases on women's 'subsistence perspective' and innate link with nature reflect the current popularity of radical populist ideology in both environment and development thinking. Much support for this has been gained from populist eco-feminist writers seeking to challenge what they see as the Enlightenment project's dominating patriarchal world view and its exploitation of both women and nature (Merchant, 1989). Vandana Shiva takes this critique a stage further to include the 'destructive' scientific knowledge that underlies most modern development planning. In contrast, she argues that the traditional dependence of rural people (especially women) upon local environments has resulted in the generation of highly developed environmental knowledge and management systems that are capable of 'satisfying basic needs through an integrated ecosystem managed for multipurpose utilisation' (Shiva, 1988, p. 94). Along with a number of other romantic populist writers, Shiva looks primarily to traditional systems of environmental management for solutions to current agro-ecological problems. In many ways, women as guardians of nature and indigenous knowledges and rural people generally 'have come to symbolise the aspirations of those wishing to return to an earlier, less complicated, more ecological state of existence' (Agrawal and Sivaramakrishnan, 2000b, p. 8).

As Bina Agarwal points out, however, 'by romanticising specific periods of the past, eco-feminism has tended to obscure, rather than grapple with, the political economy of factors underlying women's subordination, nature's degradation, and their interlinks' (Agarwal, 1998, pp. 56-7).[6] And unlike WID which has strong academic roots, much WED discourse is based on anecdotal evidence of women's involvement in environmental protection and management which has been translated into rather simplistic assumptions about a special women-environment link. While recognising the gender divisions of labour which make women responsible for 'subsistence production' and the collection of forest products, fodder and water, WED does not seek to analyse the socially constructed nature of these responsibilities (Jackson, 1993a; 1993b).

So although the adoption of eco-feminist/WED discourse by major development policy-makers marks an important step forward, there is a danger that the mainstreaming of gender issues may conceal the difficulties that many women face in developing agro-ecological knowledge and participating in development initiatives (Razavi and Miller, 1995). Indeed, instrumental WED approaches have received much criticism for their tokenism, failure to differentiate women's experiences and simplistic conceptualisation of a special women-environment link (Agarwal, 1992; 2001; Jackson, 1993a; Braidotti et al, 1994; Humble, 1998; Locke, 1999). Drawing attention to the 'striking neglect of a gender perspective on who participates, what effects this has, and what factors constrain participation' (Agarwal, 2001, p. 1624) in the mainstream literature on participation, Bina Agarwal identifies four levels of participation ranging from nominal (i.e. membership of a group) through to interactive (empowering) participation characterised by having an input into a group's decisions. Finding that women's so-called participation in community-based forest management initiatives within India rarely goes beyond nominal group membership, she recommends the use of bargaining frameworks as analytical tools for examining the possibility of positive change at the state, community and individual family levels.

Echoing a wider disenchantment with simplistic participatory and community-oriented discourses, many academics have cautioned against the special link idea, emphasising instead the need to examine the underlying socio-cultural factors and political ecologies that regulate women's access to and control over environmental resources (Agarwal, 1992; 1998; 2001; Jackson, 1993a; 1993b; 1994; Braidotti et al, 1994; Leach, Joekes and Green, 1995; Cornwall, 1998). Cecile Jackson argues that 'there cannot be a special relationship between women and environments because women are not a unitary category, and their environmental relations reflect not only divisions among women but also gender relations and the dynamics of political economies and agroecosystems' (Jackson, 1993b, p. 1950). In a similar vein, Bina Agarwal criticises eco-feminist discourse for its failure to consider the 'possible gap between women having an interest in environmental protection, and their ability to translate that interest into effective action' (Agarwal, 1998, p. 72).

Whilst many women from the South 'find identifying with nature less difficult and hence use this argument as a basis for their struggles' (Braidotti et al, 1994, pp. 98-99), Northern feminists have traditionally resisted attempts to 'naturalise' women in this way. In particular, there has been concern about the tendency of such traditional male/female dichotomisations to simultaneously universalise women's experiences and idealise stereotypical behaviour patterns that allow (if not invite) female subordination (Braidotti et al, 1994; de Beauvoir, 1988; Jackson, 1993a; 1994; Nanda, 1991). Indeed, many feminist scholars see eco-feminism as an essentially 'retrogressive and contradictory expression of feminism' (Joekes et al, 1994, p. 138) and question its narrow yet 'pervasive perception that women are closer to nature' (Jackson, 1993a, p. 399) than men.

In particular, the forms of eco-feminism that relate women's experience to their biology have been criticised for falling into the essentialist trap of universalising women's experiences. They also ignore the fact that perceptions of nature, culture and gender are socially constructed and culturally relative (Agarwal, 1992; Braidotti et al, 1994; Kaul Shah and Shah, 1995) as well as being subject to constant re-definition and bargaining within and between genders and through time (Leach, 1991; Jackson, 1995; Cornwall, 1998; Locke, 1999; Agarwal, 2001). Indeed, by identifying and homogenising women as 'other', these eco-feminist discourses help to prevent a more informed gender analysis that could differentiate between women of different economic classes, ethnic groups, religious affiliations, geographical areas and so on (Agarwal, 1992; 1998; 2001; Jackson, 1993a; Mayoux, 1995; Agrawal and Sivaramakrishnan, 2000a). They also discourage an investigation of women's exploitation by other women on the basis of factors like age, marital status and social standing. Cornwall suggests, for example, that 'although it may sit uneasily with a feminist outlook, there are women who are just as 'patriarchal' as the patriarchs themselves. These are women who use whatever strategies they can to enhance their position and power, acting at times as if they were powerful men and alongside certain kinds of men at others ... Women like these disappear quietly in representations of woman-as-oppressed' (Cornwall, 1998, p. 47).

Perhaps the most outspoken attack on the essentialist women-environment link, however, is Meera Nanda's damning critique of Vandana Shiva's brand of populist eco-feminism which she accuses of 're-creating a stereotypical feminine mystique' which is in danger of embracing 'all the age-old misogynist myths - only this time around they come with a positive valence' (Nanda, 1991, p. 47). Nanda goes on to emphasise the danger in highlighting women's 'difference' before their equality (in the form of access to education, jobs or political participation through to basic needs such as calorie intake and health care) has been achieved. She warns that without 'an improvement in gender equality, women will be forced to continue to do the "invisible" work of nature' (Nanda, 1991, p. 48): a situation that is unlikely to be accompanied by an increase in their 'economic value' (Boserup, 1970) and associated political power. Indeed, by romanticising women's 'subsistence perspective' and 'invisible' labour, populist eco-feminist discourse tends to conceal the repressive social structures that make women responsible for this work.

These critiques are relevant and link through to wider debates over gender divisions of labour and the influence that these have on the development of environmental knowledges. For example, it is widely accepted that most societies have fairly well-defined divisions of responsibility between men and women. Some tasks are fairly loosely identified as women's or men's work whereas others are more strictly gendered and may even be enforced by taboos which restrict women's (and very rarely men's) activities. Perhaps the best known examples of this are the South Asian taboos against women hunting or undertaking key agricultural activities like ploughing (Roy, 1915; 1928; Blaikie, 1985; Henshall Momsen, 1991; Sontheimer, 1991; Agrawal, 1995). As a result, simply highlighting the gender divisions of labour which give women responsibility for 'subsistence production' can conceal the socio-political nature of task allocation as well as its frequent re-negotiation in response to changing socio-economic status, intra-household power structures, etc. (Jackson, 1993a; Braidotti et al, 1994; Agarwal, 2001).

Not surprisingly, these factors have a substantial bearing on the environmental knowledges that women can actually obtain (or articulate) and suggest a rather different picture to the romanticised eco-feminist view of the 'feminine principle' as an alternative, environmentally sensitive world view or approach to development. So, although many women may indeed possess detailed environmental knowledges, this usually has less to do with an innate link with nature than with wider gender divisions of labour plus a whole series of cultural factors which determine the extent to which women are in contact with environmental resources. A study of !Ko bush people (Heinz and Maguire, n.d.), for example, initially expected women to have better developed botanical knowledges than men as they have primary responsibility for gathering activities. During fieldwork, however, it was found that the men's knowledges were equal to those of the women due to the greater areas travelled by men during hunts and other excursions.

Elsewhere, cultural restrictions on women's mobility outside the home can influence their intra-community bargaining positions as well as the extent to which they come into contact with environmental resources (Agarwal, 2001). Local inheritance systems and marital residence patterns, meanwhile, have a fundamental impact on land management and control. In societies where inheritance is patrilineal, for instance, the management of land and environmental resources is often strongly male dominated. Patrilocal residence patterns often reinforce this tendency as women frequently forego their share of their parents' property when they marry, often receiving a dowry instead: a situation that can exacerbate their sense of 'environmental dislocation' (Jackson, 1993a, p. 410) and undermine their intra-household bargaining position (Kandiyoti, 1988; Agarwal, 2001). In addition to having a lower degree of familiarity with and knowledge about local environments than men (Agarwal, 1997a), such women often see themselves as outsiders in their marital places (Jackson, 1993a).

Another important issue that has been largely ignored in most eco-feminist/WED discourse is the fact that women are sometimes responsible for environmental degradation (Jackson, 1993a; 1993b; 1994; 1995; Sarin, 1995; Locke, 1999). Fuel wood collection, in particular, is an activity that has caused immense ecological damage in many areas of the developing world and although most of the wood is gathered for subsistence (cooking and heating) purposes, significant amounts are also sold or used for commercial purposes (Jackson, 1993a; 1994; Joekes et al, 1994). In much of southern Africa, for example, beer brewing and pot firing are important income earning activities for women: both of which use a lot of fuel wood and contribute to local forest decline (Jackson, 1993a; 1994). And while fuel wood collection in many parts of Africa is seen primarily as a women's task, tree planting is largely carried out by men as it tends to be associated with land ownership. Indeed, in Kenya, tree planting by women is considered a very subversive activity because it represents a claim on men's property (Jackson, 1993a; Joekes et al, 1994).

Rather than having an innate or special link with nature, therefore, it could be argued that women's environmental interest is more often 'rooted in material reality - in their dependence on and actual use of natural resources for survival' (Agarwal, 1992, p. 149). And if this is so, the celebration of women's environmental work by eco-feminists and proponents of the WED approach is in danger of supporting the existing social structures that restrict women to '"invisible" work (along with earthworms and such) in "partnership with nature"' (Nanda, 1991, p. 51). It is also running the risk of overrating women's environmental interest while simultaneously 'invisibilising' men's agro-ecological knowledge and role in environmental management (Braidotti et al, 1994; Kaul Shah and Shah, 1995; Leach, Joekes and Green, 1995; Locke, 1999; Agarwal, 1998; Gujit and Kaul Shah, 1998a; Mawdsley, 1998).

Conclusion

In many ways, the shifts that have taken place in gender-environment discourse over the last thirty years reflect changes in the wider political ecology of development thinking over this period (Blaikie and Brookfield, 1987; Bryant, 1992; 1997; Rocheleau et al, 1996; Bryant and Bailey, 1997; Keil et al, 1998). Responding to the traditionally male-dominated nature of development planning and policy-making, WID set out in the early 1970s to address the invisibility of women's work and highlight gender inequalities (Escobar, 1995). Concern about fuel wood shortages and the tendency for women to bear the brunt of environmental degradation followed in the wake of the oil crisis. Interest in women's role as environmental custodians, meanwhile, linked through to wider debates about sustainable development. As eco-feminism gained popularity in tandem with radical environmentalist and populist thinking, the idea of a special women-environment link started to be adopted into WED approaches and thereafter into development policy-making. While on many counts, this has been a very positive shift, there has been concern about the theoretical inconsistencies and practical implications of overly simplistic portrayals of gender-environment relations.

In theoretical terms, attempts by radical eco-feminists to reject 'reductionist science' in favour of a more holistic, feminine epistemology are problematic because by placing themselves in opposition to science and Enlightenment values more generally, the critiques become locked into that same conceptual order.[7] So although eco-feminists are unwilling to accept the universality of science they are happy to both reinforce the fundamental male/female, culture/nature dichotomies that they associate with the Enlightenment project and to claim that an innate closeness to nature is common to all women. In other words, they have responded to an essentialised association of men with science, reason and culture, by creating 'an equally essentialised female way of knowing' (Nanda, 1991, p. 46).[8] Indeed, it is hard to disagree with Nanda when she asks how 'the science critique that began by questioning the legitimacy of all universal knowledge claims ended up proclaiming itself sovereign?' (Nanda, 1991, p. 46).

Also theoretically problematic are attempts by radical eco-feminists to replace reductionist patriarchal science (and its assumed violence to women and nature) with a more holistic feminine world view. A good example of this is Shiva's 'subsistence perspective' (Mies and Shiva, 1993) which is essentially a radical eco-feminist variation on a romantic populist theme concerning traditional societies that are supposedly uncontaminated by Western scientific knowledge. For Shiva, the only route away from the violence of reductionist science and science-led development is through the recovery of a 'trans-gender', 'feminine principle' which enables rural people to live in organic harmony with their environments: a vision she sets up in opposition to the neo-colonial development models informed by patriarchal (Western) science that she believes to be responsible for destroying nature and oppressing women (Agarwal, 1994a).

As Kelkar and Nathan point out, however, most societies, during the transition from hunting and gathering to settled agriculture, have become significantly less egalitarian and more patriarchal with women's traditional rights over the products of their labour becoming eroded by men. Similarly, Agarwal argues that 'the assumption in certain eco-feminist writing (e.g. Birkeland, 1993; Shiva, 1988) that tribal cultures in non-Western societies are gender equal is questionable, as is the assumption that at some historical point in time when an organic world view of nature is presumed to have prevailed, gender equality also prevailed' (Agarwal, 1998, p. 70). In light of these critiques and given the existence of highly pronounced caste- and gender-based divisions of labour throughout India's settled agricultural areas, there would appear to be a case for reassessing Shiva's belief that nature and women 'have historically been the primary food providers in natural farming, based on sustainable flows of fertility from forests and farm animals to croplands' (Shiva, 1988, p. 96).

From a more practical perspective, there is growing concern about the burden that can be placed on women from the adoption, by development agencies, of WED/eco-feminist emphases on a special women-environment link. Significantly, a number of recent empirical studies have suggested that environmentally-oriented development programmes can increase gender tensions and women's drudgery by expecting them to participate in environmental conservation programmes (Jackson, 1993a; 1993b; Braidotti et al, 1994; Sarin, 1995; Kaul Shah and Shah, 1995; Agarwal, 1997a; 1997b; 1998; 2001; Kaul Shah, 1998; Sarin, 1998a; Locke, 1999). There is also evidence that stereotypical gender concerns (especially short-term participatory efforts aimed at empowering women) can both alienate and provoke resistance from men (Cornwall, 1998; Gujit and Kaul Shah, 1998a; Crawley, 1998). Bina Agarwal's and Catherine Locke's research, meanwhile, highlights the problems of both assuming 'a necessary congruity' between women's and environmental interests and viewing women instrumentally as 'fully fledged agents', able to participate easily in environmentally-oriented development projects.

As Cecile Jackson (1994) points out, the emphasis on women as the 'key' to sustainable development (coupled with the implicit WED view that women *should* participate in environmental conservation programmes for the benefit of the wider community) is likely to deepen the exploitation of women's labour (Jackson, 1993a; 1993b; Moser, 1993).[9] Although the identification of women and indigenous communities with nature has been useful for drumming up support amongst activists and policy-makers, it is also likely to 'reduce complicated social and historical dynamics and the fraught nature of social identities to mere caricatures' (Agrawal and Sivaramakrishnan, 2000b, p. 9). A related problem with WED approaches is that they have encouraged the establishment of environmentally-oriented development programmes which are expected (because of the assumed women-environment link) to be in women's interests but which instead frequently benefit (and get taken over by) men. According to Jackson (1993a; 1993b) the problem often lies in the failure of many WED writers to distinguish between environmental work (which brings women into contact with environmental resources) and environmental management (which is usually a male domain).

For poor people in developing countries, such misunderstandings can be crucial. Yet in spite of frequent calls for context-specific analyses of the relationships between gender (instead of just women) and the environment and concern about the adoption of romantic eco-feminist/WED discourse into development policy-making, little effort has yet been made to take the debates forward on the basis of in-depth empirical studies. Similarly, although numerous theoretical discussions have highlighted the need to correct WED's undifferentiated approach and 'inflexible and narrow conceptualization of social relations' (Joekes et al, 1994, p. 138), relatively few detailed field-based studies have attempted to investigate the embedded socio-political relations that influence women's access to local environments and therefore their power to develop environmental knowledges and articulate environmental management strategies (Jackson, 1993a; 1994; Agarwal, 1998). The main exceptions to this are Bina Agarwal's work on gender, land rights and forests in India (Agarwal, 1994a; 1994b; 1997a; 1997b; 1998; 2001), Cecile Jackson's (1995) research on conjugal contracts in Zimbabwe; Melissa Leach's (1991) work on resource management in the West African forest zone and the studies carried out by Govind Kelkar and Dev Nathan (1991); Madhu Sarin (1993; 1995; 1998a), Meera Kaul Shah and Parmesh Shah (1995), Mary Hobley (1996) and Catherine Locke (1995; 1999) on gender issues in forest use and management in South Asia.

In Chapters Nine and Twelve, an attempt is made to add to this body of work by investigating the micro-political ecology of gender variations in agricultural and silvicultural knowledge possession and articulation. Spatially and temporally specific empirical data from Jharkhand on the socio-cultural, spatial and political economy contexts of local environmental problems will be drawn upon to examine the contribution that eco-feminist/WED discourse has made to both development policy-making and practical project implementation. Particular note will be taken of 'feminist environmentalism' (Agarwal, 1992) and gender analysis (Jackson 1993a; 1993b) perspectives which challenge prevailing representations of women and men as unitary categories and stress the significance of place and agency working within wider structural constraints (Agarwal, 1992; Kapadia, 1995). Chapters Four and Five attempt to trace the integration of mainstream development and eco-feminist/WED discourse (and their critiques) into India's agricultural and forestry policy-making. They also attempt to illustrate the importance of empirical research for informing and challenging the ways in which environmentally-oriented and gender-related development projects are implemented.

Notes:

[1] The World Bank established a WID division in 1987, but according to Escobar, early policy formations 'echo old home economics conceptions, although this time couched in the language of economic efficiency, motivated by the fact that "investments in human capital for women have a high payoff"' (Escobar, 1995, p. 178). Mueller, meanwhile, criticises WID discourse as 'a set of strategies for managing the problem which women represent to the functioning of development agencies in the Third World' (Adele Mueller, 1987, p. 4 quoted in Escobar, 1995, p. 112).

[2] Theoretically, Shiva's arguments reflect and support those of the man responsible for the formalisation of Gandhian economics, J.C. Kumarappa, who viewed women as 'custodians of culture and the family' (Guha, 1992, p. 58) and associated them with more altruistic, community-oriented and decentralised 'herd type'

societies. By contrast, he saw men as belonging to more predatory, centralised and individuali.tic 'pack type' societies (Guha, 1992).

3 In contrast to WED, which views gender inequalities as beyond its remit, GAD does not subscribe to the idea of a special, unvarying women-environment link rooted either in biology or gender divisions of labour. Instead, it focuses more on intra-household and intra-gender inequalities in access to resources and task allocation and views environmental relations 'merely as part of general entitlements and capabilities ascribed to individuals by social relations of gender, class and so on' (Joekes et al, 1994, p. 139). According to Humble, 'GAD theory recognises that women's lives are strongly shaped by men, hence efforts to promote women's empowerment must involve men and women alike. Gender equity is only possible by addressing gender relations and rethinking development practice' (Humble, 1998, p. 35).

4 By doing so, DfID endorsed Principle 20 of the Rio Declaration, signed at the 1992 United Nations Conference on Environment and Development (UNCED, 1992).

5 The more economistic streams of feminist thought link gender divisions of labour to patriarchal traditions that have historically allocated economically productive roles to men and social reproduction to women (Braidotti et al, 1994).

6 Meera Nanda makes the excellent point that while 'western xenophiles seeking comforting myths in the spiritual east may find the interconnections between fallow fields and menstruating women most romantic, the fact remains that in actual life, linking of the two serves as a powerful justification for treating women as polluted beings, tied to their menstruating wombs and burdened with obligations to behave as the fecund mother earth' (Nanda, 2001a, p. 2562).

7 Meera Nanda asks of radical eco-feminists: 'if all knowledge is value-laden and context-dependent, how can their own critique be anything but value-laden and motivated by their own interests?' (Nanda, 1991, pp. 43-44).

8 In a recent paper, Nanda critiques the 'indigenist intellectual dalliance with postmodernism' (Nanda, 2001a, p. 2564) within India, making the daring argument that the 'misguided, uneducated and pseudo-radical critique of critical reason and modern science, borrowed from the fashionable nonsense that prevails in some segments of the western academe, has prevented the full flowering of a radical Enlightenment-style movement in India. The culturalist understanding of science and reason has diverted the people's science movements into indigenist and economistic critiques of imperialism, rather than a self-critique of dominant and oppressive traditions. The 'reactionary modernism' of Hindutva is a direct beneficiary of the left's slide into irrationalism' (Nanda, 2001a, p. 2555).

9 Even so, women have often resisted such pressures. During the 1940s in Kenya's Murang'a district, women resisted being coerced into constructing structures to control soil erosion (Mackenzie, 1995).

4 Environmental Management and Forest Policy in India

Introduction

By the late 1970s, forest loss resulting largely from competition between industrial demand for timber, the construction requirements of rapidly expanding urban areas and the rural population's need for fuel wood and agricultural land (converted from forest land) was recognised as one of India's most serious environmental problems (Eckholm, 1984). It was not until the late 1980s, however, that the Indian government started to accept that the forest policies pursued since Independence were failing to resolve this problem. The government has been equally slow in responding to criticism over its treatment of India's large (and growing) forest-dependent population.

Although forest conservation policies managed to slow the rate of forest loss from around 1,500,000 hectares per year in the early 1980s to an estimated 47,300 hectares per year in the early 1990s, over 37,000,000 hectares of India's 75,000,000 hectares of so-called 'forest land' is classed as degraded (Palit, 1993; Shah, 1995). Out of this total, over 24,000,000 hectares require afforestation, and significantly more must be planted with trees if the Indian government is to meet its aim of achieving 33 per cent forest cover (Indurkar, 1992; Palit, 1993; Chaturvedi, 1996). Nevertheless, the FAO estimates a currently yearly loss of 3.37 million hectares (Karlsson, 2000).

For the Indian state, forest decline has become a major area of concern for political and economic as well as environmental reasons. India's timber industry which used to be a major contributor to the national economy with considerable political clout, is now unable to meet the growing demand for wood products. To overcome this problem, some of India's major timber users have had to find alternative raw materials. Many of the more indirect impacts of forest loss - such as increased rates of topsoil erosion (up to 6,000,000,000 tonnes per year), nutrient losses and water body siltation - also have high economic as well as environmental costs. Equally serious from a political ecology perspective are the deforestation-related losses of wildlife habitat, reductions in genetic and biodiversity plus a decrease in other 'ecological services' (Kumar, 2001, p. 2859) such as climate control, water body recharge and land fertility that have been linked to forest decline (Pathak, 1994; Rangan, 2000a).

Forest loss has also had severe implications for those that are dependent on woodlands as a source of fuel, house construction timber,

57

wood for making agricultural implements, fodder, food, medicinal herbs and land for cattle grazing or shifting cultivation. In the Himalayas, fuel wood satisfies 93 per cent of the cooking and home-heating needs of the area (*Wastelands News* XIII, 1998, p. 56). An average family is estimated to require four tonnes of wood per year, most of which (around 70 per cent) is used for cooking purposes (Society for Promotion of Wastelands Development, n.d.). Indeed, it is thought that forests meet 80 per cent of India's energy needs with 70 per cent of the rural population and 50 per cent of the urban population using fuel wood for cooking purposes. As forest degradation around urban hinterlands has increased, the collection of fuel wood for urban markets has spread much further afield. Fuel wood for the urban population of Bangalore, for example, comes from up to 140 kilometers away and for New Delhi it sometimes comes 700 kilometers by rail from Madhya Pradesh (Dankelman and Davidson, 1988; Henshall Momsen, 1991; Pathak, 1994). During 1976-1977, however, India's forests were too degraded to meet more than 7 per cent of the population's requirements for fuel wood. Shortages were most acute in Karnataka, Assam, Madhya Pradesh, Himachal Pradesh and Maharashtra (Ghate, 1992).

In areas where good forest cover still exists, the sale of non-timber forest products (NTFPs) such as fruit, seeds, leaves, grasses, leaf plates, bamboo baskets and medicinal herbs can provide significant incomes; sometimes half the annual income of a village (Nathan, 1988; Poffenberger, 1996; Jeffery, 1998a; Cooke, 1999). Food derived from the forest, meanwhile, provides an important source of vitamins and protein and can help to boost calorific intakes during times of shortage: particularly in the pre-harvest period. In *adivasi* areas, the edible leaves, roots, tubers, seeds, fruit, flowers, mushrooms and game that can be found in forests contribute substantially to staple diets (Saxena, 1995). Elsewhere, such food is particularly beneficial to the poor and landless who are forced to depend on the gathering of forest produce to supplement their food stocks (Arnold, 1991; Shah, 1995).

Forests are also inextricably linked to agricultural production as they provide not only compost, in the form of leaf mulch, but also a source of fodder for plough and milk cattle which in turn produce manure which can be used on farm land as a fertiliser (Arnold, 1991; Guha, 1993). In some areas, forests provide an additional source of land which can be used for agriculture, although the illegal conversion of forest land to agriculture has long been a source of conflict between forest-dwelling peasants and the Forest Department (Pathak, 1994; Jeffery, 1998a; Rangan, 2000a).[1]

In addition, forests are, for many Indians (particularly *adivasis*), places of great socio-cultural significance (Guha, 1989; 1993; Chandrakanth et al, 1990; Imam, 1990; Corbridge, 1991a; Gadgil and Subash Chandran, 1992; Pereira, 1992; Ramakrishnan and Patnaik, 1992; Vatsyayan, 1992; Pathak, 1994; Jeffery, 1998b; Bhattacharaya, 1998). Indeed, Shiva goes as far as to say that forests 'have always been central to Indian civilisation. They have been worshipped as Aranyani, the Goddess of the Forest, the primary source of life and fertility, and the forest as a community has been viewed as a model for societal and civilisational evolution' (Shiva, 1988, p. 55).

Changing Attitudes towards Forest Protection

Although the pressure of increasing human and animal populations on declining forest areas has undoubtedly exacerbated deforestation by forcing people to use these resources unsustainably, this 'official' view on the cause of forest decline is not the only explanation. Contrasting India's rural, environment-dependent 'ecosystem people' with the country's resource-wasteful, urban-industrial 'omnivores' (Gadgil and Guha, 1995), Guha seeks to transfer the blame back to India's misguided emphasis on resource-demanding industrialisation which, he argues, represents 'a gross caricature of European capitalism, reproducing and intensifying its worst features without holding out the promise of a better tomorrow' (Guha, 1989, p. 195).

In his excellent monograph, 'The Unquiet Woods', Guha suggests that local people lost interest in what they had traditionally seen as 'their' forests after forest reservation for commercial production was started by the colonial state in 1878. In many cases, local people's progressive loss of control over forests was accompanied by the spread of a 'free for all psychology' (Bahaguna, 1991, p. 1) and rapid environmental degradation (Guha, 1989; 1993; Bandyopadhyay, 1992; Alvarez, 1994). Villagers who continued to use Reserved Forests, meanwhile, were often treated as criminals and punished with some force (Corbridge and Jewitt, 1997). As Karlsson notes, forest officers didn't always bother to 'establish villagers viewpoints before sending in armed troops and arresting all those who didn't escape to the forest' (Karlsson, 2000, p. 55). In response, villagers often refused to protect and nurture forests that were being scientifically managed and policed by Forest Department employees.

Illustrating the complexity of people-forest and state-citizen relations, however, some communities reacted differently to commercial forest management, establishing their own village-based committees to protect local jungle areas (Mehrotra and Kishore, 1990; Kant et al, 1991; Ghate, 1998; Singh, 1999). A number of these were set up soon after Independence and have been operating successfully since then. There have also been occasional instances when Forest Department officers (and sometimes NGOs) tried to involve local villagers in forest management. Some of the most long-standing examples of this are the Van Panchayats (village councils) in the Uttar Pradesh hills and the Village Co-operative Forest Societies in Kangra District, Himachal Pradesh which were set up during the 1930s and 1940s (Argawal and Singh, 1995; Singh, 1999).

Like many joint forest management schemes today, the Kangra Co-operative Forest Societies encouraged local communities to manage forests in exchange for a share of the profits from timber sales. Between 1941 and 1953, 70 Co-operative Forest Societies were formed and the scheme worked well. After 1960, however, the state discontinued its annual grant-in-aid payment of 50,000 rupees to the Co-operative Societies and in 1973, the scheme was effectively discontinued (Argawal and Singh, 1995). In spite of the removal of villagers' financial stake in and effective control over local forests, however, a number of Co-operative Forest Societies still exist in Kangra: a situation that has helped to stimulate wider interest in community-based forest protection (Potter, 1998).

Additional factors that were responsible for the recent interest in local forest management systems and the shift towards joint forest management by the Indian government will be discussed in more detail below. First of all, it is helpful to examine the changes in forest management and forest policies from the pre-British colonial to the post-Independence periods; particularly the development of forest conservation, social forestry and community forestry programmes in the 1980s.

Forest Use and Management in the Pre-colonial Period

Reflecting global interest in sustainable development and alternative resource management systems during the 1980s, there are many romantic populist descriptions of a so called forest 'golden age' in India prior to the period of British colonial rule when local people apparently enjoyed unrestricted access to and use of forests (Freeman, 1998; Jeffery, 1998b; Linkenbach, 1998; Karlsson, 2000; Rangan, 2000a). Although the *de jure* ownership of most forests rested with various local rulers, many were supposedly managed *de facto* as common property resources by villagers who had a strong sense of community and a high level of concern for environmental resources (see, for example: Guha, 1983; 1989; Tiwari, 1983; Prabhu, 1983; Kulkarni, 1983; Shiva, 1988; Mehrotra and Kishore, 1990; Kant et al, 1991; Ghate, 1992). Indeed, according to Singh (1986), more than 80 per cent of India's natural resources were managed as common property prior to the end of the last century.

For tribal societies, in particular, forests have traditionally been so important for subsistence, cultural and religious purposes that many *adivasis* consider themselves to be the true owners of local woodlands. Ghate, for example, argues that from 'times immemorial the tribal people have enjoyed freedom to use the forest and hunt its animals, and this has given them a conviction, that remains even today deep in their hearts, that the forests belong to them' (Ghate, 1992, p. 17). Many accounts of the pre-British period also emphasise the religious value of forests for India's hill and tribal people who have a strong tradition of worshipping forest deities and protecting sacred groves (Roy, 1912; 1928; Guha, 1989; 1993; Shiva, 1988; Chandrakanth et al, 1990; Gadgil and Subash Chandran, 1992; Pereira, 1992; Ramakrishnan and Patnaik, 1992; Vatsyayan, 1992; Pathak, 1994; Jeffery, 1998a). In addition to the importance of forests to Animist faiths, Hindu theology has a strong tradition of tree worship and the existence of temple forests is thought to have helped to protect substantial areas of woodland (Guha, 1989; Khiewtam and Ramakrishnan, 1989; Chandrakanth et al, 1990; Gadgil and Subash Chandran, 1992; Pereira, 1992; Ramakrishnan and Patnaik, 1992; Vatsyayan, 1992).[2]

Writing of Uttarakhand prior to the reservation of forests by the British, Guha notes how through 'religion, folklore and tradition the village communities had drawn a protective ring around the forests' (Guha, 1989, p. 29). Hilltops and sacred groves were often dedicated to local deities and these areas helped to stabilise water flows and prevent landslides as well as providing additional nutritional security. Practical protection of forests and farm or household trees was secured through 'customary limitations on users' (Guha, 1989, p. 31) and the co-operation of hill villagers in managing their common lands. In Jodhpur District, Rajasthan, meanwhile, religious and moral sanctions on resource use were used to promote 'desirable collective behaviour' (Gupta, 1986, p. 313).

According to Gadgil and Iyer (1989) and Ghate (1992), these limitations included restrictions on the amount of forest produce that each household could take, the species that could be exploited, the areas from which they could be taken and the time during which resource harvesting could take place. Due to the relatively small number of communities dependent upon what were then quite vast areas of forest, these traditional conservation and management systems are believed to have worked quite well. How long such seemingly idyllic forest management systems could have continued to be sustainable is impossible to say, however, as most of them were destroyed by the onset of Britain's systematic and commercially-oriented exploitation and scientific management of India's forests.

British Colonial Forest Policy

Although at first Britain did little in the way of managing what it considered to be India's 'inexhaustible' forest resources, the realisation that the 'safety of the empire depended on its wooden walls' (Ghate, 1992, p. 31) was probably a major political factor behind the reservation of teak forests in Malabar in 1806. By the mid-nineteenth century, concern about declining forest resources, climatic change, soil erosion, falling agricultural productivity and the need to secure future supplies of timber for railway construction, ship building and military purposes, resulted in proposals for forest conservation (Guha, 1989; Pathak, 1994). The Imperial Forest Department was created in 1864, and in 1865, the first Indian Forest Act was passed. Simultaneously, 'scientific forestry' based on European silvicultural principles and focused on the production of sustained revenues and timber yields (often through the creation of timber monocultures) was introduced (Pathak, 1994).

The reservation of certain forests for commercial purposes followed soon afterwards with the passing of the Indian Forest Act of 1878. By empowering the government to 'declare any land covered with trees, brushwood or jungle as government forest by notification' (Ghate, 1992, p. 33), this Act established the colonial state's monopoly right over India's forest lands which were to be divided thenceforth into four categories: i) Reserved Forests, ii) Protected Forests, iii) Private Forests, iv) Village Forests. Of these, only Village Forests were available for unrestricted use by local people. Reserved Forests were out of bounds to local people except for the collection of certain NTFPs such as fruit. In Protected Forests, villagers were allowed to collect wood for *bona fide* domestic needs and (sometimes) graze their cattle.

Despite these measures, concern about the need to maintain good forest cover to reduce the impact of soil erosion on taxable agricultural land was expressed by the agriculturalist, Dr. J.A. Voelcker. In 1894, his recommendations were taken up in Resolution 22F on Forest Policy in India (often known as the Voelcker Resolution) which was applied throughout British India. The Resolution classified forests according to their suitability for soil protection (usually forests situated on steep slopes), commercial timber production (the most valuable forests) and the satisfaction of local people's subsistence requirements (the least valuable forests). It also established further restrictions on the rights of local people to use forests. For a number of years, however, these divisions caused tension amongst the government agencies responsible for different types of forest management; notably between the Forest Service which had prime responsibility for commercial forest management and conservation, and the Revenue Department which was in charge of watershed management and the prevention of soil erosion (Sivaramakrishnan, 1999; Rangan, 2000; Saberwal, 2000).

One of the main effects of these pieces of legislation was to convert villagers' customary forest rights into 'privileges', granted by the Forest Department through the privatisation of what were often, effectively, former common property resources. Forest reservation, although carried out, in theory, for the 'public benefit', was primarily a means of maximising timber yields and profits at the expense of other claims (notably the subsistence requirements of local people) to forest resources.[3]

As was often the case elsewhere (Shepherd, 1992b), the Indian Forest Department attempted to justify its expansion of scientific forestry by ridiculing local systems of agricultural and forest management, claiming that they were inefficient or environmentally damaging. Shifting cultivation, in particular, came in for attack by the colonial Forest Department which regarded it as 'a pernicious system ... probably as destructive to forests as any other act of man' (Stebbing, 1922, p. 31)[4] and made a whole series of attempts to discourage it

(Elwin, 1964; von Furer-Haimendorf, 1989). The main exception to this was in areas of poor quality forest where greater revenues could be obtained by allowing (or even encouraging) local people to convert forests into taxable agricultural land (Hardiman, 1987).

The full extent of the disruption that the imposition of colonial forest policy brought to the lives of local, forest-dependent (particularly *adivasi*) communities is hard to imagine. Many people had to cope with the sudden loss of important forest-based religious sites and traditional festival grounds while others were forced to undertake a form of settled agriculture that was completely alien to them. Talking of the Juang tribe of Orissa, for example, Elwin describes the 'complete cultural and religious collapse' (Elwin, 1964, p. 173) that followed the tribe's loss of access to nearby forests.

Subsistence production was also severely disrupted by forest reservation as although Village Forests were set aside for use by local people, many communities were faced with substantial limitations on the woodland areas available to them (Guha, 1989). Traditional practices such as the collection of fuel wood, fodder, twigs, leaves, fruit, seeds, honey, grass and other items of forest produce were greatly restricted in many areas and the possibility of converting forests into agricultural land was suddenly curtailed. Claiming compensation for these losses, however, was a very difficult process. This was due both to the failure of the colonial government to recognise the traditional occupancy rights of local communities (Shepherd, 1992b; 1993a) and the inability of local people (many of whom were illiterate and unfamiliar with the concept of private property rights) to register their appeal with the relevant forest settlement officer (Singh, 1986; Karlsson, 2000). Even today, many communities remain unaware of their full rights to local forests and forest produce.

In addition to this, a whole series of new rules were imposed on local people's access to forests by a highly bureaucratic Forest Department that was willing to punish 'forest crimes' with some force. In a majority of cases, it was not even possible for villagers to maintain their links with local forests by working for the Forest Department as, according to Stebbing 'the right stamp of man did not present himself for competition with the European applicants' (Stebbing, 1923, p. 31). Those that did succeed in getting jobs as forest labourers were paid a pittance (Ghate, 1992).

According to Singh, 'if the natural resources of rural and tribal people are to be usurped and no alternative made available to them, the courses of action for such people are few: they either steal, die of poverty, revolt or are forced to migrate' (Singh, 1986, p. 5). In areas where large scale forest reservation took place, many villagers resorted to 'everyday forms of peasant resistance' (Scott, 1985) such as the illegal collection of forest products as a statement of their right to satisfy their subsistence requirements from these woodlands (Blaikie, 1985; Guha, 1989; Pathak, 1994; Linkenbach, 1998; Rangan, 2000a). Others felt that there was no option but to move away (Guha, 1989). Any inclination by villagers to protect and regenerate local Village Forests were, on the other hand, often tempered by the view that this would only encourage the Forest Department to seize control over them. Speaking of Uttarakhand, Guha suggests that this attitude makes sense if one considers that 'the loss of community ownership had effectively broken the link between humans and the forest' (Guha, 1989, p. 55).

After the appropriation of local forests by the Forest Department, many communities lost interest in them, some even becoming violent towards them.[5] In Uttarakhand, for example, there were a number of incidents in 1916 when villagers deliberately set fire to newly reserved forests (Guha, 1983; 1989). Despite the introduction of increased penalties for forest crimes in the Indian Forest Act of 1927 and the right of forest officers to arrest suspected offenders without proof, however, many forest dwellers continued to ignore forest rules (Blaikie, 1985; Pathak, 1994). This usually took the form of pilfering wood and other forest products for subsistence purposes (and occasionally for sale) although resistance strategies such as the removal of iron forest boundary posts to make arrow heads were not unknown (Pathak, 1994).

In many ways, these responses reflected local people's resentment at the 'catch 22' situation in which they found themselves with regard to forest use. If on the one hand they tried to protect forests, they ran the risk that they would encourage the Forest Department to reserve them for commercial exploitation (Guha, 1989). If on the other hand they deliberately over-exploited local forests, the Forest Department could use this as justification for the need to protect them by re-classifying them as Reserved Forests (Gadgil and Iyer, 1989).

Local people's opposition to colonial forest policy was aimed not only at restrictions on access to forests but also at the changes in woodland composition brought by scientific forestry. In many cases, all-aged, mixed forests were converted into even-aged timber monocultures that had little socio-cultural or subsistence value. Nor did the new forests have much ecological value as the Forest Department's emphasis on commercially valuable timber meant that it frequently neglected 'the complex relationship within the forest community and between plant life and other resources like soil and water' (Shiva, 1988, p. 63). Significantly, almost all of the planned forest fires in Uttarakhand in 1916 occurred in newly reserved *chir* pine (*Pinus longifolia*) forest rather than in mixed oak woodlands. *Chir* pine was favoured by the colonial Forest Department for timber and resin production but local people had traditionally fired these forests as a means of obtaining new grass crops for their cattle (Guha, 1989).

Despite the highly commercial nature of British forestry policy and the forceful way in which resistance to its implementation was usually dealt with, it is important to question simplistic state-citizen dichotomies and popular representations of colonial power as inflexible, unchanging and insensitive to the plight of local people. According to Rangan, most narratives about the colonial state in India are the 'stuff of high drama, the monolithic object against which the colonised relentlessly engaged in a Manichean struggle ... the pivot around which oppositions are articulated: precolonial harmony against colonial discord, tradition against modernity, the 'periphery' against the 'core', a hallowed and distant paradise transformed into a wasteland by colonial marauders and capitalists' (Rangan, 2000a, p. 75). Yet the reality was often very different and extremely complex.

Most surprising, perhaps, is the fact that many local people look back with fondness on British forest guards and officers, regarding them as fairer, less corrupt and more respectful towards local customs than their post-colonial successors. Karlsson (2000), for example, recounts how in northern West Bengal, Rabha communities regard the British as better (and fairer) foresters and their narratives about the 'just' British rule challenge the 'unjust' rule of the present *Bangla sahib* (Bengali Government).

There are also many examples of forest management being adapted to fit the needs and knowledges of local people. Sivaramakrishnan (1999), for example, argues that forest conservancy was 'always tempered by the demands of rural livelihoods that had to remain viable if scientific forest management was to expand its domain' (p. 201). Similarly in Garhwal and Kumaon, scientific forestry did not consistently override the needs of local populations. Contrary to Guha's emphasis on the role of colonial officers as cold, unemotional scientific managers, many were upset by the disruption caused to local people and 'recorded their exasperation at being besieged by litigations against regulations and boundary disputes within their jurisdictions' (Rangan, 2000a, p. 130). The main reason behind such sentiments was an awareness by British forest officers that they were too few in number to cope with a serious case of 'unrest' over forest issues (Guha, 1989).

In addition, there are several sources which document the admiration of many colonial administrators for indigenous medical systems, agricultural practices and silvicultural knowledges. The agriculturalist, Dr. J.A. Voelcker, for example, was full of praise for Indian agriculture, stating that 'Nowhere would one find better instances of keeping land scrupulously clean from weeds, or ingenuity in device of water raising appliances, of knowledges of soils and their capabilities as well as the exact time to sow and reap, as one would in Indian agriculture ... Certain it is that I, at least, have never seen a more perfect picture of careful cultivation, combined with hard labour, perseverance and fertility of resources' (quoted in Dogra, 1983). In eastern India, meanwhile, foresters in Bengal frequently 'learnt the political pertinence of local knowledge' (p. 225). A classic example of this is when colonial Forest Department's pejorative views on fire in forest ecosystems underwent a complete *volte face* in response to evidence from local people that fire greatly assisted *sal* (*Shorea robusta*) forest regeneration (Sivaramakrishnan, 1999; Karlsson, 2000).

Some colonial administrators and 'sympathetic outsiders' became sufficiently concerned about the plight of India's forest-dependent populations that they attempted to develop more sensitive policies (Shah, 1995; Karlsson, 2000). One of the most influential of these was the Oxford-educated ex-missionary, Verrier Elwin, who criticised the restrictions that forest reservation placed upon tribal lifestyles, arguing instead for the need to involve *adivasi* people in forest management. (Guha, 1992; Jewitt, 1995a). Elwin was particularly concerned about British attempts to protect commercially valuable forests by forcing *adivasi* groups such as the Baiga of Orissa to give up shifting cultivation in favour of settled (and taxable) agriculture (Elwin, 1939; 1958; 1964). He argued that as the Baiga had strong cultural reservations about plough cultivation, they could no more 'be forced to a plough than a weaver can be forced to service in a cotton mill' (Elwin, 1964, p. 147). Elwin was also prescient in his defence of the Baiga system of shifting cultivation, arguing that it caused far less ecological damage than many foresters claimed.

Although Elwin's arguments went largely unheeded initially,[6] Collins (1921), who had observed the extremely well maintained state of many community-managed forests, successfully argued for the inclusion of provisions for the formation of village forest councils in the Indian Forest Act of 1924 (Gadgil and Iyer, 1989). Others, including Stebbing, Dietrich Brandis (the first Inspector of Forests) and Voelcker showed concern about the access of local communities to fuel wood and tried hard to involve local people in schemes for its provision.

Post-colonial Forest Policy

Instead of building on these admittedly rather feeble attempts by British administrators to be sensitive to the needs of forest-dependent villagers, the post-colonial Forest Department maintained a surprising degree of continuity with its colonial predecessor (Guha, 1989; Pathak, 1994). Indeed, while the structure and objectives of colonial and post-colonial forest policies remained very similar,[7] the greater strength of the post-colonial state meant that it was better able to enforce modernising forest management practices: a situation that resulted in the further criminalisation of forest-dependent populations (Jewitt, 1995a; Corbridge and Jewitt, 1997; Karlsson, 2000).

The Indian Forest Act, 1952, for example, endorsed scientific forestry, forest privatisation and forest reservation, attempting, according to Singh to enforce the British system of forest classification 'with greater zeal' (Singh, 1986, p. 3). The Act also emphasised the importance of the 'national interest' over and above the needs of local people, so as to ensure that India as a whole was not 'deprived of a "national asset" by the "mere accident of a village being situated close to a forest"' (Guha, 1983, p. 1888). The main reason for this was to satisfy India's large and growing industrial, commercial, communications and defence requirements.

To meet these demands, the Act's provisions for forest conservation gave way, more often than not, to the exploitation of timber and other forest produce on a vast scale (Guha, 1983; Ghate, 1992; Pathak, 1994). Between 1956 and 1967, the extraction of industrial wood increased from 4,460,000 cubic metres to 9,260,000 cubic metres and between 1948 and 1970, the consumption of printing and writing paper increased from 100,000 tonnes to over 400,000 tonnes (Pathak, 1994). Revenues also grew rapidly, multiplying fivefold between 1951-2 and 1970-1 (Pathak, 1994).

But in many cases, these increases in revenue were made at the expense of local people's traditional rights to forest produce. The Forest Department tried to insist that local people should buy such items from forest depots at concessional rates, but these depots were rarely well stocked as local industry often got preference, despite the fact that it paid lower prices than local people. Bamboo, for example, was often sold to paper mills at prices ranging from 37 paise to 22 rupees per tonne, whereas local artisans had to pay up to 1200 rupees per tonne on the open market and had problems selling their wares at prices that would cover their costs. In some areas, local people were unable to claim their customary allowance of bamboo as the paper mills left none for them to take (Pathak, 1994; Saxena, 1995).

The extent of this exploitation had severe implications not only for forest stability, but for the livelihoods and culture of forest-dependent populations whose control over remaining common property resources was further eroded and whose role in the management of state forests (with the exception of Village Forests) remained minimal. At the same time, growing rural populations put further pressure on declining areas of Village and Protected Forest in their attempts to meet subsistence needs.

Many forest dwellers (particularly *adivasis*) were finding it increasingly difficult to obtain fuel wood for cooking, fodder for their cattle and raw materials for their handicrafts. Poor people often had to migrate, either seasonally or permanently. Others tried to find wage labour (often selling NTFPs) to get cash to buy food (Pathak, 1994). Yet provisions for helping *adivasis* are conspicuous by their absence in the Indian Forest Act, 1952: a situation that was deplored by the 1960 Dhebar Commission which recommended (to no avail) that *adivasis* should be given greater control over natural resource management (Mishra, 1993).

As was the case during the colonial period, this loss of control over forests often discouraged local people from protecting them. The widespread occurrence of illegal fellings by Forest Department employees (contract cutters, in particular), further exacerbated many villagers' sense of detachment from forests (Guha, 1989; Shepherd, 1992b; 1993a). It also made them even more hesitant about re-establishing community-based forest protection or management, the benefits of which they felt sure would not be reaped by them. Instead, many villagers took the view that they would be foolish not to follow the contractors' example and help themselves to the state's property (Berkes and Taghi Farvar, 1989; Karlsson, 2000). This mainly involved satisfying their subsistence requirements by illegally collecting forest produce or encroaching on forest land, although some villagers also boosted their household cash incomes by selling fuel wood to urban markets (Eckholm, 1984; Pathak, 1994).

This response exacerbated deforestation, increased villagers' conflict with the Forest Department and, according to Guha, heightened local communities' sense of alienation from forests (Guha, 1983; 1989). Pathak (1994), however, points to a more complex political ecology than Guha's simple binary opposition, based on ideas of a moral economy between local people and the state. Instead, he argues that the Forest Department is composed of a hierarchy of officials, at the cutting edge of which are the Beat Guards whose jobs fill up 'the vacuum created by the disruption of traditional power relations' (Pathak, 1994, p. 15). As a result, they understand and can often be persuaded to ameliorate (for a fee) the problems faced by local people in obtaining their subsistence needs.[8]

At the same time, these Beat Guards (who are usually non tribal) are in a position of power which, when allied with that of influential village leaders, has enabled them to dominate many aspects of local people's lives. Threats that court cases will be taken out against villagers who commit (real or imagined) forest crimes have been a commonplace means for Beat Guards to keep local people 'under control' (Pathak, 1994; Karlsson, 2000). This in turn has often created intra-community resentment over who is in the Beat Guard's good books regarding access to forest produce (or land) and punishments for forest offences.

In some areas, forest communities became so angry about these problems and about the commercial exploitation of forest lands more generally that they took measures to resist the further imposition of Forest Department activities. The Chipko movement, for example, has been extremely successful in bringing local forest cutting to a halt in Uttarakhand (Shiva, 1988; Guha, 1989; 1993; Bandyopadhyay, 1992; Colchester, 1994; Pathak, 1994; Rangan, 2000a). In Singhbhum District, Jharkhand, meanwhile, local opposition (supported by the Jharkhand movement) to the nationalisation of NTFPs and the clear felling of mixed forests became so violent in the late 1970s and early 1980s that forest officers temporarily had to withdraw (Corbridge, 1991a; Corbridge and Jewitt, 1997; Alvarez, 1994). On many other occasions, however, such actions have attracted a repressive response by the state including greater levels of vigilance and policing to 'protect' state forests from local people and, in some cases, the use of arms to put down 'unrest' (Anderson and Huber, 1989; Corbridge, 1991a; Palit, 1993; Corbridge and Jewitt, 1997; Karlsson, 2000).

The State's Response to Forest Decline

The ways in which the Indian government tried to address forest decline reflect a number of conflicting political ecologies. Concern about the poor contribution of the forest sector to GNP often elicited an authoritarian response towards forest-dependent communities (particularly *adivasis*) which, particularly during the 1970s, were given much of the blame for deforestation (Ghate, 1992). This was in spite of reports suggesting that illegal trade in forest produce by forest contractors played a much more significant role.

On the other hand, wider environmental concerns during the early 1970s brought to light India's 'joint crisis of unsatisfied basic needs and ecological instability' (Shiva et al, 1986, p. 239) and raised neo-Malthusian concerns about population pressure coupled with fears of a 'tragedy of the commons' type situation. The Chipko movement was particularly important in highlighting the environmental role of forests; representing them as a means of preventing soil erosion which degraded agricultural land and silted up reservoirs (Pathak, 1994). Coming as it did in the wake of the 1972 United Nations Conference on the Human Environment in Stockholm, Chipko offered an alternative, locally empowering environmental management model (Guha, 1989) as well as fitting neatly into the Indian state's new emphasis on the need to promote forest conservation. Significantly, the movement's roots as an assertion of local people's traditional rights to use forests were largely ignored (Rangan, 2000a). According to Pathak, however, the fact that the Chipko protesters were mostly Hindus gave the movement greater legitimacy and effectively prevented it from being 'crushed with brutality' (Pathak, 1994, p. 49) as probably would have been the case with a tribal agitation.

With the onset of the oil crisis, the compatibility of industrial development and environmental protection was brought even further into question and Indira Gandhi decided to make a strong personal commitment to the environment (Pathak, 1994). In an attempt to gain greater control over India's resource management problems, environment-related legislative power was shifted from the States to the Centre and forests were transferred from the State List to the Concurrent list in 1976 (Pathak, 1994; Shah, 1995). Thenceforward, the central government was forced to arbitrate between three conflicting forest uses: for commercial timber production, for ecological conservation and for meeting the subsistence needs of forest-dependent populations.

Increasing Commercial Timber Production With regard to increasing the productivity of forests for future industrial use, the Indian state emphasised the need to improve the quality of existing forests through re-planting initiatives. As the Department of Forests was part of the Ministry of Agriculture at the Centre, it was the National Commission on Agriculture (NCA) which set the new agenda for forestry. The 1976 Report of the NCA emphasised the need to promote soil conservation (it was particularly concerned that cattle dung was being used as a fuel rather than as a fertiliser) and to increase commercial timber production. Its solution was to clear fell existing mixed forests and re-plant them with commercially valuable timber species. It also recommended the removal of the remaining rights and concessions to forest produce that were available to local communities.[9]

The piece of legislation that aimed to support these goals came in the form of the Draft Forest Bill of 1981 which is perhaps the best testimony to the way in which the commercial bias of British forest policy was not only taken on board but extended after India's Independence. Retaining 81 out of 84 of the provisions of the 1878 Indian Forest Act, the Bill proposed further restrictions on forest users and more severe penalties for forest offences. It also aimed to increase the power of forest officers to implement these restrictions and to control illegal fellings, despite the consistent misuse of such powers in the past (Guha, 1983; Alvarez, 1994). Essentially the Bill aimed to negate customary access to forests and to tighten the state's monopoly over them (Pathak, 1994).

Fortunately for India's forest dwellers, the Bill was eventually abandoned due to heavy criticism from various academics and organisations representing forest-dependent communities. A number of populist 'forest intellectuals' were particularly influential in critiquing the prevailing ideology of scientific forestry and modernisation more generally (see, for example: D'Abreo, 1982; Fernandes and Kulkarni, 1983; Kulkarni, 1983; Guha, 1983; 1989; Shiva, 1988; Alvarez, 1994; Gadgil and Guha, 1995). Using rather simplistic state-citizen dichotomisations to promote their views on India's forest struggles to an international audience, they helped to stimulate a shift towards more participatory approaches to. forest management. Guha, in particular, has highlighted the virtues of a moral economy of forest management by knowledgeable and active forest communities. He has also sought to challenge representations of the Chipko movement as backward looking or defensive, characterising it instead as a forward looking attempt to 'escape the tentacles of the commercial economy and the centralising state' (Guha, 1989, p. 196) and develop 'a way of life more harmoniously adjusted with natural processes' (ibid.).

Another factor that helped to undermine the Bill stemmed from Indira Gandhi's decision to portray Chipko as an environmental (forest protection) movement which made it politically embarrassing to then lay the blame for forest decline on local people; especially given the national and international support that Chipko had attracted (Pathak, 1994). An additional, if unexpected, ally fighting on behalf of forest-dependent populations was the judiciary which occasionally prevented the felling of natural forests on the grounds that it deprived forest dwellers of their customary rights and livelihoods and caused ecological degradation (Pathak, 1994). As a result, the goals of commercial forest production and forest conservation remained in confusion until the late 1980s.

Forest Conservation In principle, the role of forest conservation for ecological purposes received an important boost in 1980 when the Department of Environment was created within the Ministry of Science and Technology. In practice, however, the situation of the Department of Forests within the Ministry of Agriculture at the Centre created a power struggle between the two ministries regarding the use to which forests should be put. Initially, as the 1981 Draft Bill indicated, the goal of commercial forest exploitation seemed to be the dominant force (Pathak, 1994).

The main piece of legislation supporting the protection of forest ecosystems and habitats was the Forest Conservation Act, 1980. This Act required State governments to seek permission from the central government before 'de-reserving' forests or converting them to non-forest use (especially for agricultural use by local communities). In addition, the Sixth Five Year Plan (1979-1984) emphasised the need to protect forests to help sustain agriculture and significant amounts of money were put into forest conservation, afforestation, fuel wood creation and the establishment of environmental stability. The Plan also reflected wider environmental concern for genetic diversity in its emphasis on the need for the complete protection of representative samples of land, water, flora and fauna in sanctuaries and biosphere reserves. Project Tiger had been set up in 1973 and between 1970 and 1988, the area under wildlife reserves increased from two point three per cent to four per cent of India's land area (almost twenty per cent of the total forest area).[10]

In 1985, the Forest Department was moved from the Ministry of Agriculture to the Ministry of Environment to form a separate Ministry of Environment and Forests. The inherent conflict between conservation, development and local people's use of forests still needed to be resolved, however, and in 1988 the Forest Conservation Act, 1980 was amended to make it necessary for States to seek the central government's permission for clear felling and the sale, lease or transfer of forest land to other uses. The clause on clear felling, of course, reversed the NCA's policy of clear felling mixed forests to raise industrial timber plantations and the balance of forest use at last started to swing away from purely commercial goals. This was consolidated with the National Forest Policy 1988 which, while recognising the production role of forests and maintaining the Forest Department's monopoly *vis-à-vis* local rights, reflected both a new climate of environmental awareness plus pressure within the World Bank and other donor agencies to concentrate on 'green issues' in developing countries (Ghate, 1992).

Meeting the Needs of Local People In 1976, the NCA's solution for the satisfaction of villagers' subsistence needs for forest produce had been for local people to procure their timber, fuel, and fodder requirements from 'Social Forests'. These were to be planted on commons, Panchayat lands, wastelands, roadsides, canal banks and other public lands and would act as substitute mixed forests: a situation that took away the justification for local people to enter state forests and be tempted to pilfer from them (Pathak, 1994).[11]

At the root of the NCA's argument for social forestry was the feeling that 'rural people have not contributed much towards the maintenance or regeneration of forests. Having over-exploited the resource, they cannot in all fairness expect that someone will take the trouble of providing them with forest produce free of charge' (Government of India, 1976, p. 25). This lack of consideration for forest dwellers' interests marked a low point in the history of Indian forest policy that echoes Elwin's comment that the 'rights and privileges' granted to local people by the colonial Forest Department had been downgraded to 'rights and concessions' in 1952 and 'concessions' thereafter (Elwin, 1964).

Social forestry programmes only really started to gain momentum in the early 1980s. Notably, most attention was given to the more profit-oriented farm forestry projects rather than to community forestry, which aimed to meet the needs of the poor.[12] In addition, attempts were made to 'take forest resources to the people rather than re-link people with the forests by involving them directly in the improved management of natural forests' (Palit, 1993, p. 4). Since then there has been much criticism of social forestry, as project after project failed to fulfil its expectations (Nesmith, 1991).

One of the main reasons for this was that the Forest Department and most external funding agencies used a very top-down approach that had few, if any, provisions for involving local people in the planning of schemes that were, in theory, supposed to serve their interests. This hindered the development of understandings about the political ecologies of recipient societies and the existence of intra-village tensions and conflicts. The failure of many projects to involve women has been especially criticised (Shepherd, 1985; Noronha and Spears, 1988).

As a result, economically or socially marginalised community members were often excluded from the benefits of social forestry schemes by richer farmers, who could use their greater influence with officials to obtain free and subsidised saplings. Wealthier farmers could also afford both the necessary inputs and the time required for trees to mature. Landless people, by contrast, were often forced into losing their traditional access to village common property resources by the appropriation of village commons or 'wastes' by the state for social forestry programmes (Noronha, 1981; Shepherd, 1985; Foley and Barnard, 1985; Blaikie et al, 1986; Pathak, 1994).

The imposition of schemes developed and managed from outside frequently also caused suspicion in the minds of the farmers working within them. This was especially true where villagers did not have secure land tenure and, as a result, were unwilling to commit themselves fully to such long-term projects. In particular, they were concerned that the government might try to 'reserve' their social forestry plantations at some future date and take away the benefits from them

(Shiva et al, 1983; FAO, 1985; Shepherd, 1985; Arnold and Campbell, 1986; Falconer, 1987).

The ecological benefits of social forestry, especially its emphasis on eucalyptus were also questionable. Although eucalyptus is usually fast-growing, it is not favoured as a fuel wood and cannot be used for fodder as it is unpalatable to cattle. Many writers have also emphasised that the species has a negative effect on the water table, inhibits undergrowth and provides few non-timber benefits, (such as the provision of fruit, fodder, green manure, moisture conservation and soil binding properties) compared to native trees (Shiva et al, 1986; Shiva, 1988; Arnold, 1991; Ghate, 1992). Shiva is particularly critical of the way in which 'experts decided that indigenous knowledge was worthless and "unscientific", and proceeded to destroy the diversity of indigenous species by replacing them with row after row of eucalyptus seedlings in polythene bags, in government nurseries' (Shiva, 1988, p. 79).

Another problem was that eucalyptus is better suited to industrial (especially paper) and urban markets for poles, rather than for use as fire wood (Eckholm, 1984). In some areas it was grown as a cash crop and displaced traditional agricultural crops (Shiva et al, 1986; Foley and Barnard, 1985). Indeed, Guha argues that social forestry programmes were 'launched, under a populist guise, to meet the demands of the paper industry' (Guha, 1983, p. 1892) as deforestation in the state's Reserved and Protected forests created shortfalls in the growing demands for pulp-wood.

This view also seems to have been shared by a number of local people who revealed their 'weapons of the weak' (Scott, 1985) by opposing their lack of input into social forestry and asserting their right to have a say regarding the species being planted. In the villages of Barha and Holahalli in Tumkur District, Karnataka, for example, local people uprooted eucalyptus seedlings that had been planted as part of a social forestry project and replaced them with more ecologically suitable and useful tamarind and mango seeds (Shiva, 1988; Gadgil and Guha, 1994). A similar incident also occurred in Garhwal, where local women refused to plant exotic poplar trees given to them by the Forest Department, insisting that native fodder species should be provided instead (Shiva, 1988; Gadgil and Guha, 1994).

Many other forest-dependent communities disregarded social forestry altogether as its attempts to provide for their subsistence needs for forest produce offered too little and came too late. After such a long history of conflict with the state over access to forest resources, many forest dwellers (particularly *adivasis*) had become wary of any promises by the Forest Department to improve their situation. As a result, a number of communities countered the state's supply side populism with a demand side populism of their own, in the form of village-level forest protection and management.

Autonomous Forest Management in Bihar and Orissa

Some very interesting research on autonomous forest protection committees in Bihar and Orissa has been done by Mehrotra and Kishore (1990) and Kant et al (1991), as part of a project on community-based forest protection and management funded by ISO/Swedforest, the Swedish International Development Authority, the Indian Institute of Forest Management, Bhopal, and the Bihar Social Forestry Project. Their work suggested that forest protection committees are most likely to become established in areas affected by large scale deforestation, by communities that are unified, homogeneous and which have a high degree of dependence on and equal access to local forests.

In many of the Orissan examples, some external stimulus, such as interest in local forest management by forest officers or NGOs, was often responsible for the establishment of village-based forest protection committees (Kant et al, 1992; Sarin, 1993). In Bihar, by contrast, the Forest Department's neglect of local forests coupled with instances of illegal felling by forest contractors, encouraged a number of communities to revive (or re-invent) informal village institutions for the purpose of village-level decision-making and forest protection.

In the early 1970s, for example, a student in the predominantly *adivasi* Karandih Panchayat, Jamshedpur, mobilised local villagers' concern about forest degradation and the problems that this was causing them in terms of fuel wood shortages, local climatic change, soil erosion and water table lowering. Realising that it was pointless waiting for the state to intervene, a forest protection committee was established and villagers started to guard local forests and to limit their own use of forest produce. Forest regeneration was rapid and the committee gained the support of the Forest Department. Soon afterwards, thirty five nearby villages started to protect their forests and many committees have also taken on board wider issues such as health and education (Mishra, 1993).

The establishment of community-based forest management has been particularly common in districts with significant *adivasi* populations where there is a strong tradition of village-level administrative institutions (Gadgil and Guha, 1994). Indeed, the fact that a number of forest protection committees date from the early 1950s may reflect the strength of the Jharkhand movement's demand for a separate State at that time as well as the impact of forest 'nationalisation', increased commercial exploitation plus widespread illegal fellings by contract cutters.

In terms of their administration, Mehrotra and Kishore (1990) and Kant et al (1992) found that most autonomous forest protection committees had a set of rules governing villagers' access to forest produce and a punishment system for dealing with 'offenders' from both outside and within the community. Where resources were scarce, many committees distributed forest produce (particularly large timber) on a strict need basis. Occasionally, coupe cutting or permit systems were practised where the amount of each species that could be cut and the areas from which (or the season during which) these could be taken were specified.

With regard to punishments, regulations were found to exist in a number of cases stating that committee members who failed to attend meetings could lose both their membership and their share of the benefits from forest protection. Frequent offenders (especially where there was collaboration between the committee and the Forest Department) were often handed over to local forest officials for punishment (Sarin, 1993). In order to catch offenders, most villages practised one of three main forest protection systems. The first was where each household sent one member on a rota basis for 'guard duty', the second was where villagers paid (usually in kind) one or more guards to patrol the forest and the third was where all villagers watched the forest and reported any unauthorised fellings.

One of the main difficulties faced by autonomous forest protection committees involved overcoming the power of dominant groups within the village in order to ensure the fair and equal access of all villagers to forest produce. Also, the almost complete absence of women from these organisations sometimes hindered the satisfaction of their forest requirements (Sarin, 1993). Other problems included a gradual decline of interest in forest protection over time and a break-up of community cohesion resulting in the sudden felling of the entire protected area. From an environmental perspective, there is also a danger that although regeneration from *sal* (*Shorea robusta*) coppice has been very rapid within most protected areas, the tendency of many groups to protect only certain tree species could promote a future ecological imbalance if it resulted in the eradication of other forest trees.[13]

A More Participatory Approach

Interestingly, despite the problems that social forestry has had in meeting its twin goals of halting deforestation and satisfying local people's needs for forest products, there have been few calls for its abandonment. Instead, it has been modified repeatedly to rectify old problems and to incorporate new ideas; notably the emphasis on common property resource management by village-level institutions expressed in many of the critiques of the 1981 Draft Forest Bill (D'Abreo, 1982; Fernandes and Kulkarni, 1983; Kulkarni, 1983; Guha, 1983) as well as in the wider rural development literature (FAO, 1985; Blaikie et al, 1986; Korton, 1986; Wade, 1988; Falconer, 1987; Noronha and Spears, 1988; Berkes, 1989).

Empirical research on community-based systems of common property (particularly forest) management has also indicated that such systems could be more socially equitable and ecologically sustainable than commercially-oriented forest management strategies (Arnold and

Campbell, 1986; Blaikie et al, 1986; Singh, 1986; Shiva, 1986; Wade, 1988; Gadgil and Iyer, 1989; Ostrom, 1990; Sengupta, 1991; Shepherd, 1992a; 1992b; 1993b; Ostrom et al, 1993; Messerschmidt, 1995; Baland and Platteau, 1996; Ostrom, 1999). From a more radical perspective, meanwhile, a number of populist and anti-developmentist writers have emphasised local agro-ecological knowledges and community resource management as part of a strategy for overcoming some of the Third World's environmental and poverty-related problems (Shiva, 1988; Esteva, 1992; Escobar, 1992; Sachs, 1992; Alvarez, 1994).

Both contributing to and reflecting this interest, a number of countries have tried to re-establish systems of common property resource (particularly forest) management as a means of arresting deforestation and meeting local fuel wood and fodder needs. The Government of Nepal, for example, reversed in 1978, much of the nationalisation of forests that had taken place in 1957 and returned the authority to protect and manage forests to local communities. This move resulted, in part, from the realisation that attempts to prevent local communities from exercising their traditional use rights in privatised forests had resulted not only in conflict between local communities and the Forest Department, but also in forest decline. This stemmed mainly from the fact that local people continued to exploit forest resources, but felt no obligation to protect or control their use of forests as they had done previously (Arnold and Campbell, 1985; Ostrom, 1990; Messerschmidt, 1995).

The transfer of forest management back to local Panchayats was greeted with widespread enthusiasm, not least because the incomes from timber sales were to be shared, thenceforth, between local communities and the Forest Department. To ensure good forest growth and regeneration, local communities imposed restrictions on the amount of forest produce that villagers could take and the times during which they could use the forest. Many villages employed guards to deter or report on rule-breakers and paid them in kind for their services. Tree nurseries and planting programmes were also established (after 1980) with assistance from the Community Forestry Development Project (Arnold and Campbell, 1986; Poffenberger, 1990). According to Messerschmidt, the resulting forest management systems have been both ecologically very effective and consistent with 'acceptable preexisting patterns of social organisation, custom, and use' (Messerschmidt, 1987, p. 379).

Community Forestry in India

In India, too, there has been a lot of emphasis, particularly since the publication of the 1981 Draft Bill, on indigenous silvicultural knowledges and the importance of commons as an important 'survival base for rural India' (Shiva, 1988, p. 83). In contrast to the 'western tradition of forest management, which views trees primarily in terms of their woody biomass' (Shiva, 1988, p. 58), ethnobotanical work amongst India's tribal and hill populations has uncovered the existence of complex systems of local knowledge about forest use (for food, fodder, medicine, agricultural inputs, soil conservation, household items and liquor) and management strategies.[14]

By the late 1980s, the futility of trying to protect forests without villagers' co-operation started to be recognised within India's forest planning rhetoric; particularly after the publication of the National Forest Policy 1988 which emphasised the need to involve local people in forest management. Taking on board the somewhat romanticised populist vision of active forest communities poised for the re-formation of traditional resource management institutions, policy-makers promoted community forestry as a means of both providing fuel for local villagers and promoting afforestation on communal and waste lands. The idea was for forest officers to plan community forestry schemes while local people shared responsibility with the state for managing degraded land (Indurkar, 1992; Pathak, 1994).

Unfortunately, disagreements over the management of these lands often prevented the devolution of responsibility to local communities with the result that these schemes were criticised for 'unwittingly converting common property resources into state controlled resources' (Arnold, 1991, p. 9). As some of the poorest rural families derive most of their fuel and fodder, between fifteen and twenty three per cent of their income and up to ten per cent of their nutritional requirements from commons (Shiva, 1988; Arnold, 1991), this had important implications for the livelihoods of many rural communities. Also, the reality of community resource management often had little in common with the idea of a wider moral economy of provision (Guha, 1989). Instead, intra-village elites often succeeded in making these so-called community programmes serve their own interests, rather than those of the poor and landless for whom the schemes were originally intended (Guha, 1983; Foley and Barnard, 1985; Pathak, 1994). In terms of rhetoric however, community forestry did at least make a step in the right direction with its emphasis on the need to involve local people in forest management: even if it was only tree planting on their privately owned land.

By the late 1980s, the need for people's involvement in forest management became widely accepted and it was argued that 'only people's participation on a large scale [would] give success in regaining the lost forest wealth and in preserving the present forest land and tree cover' (Indurkar, 1992, p. 48). As a result, the need to slow the rate of forest decline and to involve local communities fully in this goal started to be articulated at both national and state levels. The National Forest Policy, 1988 took on board many of the criticisms of the 1981 Draft Bill and developed a strong participatory thrust, aiming to meet 'the requirements of fuel wood, fodder, minor forest products and small timber of the rural populations' by 'creating a massive people's movement, with the involvement of women' (Government of India, 1988, p. 13).

In 1992, a major step towards participatory development was made with the passing of India's Constitution (Seventy-Third Amendment) Act which aimed to vest local government bodies known as Panchayati Raj Institutions (PRIs) with constitutional powers and make provisions for them to receive financial resources from the state. The Eleventh Schedule details the areas over which PRIs have jurisdiction, including, *inter alia*, agriculture, land improvement, watershed development, social forestry, minor forest produce, fodder and fuel and maintenance of community assets (Dhar, 1999).

A perceived step backwards was taken soon afterwards, however, with the publication of the 1994 Draft Forest Bill. In a situation somewhat reminiscent of the 1981 Draft Bill, activists deemed this Bill unacceptable in its existing form and were working with the Ministry of Environment and Forests on an alternative (Fernandes, 1998).[15] Things looked to be improving with the passing of the Panchayats (Extension of Scheduled Areas) Act (PESA) 1996 which stated that all Panchayats and *gram sabhas* were to be given ownership rights over minor forest produce, the power to exercise control over institutions and functionaries in all social sectors and the power to control local plans and resources (Shabbeer and Tankha, 1998). When the Ministry of Environment and Forests constituted an expert committee to decide mechanisms for the transfer of rights over NTFPs, however, the committee recommended that usufruct and not ownership rights be given (Mishra, 1998; Shabbeer and Tankha, 1998).

Reflecting this internal government conflict over local people's rights to natural resources, much of the initial emphasis on local participation in forest protection was linked to an attempt by the Indian state to make villagers aware of their 'duty' to protect forests: a duty that was expected in exchange for their 'privileges' and rights to forest produce (Government of India, 1990; Society for Promotion of Wastelands Development, 1993; Singh and Khare, 1993). In the National Forest Policy, 1988, for example, it is stated that 'the holders of customary rights and concessions in forest areas should be motivated to identify themselves with protection and development of forests from which they derive benefits' (quoted in Kant et al, 1992, pp. 2-3).

Nevertheless, the re-acceptance of villagers' customary forest rights was an important step forward from the 1976 Report of the National Commission on Agriculture which had advocated the withdrawal of concessions to forest produce. Indeed, the National Forest Policy, 1988 went to some lengths to emphasise the importance of both NTFPs and forests more generally to the livelihoods of local (particularly *adivasi*) communities. It also acknowledged the role of forest contractors in illegal tree felling, recommending their replacement by tribal and labour co-operatives.

In addition, the objectives of the National Wastelands Development Board, the National Forest Policy, 1988 and the Eighth Five Year Plan focused upon the need for conservation, protection and regeneration or afforestation of degraded land (Shah, 1995). The necessity of people's participation at all stages was also stressed, although from a cynical point of view, this could be interpreted as an attempt to get villagers to use their own time, resources and institutional capital to improve the state's forest resources. Nevertheless, the shift in emphasis from a 'beneficiary approach' (where land is held and managed individually) to a more participatory approach (where resources are communally managed) to forest management marked an important 'trend of transition' (Singh and Khare, 1993) for Indian forestry and environmental policy.

Joint Forest Management

Although intermittent interest has been shown since the British colonial period in the potential of community-based forest management for addressing land degradation problems (Eckholm, 1984; Pathak, 1994; Agarwal and Singh, 1995), these systems attracted little attention from policy-makers until after the publication of the 1981 Draft Forest Bill. Over the last ten years, however, the type of potentially empowering forest partnership envisaged by forest intellectuals like Guha has received significant attention; primarily as a result of the success, in a number of States, of forest officers involving local communities in forest regeneration and planting (Kant et al, 1992).

At the heart of participatory forest management is the 'care and share' principle (Singh and Khare, 1993). This assumes that villagers will manage forests carefully if they have an incentive to do so, in the form of an entitlement to the produce from the regenerated forest or a share in the final timber harvest. One of the first successful examples of this type of approach was the 'Arabari experiment' in West Bengal. This started in 1971, when Dr. A.K. Banerjee, a Divisional Forest Officer, encouraged the inhabitants of eleven villages in Arabari to protect over 1250 hectares of degraded *sal* forest (Singh and Khare, 1993; Saigal, 1995). Villagers patrolled the forest on a rota basis and in return for this, they received the right to collect fuel wood and NTFPs and were given some employment opportunities in the Forest Department. The protected area regenerated rapidly from coppice growth and in 1987, the Forest Department made an agreement to share 25 per cent of the net profits from timber sales with participating villagers (Roy, 1991; Roy and Mukherjee, 1991).

A similar experiment was conducted in 1987 in a degraded teak forest area of Surat District, Gujarat, by the Southern Circle Officer, R.S. Pathan. By talking to local people and organising rallies to make people aware of the need to protect forests from illegal felling by logging gangs, he succeeded in encouraging the establishment of about 100 forest protection committees in 1988. In return for guarding the forest from human intruders and grazing animals, participating villagers were allowed to take wood from thinnings and clearings for fire wood, free of cost. Where teak root-stock was present, villagers tried to promote coppice regeneration and where it was more limited, the committees were given free seedlings to plant by the Forest Department. In one village, local communities were able to obtain 50 per cent of their fuel wood, 40 per cent of their fodder plus assorted NTFPs (the amount of which will increase as the newly planted trees mature) from the protected area. In another, the employment generated by afforestation activities was sufficient to prevent villagers from having to migrate to find work. Indeed, when the Chief Minister of Gujarat visited the project, he was sufficiently impressed with its success that he ordered the commitment of a 25 per cent share of timber sale profits to the forest protection committees (Pathan, n.d.; Arul and Poffenberger, n.d.).

   •   In 1990, the Government of Haryana established the 'Rehabilitation of Common Lands in Aravalli Hills' project as a means of restoring sustainable village-based common property management for the production of fuel wood and fodder. 'Village Forest Committees' were established under the project and were promised a share of any timber that they succeeded in growing as an incentive to participate (Srivastava and Kaul, 1994). In addition, the Indian Institute of Forest Management, Bhopal, the Ministry of Environment and Forests and the National Wastelands Development Board have all collaborated in their recognition of the importance of village commons and their concern over the impacts associated with the privatisation of common property resources. In 1991, a national workshop entitled 'Managing Common Lands for the Sustainable Development of Our Villages' was held at Bhubaneshwar. The participants emphasised the need for increased community participation in common property management (through the establishment of forest protection and pasture management committees) to ensure that the potential of India's common lands for producing fuel wood, fodder, timber and fruit be realised (Singh, 1991a).

   By the late 1980s, interest in the establishment of such committees was being expressed by the central government and a number of State governments: not only for the management of village commons, but also for the protection and management of certain state-owned forests (Government of India, 1990; Government of Orissa, 1990; Bahaguna, 1991; Kumar, 1991; Singh, 1991a; Palit, 1993; Sarin, 1993). Indeed, financial incentives for villagers' long-term commitment to these joint forest management (JFM) schemes soon became quite widespread; usually taking the form of a share in the final resource harvest. The success of the Arabari experiment, in particular, attracted attention from international donor agencies including the World Bank and the Ford Foundation.

   As a result, efforts to encourage the co-operation of local people in the regeneration of degraded forest areas increased significantly: notably through a growing emphasis during the late 1980s and early 1990s, on the development of 'partnerships between local institutions and [Forest Departments] for the sustainable management of forest areas on the basis of trust and mutually defined rights and responsibilities of both parties' (Sarin, 1993, p. 2). It is difficult, however, to resist the more cynical view that part of JFM's appeal lies in its potential for getting local people to use their own time and energy to promote forest regeneration where the Forest Department has failed.

Nevertheless, JFM did represent a significant change in attitudes towards both rural development and rural people in forest areas as local communities started to be seen as part of the solution to deforestation, rather than as the main problem. Many government officials and development workers quickly became enthusiastic about the potential of JFM to tackle forest decline in a more sustainable, equitable and bottom-up manner. The Society for Promotion of Wastelands Development went as far as to suggest that the new trend towards a decentralisation of political power 'must ultimately end up in solving the most vital problem facing India, today i.e. regeneration and restoration of the environment, specially the ecologically fragile regions' (SPWD, n.d., p. 1).

This shift in approach was endorsed formally in August 1988 and July 1989, when the Governments of Orissa and West Bengal issued orders to involve local people in forest management and protection. The Government of Orissa's Resolution was strengthened in December 1990 to give Divisional Forest Officers the power to constitute forest protection committees. These committees consist of ten to fifteen members including the Panchayat leader (*Sarpanch*), the village Ward Member(s), the Forester, the Forest Guard, a nominee from a concerned non-governmental organisation and six to eight village representatives (of whom at least three should be women) elected by the committee (Government of Orissa, 1993). Their duties are to undertake the 'proper upkeep and maintenance of the assigned Reserved Forest and Protected Forest areas' (Government of Orissa, 1990, pp. 1-2) and to protect these forests from fire, illegal felling and grazing animals. In return for this, they are entitled to collect certain NTFPs and to 'obtain their bonafide requirement of small timber and firewood for household consumption only, and not for sale or barter' (ibid., p. 2).

In contrast to this, the West Bengal model has provisions for 25 per cent of the sale value of timber grown in Protected Forests to be given back to the forest protection committees after harvesting. The idea behind this was to provide local communities (particularly low income groups) with a powerful incentive to undertake long-term forest management and protection (Roy, 1991).[16] This innovation was subsequently taken on board by a number of other State Governments. It was also endorsed by the Government of India which states in its Resolution of June 1990, that: 'beneficiaries should be given usufructs like grasses, lops and tops of branches and minor forest produce' and if they 'successfully protect the forest, they may be given a portion of the proceeds from the sale of the trees when they mature' (Government of India, 1990, p. 1).

Since the West Bengal Resolution on community-based forest protection was passed, JFM has spread rapidly and has achieved some spectacular successes (Kishwan, 1996; Shabbeer and Tankha, 1998).

Although, relatively few investigations have been made into the individual institutional ecologies of successful community-based forest management systems, the best results have tended to be found in small, relatively homogeneous villages with a high proportion of *adivasi* inhabitants and a recent history of community-based forest management (Mehrotra and Kishore, 1990; Kant et al, 1991; Sarin, 1993; Singh, 1999). The involvement of NGOs can assist in getting JFM off the ground (Dhar, 1999); often helping to ensure that different village groups (notably women) are represented on the executive committee and assisting villagers in challenging forest officers who are resistant to power sharing (Arora, 1997; Potter, 1998). Some NGOs have even helped villagers to secure better NTFP sale prices (Tankha, 1998). Indeed, the importance of NGO contributions to JFM has been officially recognised in both the 1990 government order on JFM as well as the most recent central government JFM Resolution No. 22-8/2000-JFM (FPD) (Government of India, 2000).

The most common teething problems with JFM have included an imbalance of power between villagers and the Forest Department,[17] the development of vested interests and the under-representation of women in forest protection committees which still need to be addressed before 'these institutions get well-established on democratic lines' (Singh and Khare, 1993, p. 37). Additional difficulties have centred around villagers' suspicion about long-term forest access (Reddy, 1999), loss of interest during the long lag time between JFM initiation and benefit sharing (Kishwan, 1996), the heavy social costs of sustaining community participation (Chaturvedi, 1996), resistance to power sharing by lower-level Forest Department staff (Karlsson, 2000) and the perennial conflict between short term livelihood needs and a desire for long-term forest protection (Rastogi, 1998). In heterogeneous communities, there have also been problems with wealthier villagers using JFM 'as a means of gaining control over additional forest resources (thereby increasing their own political and economic power), while further reducing the access of marginalised groups who depend on forest resources to meet their basic needs' (Saxena et al, 1997).

In spite of these problems, the success of forest protection committees in West Bengal and Orissa encouraged twenty other States to develop their own JFM initiatives. By January 2000, over 10.24 million hectares of India's 76.5 million hectares of forest land were being managed under JFM programmes by 36,075 committees (Government of India, 2000). In some States, the scope of JFM has been extended to include wider development issues. In the Harda Forest Division of Maharashtra, for example, some forest protection committees have established 'social banks' (funded by local people's earnings from forest work and fines for illegal felling) to provide loans to needy villages.

In an attempt to build on this type of success story and combat the most serious problems facing exiting JFM programmes, the Indian government published an updated Resolution on JFM (Resolution No. 22-8/2000-JFM (FPD). Reflecting the effectiveness of donors and NGOs in bargaining with the state over JFM membership (Agarwal, 2001), the Resolution seeks to address the exclusion of women and other frequently marginalised groups from JFM decision-making by recommending that 'All adults of the village should become eligible to become members of the JFM committees' (Government of India, 2000, p. 1). It also aims to give legal backing to self-initiated forest protection committees by recommending that they be formally registered under the Societies Registration Act 1860 and given opportunities for JFM benefit sharing.[18] The development of conflict resolution mechanisms that involve all stakeholders (including NGOs) is also encouraged. Updated recommendations on benefit sharing state that 25 per cent of the village community's share should be deposited 'in the village development fund for meeting the conservation and development needs of the forests. A matching contribution may be made by the forest department from its share of such sales' (Government of India, 2000, p. 3).

With regard to forest management more specifically, the Resolution proposes (with a number of caveats) the extension of JFM into well stocked forest areas.[19] Maintaining quite a strong participatory emphasis, it is recommended that working plans should be prepared by Forest Officers and Village Forest Protection Committees after detailed PRA and must reflect the 'consumption and livelihood needs of the local communities as well as provisions for meeting the same sustainably' (Government of India, 2000, p. 2). In addition, the working plan 'should aim to utilise locally available knowledge as well as aim to strengthen the local institutions. It should also take into account marketing linkages for better return of NTFPs to the gatherers and should also reflect the needs of local industries/markets' (Government of India, 2000, p. 2). Significantly, the document permits deviations from existing working plans to allow villager-oriented changes (such as a shift to greater NTFP harvesting) to occur.

The Bihar JFM Programme The Bihar JFM programme was adopted in August 1990 and claims to return 80 per cent of the profits raised from forest protection back to forest protection committees.[20] The remaining 20 per cent is kept by the Forest Department as a 'royalty' thus acknowledging the state's monopoly rights over forests. Officially, the programme currently applies only to Bihar's 'degraded' Protected Forests, of which there are 14,000 square acres.[21] The committees established under the programme are known as 'village forest protection and management committees' (VFPMCs) and are formed on the basis of the revenue village. Where a forest serves more than one village, it is divided up into separate areas, each of which serves a different village. The Bihar Forest Department bears the expense of establishing VFPMCs and has the power, should the need arise, to dissolve them.

The VFPMCs are made up of one member from each household and in larger villages where this system would present administrative problems, an 'executive committee' is established. This is made up of fifteen to eighteen people including the current (and if alive, the former) Panchayat leader (*Mukhia*), the Panchayat committee leader (*Sarpanch*), the village religious leader, a local school teacher, four elected Scheduled Caste and/or Tribe representatives and between three and five women.[22] The Forester holds the position of Member Secretary to the Executive Board and co-ordinates between the villagers and the Forest Department. The forest guard is classed as a 'specially invited member', but is expected to attend the monthly committee meetings.

The programme draws the rights and responsibilities of forest users together by giving the VFPMC the duties of preventing further forest decline within the protected area, reporting and punishing illegal tree felling and formulating detailed management plans for the forest. These plans are supposed to promote forest regeneration and address other local needs and must be submitted to the Forest Department for approval before being implemented. The Forest Department then provides the committee with the materials and financial help necessary for their implementation. There is also a provision within the programme for other development needs such as roads, wells and small dams, although the Forest Department's funds for these are small.

In return for protecting and managing the forest, VFPMC members can take fuel wood and NTFPs for *bona fide* domestic use. If they succeed in protecting the forest until the trees within it mature, they are entitled to a share of the final harvest (in addition to wood from periodic coupes and thinnings) according to their customary rights for large timber. The committee can sell any surplus timber, either to local villagers or on the open market with the resulting income (minus the Forest Department's royalty) being divided into three parts: one part for the future development of the forest, one part for the development of the village, and the remainder for the VFPMC's executive fund.[23]

Conditions for the Success of JFM In spite of populist emphases on the role that India's 'ecosystem people' could and should play in resource management, many forest intellectuals would no doubt accept that this could not occur without the spatially sensitive and socio-politically astute development of new institutional arrangements (Corbridge and Jewitt, 1997). Guha's work in conjunction with Gadgil (Gadgil and Guha, 1995), for example, indicates a desire to move beyond the simplistic binary accounts of opposition between the India's forest citizens and the state that brought India's forests onto the international environmental agenda. But although they hint at their disenchantment with romantic populist and eco-feminist accounts that idealise traditional forest management systems, they offer no practical suggestions for promoting successful modern equivalents.

Much of the locally-based empirical research that has been carried out on JFM to date suggests that the most important general pre-conditions for success are a strong cultural or subsistence-related attachment to forests by local communities accompanied by a widely-held desire to address the problem of forest decline (Kant et al, 1991;

Kumar, 1991; Palit, 1993; Sarin, 1993; Jewitt, 1998; MYRADA, 1998; Mukherji, 1998). Small, socio-economically homogeneous villages have been found to be most receptive to the idea of forest protection as forest decline is more likely to be a shared concern (Mehrotra and Kishore, 1990; Kant et al, 1991; Sarin, 1993; Singh, 1999). A strong leader who represents the interests of the community as a whole rather than those of the wealthier villagers can also be important in initiating and maintaining support for JFM (Mehrotra and Kishore, 1990; Sarin, 1993). According to Mehrotra and Kishore (1990), such a person is also often in a good position to assert social pressure on villagers who refuse to participate fully or who habitually break the rules of forest exploitation. Nevertheless, schemes that have provisions for reducing the 'lead time' before benefits are available to local people can help to reduce this problem (Sarin, 1993). Programmes that provide local employment and/or additional income sources can be an important means of both cushioning villagers from restrictions on their own use of forests and maintaining their interest in the scheme until large-scale benefit sharing takes place (Kishwan, 1996; Kumar and Rangan, 1996; Corbridge and Jewitt, 1997; Lal, 1998; Mukherji, 1998; Pathan, 1998; Cooke, 1999; Sitling, 1999). The planting of grasses, bushes and shrubs as an under-storey to quick-growing fuel wood and fodder tree species, for example, allows villagers to collect some NTFPs almost immediately (Sengupta, 1993). The rapid growth of shrubs, bushes and fuel wood trees, meanwhile, sustains (in theory at least) the community's interest in the project.

A pivotal role in most participatory forest initiatives is held by State Forest Departments and success depends, to a large extent, on whether individual villages receive long-term commitment and support from Forest Department staff (Menon, 1995; Jewitt, 1998). A willingness to be flexible and share the responsibility for forest management with local people is also crucial (Sarin, 1993; Agarwal and Singh, 1995; Saigal, 1995). Often, where assistance from forest officers is forthcoming, the Department can act as a 'buffer' between conflicting intra- or inter-village interests. Mehrotra and Kishore (1990) and Kant et al (1991) suggest that the presence of outsiders on executive committees may stimulate the solution of problems that cannot be tackled internally and should, in theory, help to ensure that vested interest groups do not take over the committee.

Conclusion

Although India's populist writers and forest intellectuals have been extremely successful in challenging the criminalisation of local forest users and helping to pave the way for participatory forest management, few have made suggestions regarding the practicalities of how this might work.[24] Indeed their portrayal of active communities eager to re-establish traditional resource management systems misrepresent many local people's development aspirations and gloss over the skill and flexibility needed to stimulate successful community-based resource management in less than ideal conditions. As Gujit and Kaul Shah point out, the tendency to view 'all local people as insiders clearly perpetuates a simplification of intra-communal differences, and hides the reality of high levels of participation by some groups and none by others' (Gujit and Kaul Shah, 1998b, p. 10).

At grassroots level, the development of participatory forest management would benefit greatly from a combination of wider resource management theories with detailed empirical research into the diverse stakeholder interests and dynamic socio-political structures governing resource use in both forest and agricultural ecosystems. This would be particularly valuable in view of the enthusiasm that JFM has generated for a more sustainable form of forest management in India that could promote forest regeneration, watershed protection and other development benefits. Also attractive is its potential for linking indigenous and scientific silvicultural knowledges and for improving understandings of local people-forest relations and wider development priorities. From an institutional ecology perspective, meanwhile, it is important to recognise JFM's potential for helping to overcome fundamental conflicts over the use of forests for conservation, commercial exploitation and rural people's subsistence requirements.

With the exception of Mehrotra and Kishore's (1990) and Kant et al's (1991) work in Bihar and Orissa, however, there have been few local level studies of forest management in tribal India that have tried to reconstruct local forest management histories or to investigate the practical potential of participatory (or even autonomous) forest management in promoting forest protection and regeneration. Using detailed empirical data from two *adivasi*-dominated villages in Jharkhand, my own research has tried to close this gap somewhat by focusing on the political ecology of local resource use and management as well as the main factors influencing the success (or otherwise) of autonomous and joint forest management. Particular attention has been placed on gender-environmental relations and how these influence control over natural resources at the agriculture-forest interface. The wider policy context for this research is provided in the following Chapter.

Notes:

[1] Pathak draws attention to the 'subsistence ethic' (Scott, 1976) of forest-dwellers in Mathwad forest range, Madhya Pradesh. He likens them to 'Chayanovian peasants ... resolutely struggling to cultivate on shallow degraded forest soils' (Pathak, 1994, p. 12) with the help of bribe-taking lower forest officials but in the face of wider opposition from the Forest Department.

[2] To many devout Hindus, actually planting a tree is considered an act of worship. Of particular importance is the *pipal* tree (*Ficus religiosa*) which is said to embody Brahma, Shiva and Vishnu. These trees are often found in villages with altars built around them because it is believed that the person who creates such a *pipal* tree

temple will be forgiven for any bad deeds that he or she has committed (Chandrakanth et al, 1990).

[3] The colonial Forest Department made a ten fold gain in revenue (from 5.6 million to 56.7 million rupees) between 1869 and 1925 (Ghate, 1992).

[4] E.P. Stebbing worked for the Indian Forest Service and later became Professor of Forestry at Edinburgh University.

[5] Guha (1989) attributes these actions to a widespread sense of 'alienation' from forests, although Pathak (1994) questions the appropriateness of using this concept with respect to natural resources.

[6] Although Elwin's influence on British forest policy was rather limited, he did play an important role in the development of state-level tribal policies after India's Independence (Guha, 1992).

[7] The Indian Forest Act, 1952 is based on the 1894 'Voelcker Resolution'.

[8] This, Pathak argues, helps to reinforce local people's dual image of the Indian state which, although not replacing the pre-capitalist 'moral economy', is seen as fundamentally 'just' and 'compassionate'. And, if this does not seem to be the case, it can be attributed 'to the inefficiency of the lower level officials' (Pathak, 1994, p. 15).

[9] This recommendation resulted from a belief that the rights and concessions available to forest dwellers acted as 'convenient tools for entering the forests for carrying out clandestine removal of forest produce' (Government of India: Ministry of Agriculture and Irrigation, 1976, pp. 32-33).

[10] Ironically, this emphasis on wildlife conservation indirectly resulted in 270 people being killed by tigers in these reserves during the 1980s (Pathak, 1994).

[11] The Panchayat system of local governance was introduced in 1949. Each Panchayat aimed to administer around 10,000 people (at the time of the 1950 Census) and was led by an elected leader and a village committee.

[12] According to Alvarez, the Forest Department 'pretends to ecological concerns but its manual of operations is based on the assumption that the Conservator of Forests is what he was in British Times, a conservator of revenue' (Alvarez, 1994, p. 117).

[13] A list of Latin and local tree names can be found in Appendix I.

[14] According to Pereira (1992) the *bhagats* or medicine men of the Warli tribe discuss and learn about a whole variety of indigenous health care methods at annual training camps.

[15] The Draft Bill was criticised for attempting to increase bureaucratic control over forest management and for advocating only limited participation by local people. In the event, it was not passed by parliament.

[16] Mukherji (1998) estimates that after 10 years, the first thinning will provide an income of at least Rs. one lakh from teak forests. Bamboo forests are estimated to yield over 10,000 rupees per hectare from the sixth year of protection and NTFPs are likely to provide even higher incomes. Tiwary (2001), by contrast, argues that the usufruct value that villagers get from JFM is often a more important incentive to participate than the share of the timber revenue, which is often rather meagre.

[17] In a number of cases, Forest Department staff have been reluctant to devolve power and responsibility for forest management to local people (Singh and Khare, 1993; Menon, 1995).

[18] Although one wonders to what extent this will benefit the villagers themselves who have presumably established their own benefit-sharing scheme that requires no dividends to be paid to the state.

[19] JFM in 'good forest' areas is recommended so long as micro- and treatment plans focus on NTFP management. To be eligible for benefit sharing of timber, villagers must have 'satisfactorily protected the good forests for at least 10 years and the sharing percentage should be kept to a maximum of 20% of the revenue of the final harvest - a certain percentage of which should be ploughed back into forest management. The amount of good forests managed under JFM must be limited to a maximum of 100 ha. and 2 km from the village boundary. Good forest JFM shall be carried out on a pilot basis and can be extended further in consultation with the Central Government' (Government of India, 2000, p. 2).

[20] It is expected that the newly-created Jharkhand State will maintain the structure of this programme in its existing form. As my research was conducted prior to the formation of Jharkhand, however, this book is restricted to an analysis of the Bihar JFM programme prior to the formation of Jharkhand in November 2000.

[21] 'Degraded' is defined as having a vegetation density of less than 0.4 as opposed to a closed canopy which has a vegetation density of 1.0.

[22] Significantly, however, Bihar's Panchayats were dissolved in 1997 as a result of overdue elections. Nevertheless, ex-Panchayat leaders continue to play a key role in VFPMCs.

[23] Further details on the implementation of Bihar's JFM programme at the local level will be given in Chapter Eleven.

[24] A significant exception is Bina Agarwal who falls firmly into the 'forest intellectual' rather than the populist camp. Her recent paper on gender and participation in community forestry (Agarwal, 2001) makes some important suggestions about how to make JFM more gender sensitive and open to participation by women.

5 Gender and Environmentally-oriented Development in India

Introduction

Detailed empirical research on gender-environment issues in South Asia started primarily as a result of WID investigations into the South's 'fuel wood crisis' which brought attention to the unequal impact of resource shortages on women (Agarwal, 1986a; 1986b; Dankelman and Davidson, 1988; Moser, 1993; Braidotti et al, 1994; Sarin et al, 1998; Sittirak, 1998). Bina Agarwal's work in India was particularly important in promoting the idea that 'women, especially those in poor rural households in India, on the one hand, are victims of environmental degradation in quite gender-specific ways. On the other hand, they have been active agents in movements of environmental protection and regeneration' (Agarwal, 1992, p. 119).

The publicity given to women's involvement in environmental protection was important in lending support to the idea of a women's 'subsistence perspective' (Mies and Shiva, 1993) which combines concern for the environmental resources on which local communities depend and a willingness to mobilise and protect them. The best known example of this is the Chipko movement where women showed their opposition to commercial forestry practices and timber auctions by hugging trees to prevent them from being felled by forest contractors (Shiva, 1988; Bandyopadhyay, 1992). They also opposed the replacement of mixed natural forest with commercially valuable pine plantations, emphasising the role of forests as life support systems that provide 'soil, water and pure air' (Shiva, 1988). Other well known cases of women opposing what they viewed to be environmentally inappropriate developments include the 'large dam' protests against the Narmada dam in central India, the Silent Valley project in Kerala, and the Inchampalli and Bhopalpatnam dams in Andhra Pradesh (Agarwal, 1997a; 1997b).

The Special Women-environment Link

From a radical eco-feminist perspective, women came to be viewed as having a special relationship with and knowledge of nature. In large part, this is reflected in the recent attention that eco-feminist/WED discourse has given to gender divisions of labour; especially women's role in forest use and subsistence production more generally. In particular, subsistence agriculture and the collection of fuel wood, fruit, seeds, mushrooms, twigs, leaves and medicinal plants have come to be regarded by many as jobs for women, particularly in *adivasi* areas (Shiva, 1988; Centre for Women's Development Studies, 1991; Roy, 1991; Kakat, 1993; MYRADA, 1998; Sarin et al, 1998).[1] Indeed, Vandana Shiva (1988) suggests that because women relate to nature through the 'feminine principle' (*prakriti*), they are much more conservation-oriented and aware of the need for ecological diversity than men who she associates with violence, domination and destruction. In India, she goes on to argue, indigenous forest management is 'largely a women's domain for producing sustenance' (ibid., p. 61). Furthermore, she emphasises that traditionally, the 'backyard of each rural home was a nursery, and each peasant woman the sylviculturalist' (ibid., p. 79).

With regard to the eco-feminist claim that women's agro-ecological knowledges are better developed than men's, the empirical evidence is less widespread although there are some South Asian examples to support this view. Chapman (1983) discovered certain groups of Bihari women to be much better than their menfolk at distinguishing between rice varieties. Similarly, the IDS Workshop (1989a) found that in subsistence households, women frequently select and store grain and therefore know more about its characteristics than do men. Vandana Shiva, meanwhile, claims that women act as 'intellectual gene pools' (Shiva, 1988, p. 48) through their emphasis on biodiversity and conservation. Citing a study of rural women in Nepal, for instance, she illustrates how women make decisions about seed selection (and, she infers, conservation) in many more cases than do men.

The Adoption of Eco-feminist/WED Discourse into Policy-making

In time, the eco-feminist/WED idea of a special women-environment link started to make headway within India's environment and development policies as well as her wider government power structures. Particularly influential was the criticism aimed at the Forest Department's failure to involve women in social forestry programmes, despite their primary role in fuel wood collection (Foley and Barnard, 1985; Shepherd, 1985; Noronha and Spears, 1988; Agarwal, 1997a; 1997b; Saxena, 1998). In 1988, India's National Forest Policy recommended a more participatory and gender-sensitive approach to forest management which has since been adopted by many state- and externally-funded forestry programmes (Jewitt and Kumar, 2000). Recognising the unequal impact that development has had on women, India's Constitution (Seventy-Third Amendment) Act (1992) specified that a third of all elected seats in Panchayati Raj Institutions (PRIs) must be reserved for women. Through this measure, it is estimated that one million women could emerge as grassroots leaders (Sarin et al, 1998).

The Eighth Five Year Plan (1992-1997), meanwhile, recommended that state governments 'allot 40 per cent of surplus land to women alone' (Sarin et al, 1998, p. 44).

Gender and Participatory Forest Management

Although many early participatory forestry programmes were criticised for their failure to encourage women to become members of forest protection committees (Sarin, 1995; 1998a; 1998b; Kaul Shah and Shah, 1995), later initiatives have responded to efforts by donors and NGOs to bargain, on women's behalf, for increased gender participation in forest management (Agarwal, 2001). Large scale, externally funded forestry programmes have helped to stimulate this process with their adoption of WED discourse about women's role as environmental guardians.[2] In West Bengal, the World Bank's Forestry Project has opened up the membership of JFM committees to two household members and has developed a number of practical measures to assist women's participation in forest management. These include the employment of a project co-ordinator for women's activities and the establishment of a women's sub-committee as well as planning teams that include women on a preferential basis. The Project has also set up a special 'women in development' cell within the Forest Department.

These types of initiative have been influential in encouraging Indian foresters to promote women's involvement in JFM, although the form that this has taken varies from State to State (see table 5.1). Although the Bihar JFM programme claims to have made the structure of its forest protection committees more representative than other states, it made only very limited provisions for women's participation; namely a minimum of three and a maximum of five women members on the executive committee. Haryana, by contrast, allowed all adults to be independently eligible for 'general body' membership of its Hill Resource Management Committees (*Wastelands News*, May-July 1993). The Government of Orissa, meanwhile, published a notification in 1993 that the *Vana Samrakshana Samiti* (forest protection committee) should be made up of two people from each household, one of whom must be a woman (*Wastelands News*, May-July 1993).

In an attempt to both address such inconsistencies and further promote women's involvement in participatory forest management, the latest central government resolution on JFM recognises the need for women's involvement in forestry and recommends guidelines to ensure the 'meaningful participation of women in JFM' (Government of India, 2000, p. 1). In particular, it specifies that women should comprise at least 50 per cent of the JFM general body membership and at least a third of the JFM Executive or Management committee. Furthermore, it states that one of the 'office bearer' posts (e.g. President, Vice President, Secretary) must be held by a woman and recommends that Executive/Management committee meetings should not be considered

Table 5.1   **Women's representation in selected Village Forest Committees and entitlement to benefit sharing in JFM resolutions**

State	General body membership per household	Minimum representation of women in the managing committee	Benefit sharing entitlement
Andhra Pradesh	1 woman, 1 man	30% women	Unspecified
Bihar	1 representative	3-5 out of 15-18 members	Committee decides
Gujarat	All adults	2	Not known
Haryana	All adults	1-3 women	Equal access to loans for all
Himachal Pradesh	1 woman, 1 man	2-3 out of 9-12 members	For all villagers
Jammu and Kashmir	1 woman, 1 man	2 out of 11 members	Institution to decide
Karnataka	1 woman, 1 man	2 out of 15 members	Among beneficiaries
Kerala	1 woman, 1 man	3 women	Not known
Madhya Pradesh	1 representative	2 women	Equally among members
Maharashtra	1 representative	1-3 women	Equally among members
Orissa	1 woman, 1 man	3 out of 11-13 members	Equally between households
Punjab	No general body	1 woman	Equally between households
Rajasthan	1 woman, 1 man	3	Equal shares for members
Tamil Nadu	1 woman, 1 man	3-5 women	Not known
Tripura	1 representative	Unspecified	Distributed among members
West Bengal	1 woman, 1 man	Not specified	Either husband or wife

Source: Sarin, 1998; Agarwal, 2001.

quorate unless a third of the women members (or at least one, whichever is more) is present (Government of India, 2000).

Critiques of the Special Women-environment Link

As is the case with populist representations of community-based resource management systems, eco-feminist/WED emphases on women's role as environmental guardians are sometimes problematic in that they are not always grounded in empirical reality. As a result, they provide a rather shaky foundation on which to build environmentally-oriented development projects. Particularly dangerous are the heavily romanticised populist and eco-feminist visions of traditional egalitarian societies living in harmony with nature which are not supported by recent research on environmental history. Bina Agarwal, for example, stresses that the 'pre-modern period was far from one of ecological stability and subsistence balance ... Nor was it a period of social harmony: caste and class divisions were deeply oppressive, especially as experienced by the poorest and lowest castes, and caste (and class)-linked sexual exploitation of and violence against women was common' (Agarwal, 1998, p. 61). In addition, pre-existing social (including gender) and economic inequalities affected people's access to resources long before British colonialism, let alone neo-colonial development was imposed (Nanda, 1991; Agarwal, 1992; 1998).

In light of this, Shiva's romantic vision of India's rural people living, through the Hindu concept of *prakriti*, in a spiritual, organic relationship with nature can be criticised for its orientalism and indeed its nationalism given its tendency to conflate the Indian with the Hindu (Nanda, 1991; Agarwal, 1992; 1998; Jackson, 1993a; 1994). More serious, however, is Shiva's failure to differentiate populist and eco-feminist fantasies from Indian realities. Meera Nanda is especially scathing of Shiva's interpretation of perceptions of nature within Indian cosmology and her failure to question 'why the same cosmology that venerates nature legitimated, and continues to condone, a caste system that utterly degrades (through the phenomenon of Untouchability) those whose work involves close contact with the elements of nature' (Nanda, 1991, p. 38).

In more practical terms, it has become clear that many attempts to translate simplistic eco-feminist/WED discourse into development rhetoric and policy have given insufficient attention to the fact that 'resource use is intertwined with social issues and the identification of 'women's interests' cannot be read off from the current allocation of resource use tasks' (Locke, 1999, p. 280). Even seemingly well planned measures like the World Bank's provision of special fora for women's participation are unlikely to fulfil expectations if they ignore the culturally constructed and socio-politically embedded nature of gender divisions of labour and fail to consider how these may influence people's abilities to acquire and communicate agro-ecological knowledge. The assumed gender neutrality of the participatory appraisal techniques used by the World Bank and other major donor agencies, meanwhile, 'may, in fact, create biases against women due to gendered power relations at the local level' (Crawley, 1998, p. 28).

In South Asia, for instance, the practice of female seclusion (*purdah*) has a significant influence on women's mobility outside the home as well as on the extent to which they interact publicly with men and play a role in decision-making (Kaul Shah, 1998; Sarin, 1998b; 1998c; Singh, 1999; Gururani, 2000; Jackson and Chattopadhyay, 2000; Agarwal, 2001). Agarwal points out that for women to 'speak out in group meetings, especially in the presence of men but also in the presence of other women, requires overcoming learned behaviour patterns that emphasise the virtue of women's silence and soft-spokenness' (Agarwal 1994a, p. 110). She also argues that Asian women sometimes feel obliged to play down their agro-ecological knowledges on occasions when voicing them could be interpreted as inappropriate or even rebellious. Similarly, Hobley documents that men's comments about 'hens crowing' often discourage women from articulating their views in public (Hobley, 1996, p. 148).

Although *purdah* varies tremendously from area to area and between socio-economic and religious groups, it tends to be most strongly observed in northern India, amongst upper-caste Hindu communities and Muslim groups, and weaker in South India and amongst lower caste and tribal communities (Agarwal, 1997a; 1997b; Murthy, 1998; Singh, 1999; Jackson and Chattopadhyay, 2000). Nevertheless, these regional, religious and socio-cultural variations are usually cross-cut by broader rural/urban and class/caste divisions which influence the extent to which women observe *purdah* and participate in household decision-making. A common generalisation is that poorer women, because they cannot afford *not* to work (and are therefore unable to observe *purdah*), have greater mobility outside the household and increased autonomy within it: a situation that often makes them more active in household and wider agro-ecological decision-making.[3] Within a household, however, the actual division of tasks is usually negotiated according to the relative bargaining positions of different household members: a situation that often changes substantially over time in response to factors such as age, education, income, seniority and inter-personal relationships (Kandiyoti, 1988; Leach, 1991; Jackson, 1993b; 1994; 1995; Mayoux, 1995; Cornwall, 1998; Locke, 1999; Agarwal, 2001).

An additional difficulty is that in a society where dowry giving predominates and girl children often become financial liabilities for their parents, restrictions on women's mobility outside the home further reinforces their low 'economic value'. Moreover, the existence of taboos which are applied primarily to women helps to constrain their activities as well as reducing their 'bargaining power both within and outside the household' (Agarwal 1997, p. 33). Taboos may also restrict women from developing their agro-ecological knowledge bases, honing their decision-making skills and making more general improvements to their economic status. The taboo on women ploughing, for instance, limits their control over agricultural production as well as their ability to obtain agricultural expertise.

Bina Agarwal's work has also shown how women's thinking on environmental resources is influenced by land ownership and inheritance customs (Agarwal, 1994a; 1994b; 1997a; 1997b). With the exception of a few scattered areas of matriliny and bilaterality, India's land inheritance systems are, *de facto*, overwhelmingly patrilineal (Agarwal 1994a; 1997). Inevitably, this discourages women from having the same concern for land as men because their interest in environmental improvement is 'filtered through perceptions of security of land tenure that differ from those of men with primary land rights' (Jackson, 1993b, p. 1953). Although these feelings may be countered to some extent by a desire to invest in land for the sake of their children, women are, on the whole, likely to have shorter-term environmental interests and less well developed agro-ecological knowledges than men. These inequalities are often intensified by the fact that divorce and widowhood tend to disadvantage women financially and/or cause them to move away, thereby further reducing their incentive to invest in environmental improvement or protection unless they have clearly defined (use or usufruct) land rights (Joekes et al, 1994; Sarin et al, 1998). Another difficulty is that although women may have legal rights to inherit land, these are often not taken up in practice (Agarwal, 1997; Jackson, 1993a); and especially not by women with limited formal education for whom the paperwork and bureaucracy needed to challenge land titles is a major problem (Development Studies Unit, Stockholm University, 1996). In many patrilineal societies, even common property resources have a strong tradition of male management, so although women may have clear use rights to such resources, their role in managing them is often limited (Agarwal, 1997b; 1998). According to Singh, institutions like Van Panchayats and *gram sabhas* in the Uttar Pradesh hills are strongly male dominated 'and if women are given a position in them it is due to the necessity to implement state guidelines and are in effect no more than token representation' (Singh, 1999, p. 35). Research by Gururani (2000) in Kumaon and Jackson and Chattopadhyay (2000) in Jharkhand, meanwhile, indicates that in spite of hill women's extensive forest-based work, men often regard them as incapable of understanding how to promote forest regeneration and blame them for forest degradation.

An additional factor that weakens the relationship between women and land is the tendency for patrilinearity to be accompanied by patrilocal residence patterns. This makes it difficult for women to assert their land inheritance claims or to manage any land that they do inherit; especially when their marital home is far away from their natal place (Agarwal, 1997a). Indeed, research carried out by Dyson and Moore (1983) and Vlassoff (1991) in India showed that there is often a negative link between the distance that women move when they marry and their autonomy within the household (Jackson, 1993b). Distant marriages tend to be most widely practised by the higher castes/classes which stress the need for strategic marriage alliances and frequently discourage women from working outside the home. According to

Jackson, such alliances are 'at least partly purchased at the cost of individual women's security and power' (Jackson, 1993a, p. 410).

The Dangers of a Superficial Gender Approach

Unfortunately, the rather superficial investigations that most development planners make into gender-environment relations tend to miss these complex political ecologies (Correa, 1997; Sarin, 1995; 1998b; Agarwal; 1997b; Gujit and Kaul Shah, 1998a; Jahan, 1998; Locke, 1999). According to Gujit and Kaul Shah 'pressure from donors to incorporate gender concerns in projects has resulted in many organisations taking up gender issues in a mechanistic fashion. They are incorporated into many programmes only to meet a requirement for resource mobilisation' (Gujit and Kaul Shah, 1998b, pp. 5-6). The same is broadly true of government policy-making in India. In spite of the state's laudable attempts to promote both participatory development and gender equality with the Constitution (Seventy-Third Amendment) Act, many of the women elected into Panchayati Raj Institutions have a minimal input into decision-making as they are 'not allowed by their husbands to attend panchayat meetings ... their husbands participate in the meetings on their behalf and take decisions' (Singh, 1998, p. 13).

With regard to participatory forest programmes more specifically, the provisions for women's involvement often show a strong element of tokenism. Many projects are characterised by very limited levels of discussion between project staff and local people; providing support for those critics who liken participation 'to a Trojan Horse that can hide manipulation and even coercion under a cloak of social palatability' (Gujit and Kaul Shah, 1998b, p. 9). Although donor agencies and NGOs all claim to be using PRA techniques, 'many continue 'grabbing the stick' from villagers' hands and putting colours on the maps themselves. They find it difficult to shift towards a role of facilitating a process that leads villagers to speak, construct, analyse and decide' (Sarin, 1998b, p. 123). During her research on an ODA-funded participatory forestry programme in southern India, for example, Catherine Locke found that the perceived need for donors to pay lip service to gender issues had been 'translated into a preoccupation with formal provisions for women's participation in local forest management institutions and with the necessity of identifying women's preferences for forest resources and, to a lesser degree, their knowledge and values about forest resources' (Locke, 1999, p. 270).

In spite of calls for a 'sophisticated analysis of gendered bargaining processes that captures the negotiation within and between households and communities over resources and their management' (Locke, 1999, p. 277), there is still great pressure to produce results quickly (Blackburn with Holland, 1998b). The failure of most donors to carry out socio-economic evaluations prior to project implementation (Correa, 1997; Sarin, 1996; 1998a; Locke, 1999) is compounded by the fact that the 'community action plans that emerge from PRA can so easily become the action plans of the powerful minority, who have co-opted the process to their own advantage' (Cornwall, 1998, p. 48). Yet according to Jahan (1998) the World Bank undertakes few measures to examine whether its 'gender sensitive' policies work in practice or if greater participation by women is associated with their empowerment and increased control over environmental resources. As a result, project impacts on villagers from different class, caste and age groups are often poorly documented, if investigated at all and misconceptions about gender divisions of labour and environmental management remain largely uncorrected.

Similarly, few projects acknowledge the fluidity and constantly re-negotiated nature of task allocation. Drawing upon empirical data from Jharkhand, for example, Kelkar and Nathan challenge the eco-feminist/WED argument that the collection of forest produce (and subsistence production more generally) is done exclusively, or even primarily by women. Instead, they argue that with respect to the gathering of forest produce 'there is a considerable sharing of jobs with some, not very rigid, division of labour. In the case of fruits, flowers and seeds, the usual practice is for men to shake the trees, while women and children collect the produce. Gathering activity engages the whole family' (Kelkar and Nathan, 1991, p. 57).

In many cases, broad gender variations in forest-based decision-making are intersected by more specific community-, class-, age- and personality-related factors. Resource management institutions in India, for example have a strong tradition of male-dominance and research on autonomous and joint forest protection initiatives suggests that men have a significantly greater role in village-based environmental decision-making than women (Kelkar and Nathan, 1991; Sarin, 1995; Kaul Shah and Shah, 1995; Agarwal, 1997a; 1997b; 1998; 2001; Kaul Shah, 1998; Pangare, 1998; Saxena, 1998; Locke, 1999; Singh, 1999; Gururani, 2000; Jackson and Chattopadhyay, 2000).

One possible interpretation of this is that it represents an instance of gender bargaining whereby women avoid the time-consuming nature of environmental decision-making by accepting its male dominance. Possibly implying a degree of support for this idea, Kaul Shah and Shah's study of Sakwa village, Baruch District shows how women set about forming a forest protection group as a means of encouraging the village men (who felt threatened by the women's initiative) to do this work themselves (Kaul Shah and Shah, 1995). Nevertheless, the fact that restrictions on women's participation in forest management exist even in the relatively gender-equal hill and tribal areas seems to point away from this type of gender bargaining as a widespread explanation. A more significant cause seems to be 'gender ideology, viz. the social constructions of acceptable female behaviour, notions about male and female spaces, and assumptions about men's and women's capabilities and appropriate roles in society' (Agarwal, 1997b, p. 26).

Indicating support for this, a number of studies carried out on JFM throughout India have shown how women often feel uncomfortable about attending mixed public meetings or making their views known when they are allowed to participate in forest management groups (Sarin, 1995; 1998b; Kaul Shah and Shah, 1995; Agarwal, 1997a; 1997b; 1998; 2001; Locke, 1999; Gururani, 2000; Jackson and Chattopadhyay, 2000). To some extent, this problem is exacerbated by the fact that many State JFM programmes specify a minimum number of women on the executive committee but not on the 'general body' committee: a situation that limits the power and confidence of female executive committee members. Other problems include the tendency of male-dominated committees to hold meetings at inconvenient times and to ignore women's views when they do attend (Kaul Shah, 1998).

Other factors that work against women include the frequency with which men in patrilineal inheritance systems control income from land resources. With regard to JFM in particular, Agarwal is concerned that a shift from usufruct (citizenship) to membership rights over forests may generate a 'system of property rights in communal land which, like existing rights in privatized land, are strongly male centred' (Agarwal, 1997b, p. 1): a situation that could increase women's economic vulnerability and dependence on men. And even where participatory approaches are successful, their attempts to encourage women's involvement sometimes creates male-female tension (Hobley, 1996; Cornwall, 1998; Crawley, 1998; Gujit and Kaul Shah, 1998a).

The most significant future cause of gender conflict over JFM, however, is likely to concern the practicalities of 'benefit sharing' amongst participating villagers (Sarin, 1995; Kaul Shah and Shah, 1995; Agarwal, 1997b; Jewitt and Kumar, 2000). The Bihar JFM programme has provisions for dividing any benefits (cash or kind) amongst the 'general body' members of which there is currently a maximum of one per household. Because most designated committee members are men, this means that in practice, few women will receive any independent financial benefits from JFM. In many villages, this has caused resentment amongst women who feel that they have suffered more than men from the restricted availability of fuel wood and are entitled to a separate share of the benefits. Others have also expressed a desire for formal membership on the 'general body' committee.

Conclusion

Many of these factors suggest that eco-feminist/WED emphases on a special women-environment link have a tendency to overestimate women's environmental concern as well as their ability to obtain and make use of agro-ecological knowledges. In some cases, the wider public appeal of women's environmental protection initiatives can overshadow the real nature of the event; often 'invisibilising' men and their environmental concern in the process (Braidotti et al, 1994; Kaul

Shah and Shah, 1995; Leach, Joekes and Green, 1995; Locke, 1999; Agarwal, 1998; Cornwall, 1998; Gujit and Kaul Shah, 1998a; Mawdsley, 1998).

A classic example of this is the Chipko movement which has been widely represented by the international media as a women's environmental movement. As Chapters Two and Four indicate, however, the movement grew out of a long history of opposition to commercial forestry practices which, by limiting local people's access to forests and forest produce, deprived them of their right to subsist (Guha, 1989; 1993; Gadgil and Guha, 1992; 1994; Pathak, 1994; Rangan, 1996; 2000a; Mawdsley, 1998). Another omission made by many popular media (and eco-feminist) accounts is the fact that it was a male dominated local voluntary organisation (the Dashauli Gram Swaraj Sangh - DGSS) led by Chandi Prasad Bhatt which popularised tree hugging in this part of Uttarakhand in early 1973. In fact, the famous incident when the women of Reni village hugged trees to prevent forest contractors from felling them was more a case of the village women standing in for their temporarily absent men folk than an example of a female-led forest protection initiative. Instead of being an eco-feminist movement which aimed to protect trees, therefore, Chipko can be viewed as a (predominantly male-led) peasant protest that sought to defend a resource base and traditional moral economy that was vital to local livelihoods (Guha, 1989; Joekes et al, 1994; Pathak, 1994; Mawdsley, 1998).[4] Indeed, despite the representation of Chipko as a women's movement, the position of women in Uttarakhand 'remains secondary in these village communities and their collective actions often remain within the permissible patriarchal boundaries' (Singh, 1999, p. 32).

As noted in Chapter Three an additional problem with popular (and increasingly policy-related) perceptions about women's environmental concern is that they draw attention away from examples of women who have little environmental interest or who are actively involved in environmental degradation. In India, for example, women's fuel wood cutting activities have contributed significantly to deforestation, but the strongly male-dominated tradition of environmental management ensures that most autonomous forest protection and management initiatives are established and run by men.

In some cases, the adoption of eco-feminist/WED ideology into policy-making has stimulated the initiation of environmentally-oriented development programmes that expect women's participation and thereby increase their workloads (Jackson, 1993a; 1993b; Braidotti et al, 1994; Kaul Shah and Shah, 1995; Agarwal, 1997a; 1997b; 1998; 2001; 2001; Kaul Shah, 1998; Sarin, 1998b; Locke, 1999). JFM, in particular, makes it more difficult for women to obtain fuel wood but does little to challenge (or investigate the reasons behind) their traditional lack of involvement in environmental decision-making. Locke, for example, found that typically the 'closure of forest areas for regeneration occurs without any consideration for how women who previously used the forest are to meet their obligations regarding the collection of forest resources' (Locke, 1999, p. 275). The women questioned by Kaul Shah in Baluch District Gujarat revealed that the forest closure associated with JFM had caused them to spend six hours per day collecting fuel wood, walking five times further than before to collect the same quantity. When asked how they felt about a local reforestation initiative, they replied sarcastically 'What forest? We don't know anything about it now. We used to go to the forest to pick fuelwood but ever since the men have started protecting it they don't even allow us to look at it' (Kaul Shah, 1998, p. 247).

In a similar way, Agarwal's research has pointed to the danger that JFM programmes could marginalise women by failing to give them a 'formal independent right in the new resource and not merely indirect benefits mediated through male members' (Agarwal, 1997b, p. 23). She also found that few programmes actively tried either to empower women or enable them to air their views and develop self-confidence by formalising their participation in local decision-making bodies (Agarwal, 1997b). To alleviate such problems and promote women's role in JFM from a nominal to an active (or even interactive) level of participation, Agarwal suggests that support from NGOs, female forest officers and women-only groups can help to women to challenge existing social norms and increase their self-confidence as well as their intra-community and intra-household bargaining power (Agarwal, 2001). She also recommends women's greater participation in rule-making as a means of reducing their tendency to break community forest management rules and argues that villagers' contributions to JFM (in the form of membership fees, protection duties etc.) would form more equitable criteria for the distribution of JFM benefits.

A rather different problem that is yet to be overcome by many JFM programmes is the need to adapt to specific institutional ecologies and local environmental management traditions. In the Jharkhand region alone, many different land tenure and environmental decision-making systems exist amongst the various tribal and non-tribal communities. While some forest management regimes or land tenure systems may fit quite easily into the standard JFM model, others require substantial adjustment. Similarly, forest management practices that are welcomed by men or better-off villagers may have negative impacts for women or the poor (Sarin, 1995; 1996). As a result, the imposition of what may be perceived to be socio-economically inappropriate or gender-insensitive programmes may be resented by different groups of local people.

Nevertheless, unlike most traditional, male-dominated forest management systems, JFM does at least offer opportunities for women to play a role in forest management (Sarin, 1993). Many programmes have also shown a willingness to respond to past problems; notably the initial failure to encourage women's participation and to acknowledge that a large number of households are female-headed (Sarin, 1993; World Bank, 1992b; 1994; 1995a; 1995b; 1998; 2001). Before examining empirical data on Jharkhandi gender-environment relations, agro-ecological knowledges and forest management systems, however, it is necessary to provide some background information on the research area and its people.

Notes:

[1] According to Linkenbach (1988), forests in Garhwal are also seen as 'a space of secrecy, joy and relaxation' where women can get away from their homes, gossip with friends and 'withdraw from social control for at least some hours' (Linkenbach, 1998, p. 92).

[2] By 1997, 15 externally funded forestry projects had been set up in 11 States at a total cost of over 27 billion rupees (Government of India, Ministry of Environment and Forests, 1997). The World Bank-assisted projects in West Bengal, Andhra Pradesh, Uttar Pradesh, Kerala and Maharashtra claim to view the participation of local people, particularly women, as both a critical criterion for the success of official programmes and a key to forest regeneration and development (World Bank, 1992b; 1995a; 1995b; 1998).

[3] According to Murthy (1998), certain forms of gender discrimination are highly caste-specific. 'Women from upper/forward castes live in severe seclusion and do not work outside their home, irrespective of their economic condition. They experience a much higher incidence of dowry than amongst other castes. Domestic work and child care are solely their responsibility. Women from backward castes engage in income-generating activities ... Women from *dalit* communities play an important economic role in the household, at times contributing more than their husbands' (Murthy, 1998, p. 83). In Jharkhand, meanwhile, Jackson and Chattopadhyay's research in Phulchi village showed that the wealthier caste Hindu Bhumihar women are subject to social restrictions on their mobility which prevent them from undertaking field-based agricultural or forest-based labour.

[4] In addition to Chandi Prasad Bhatt's DGSS, the other two other 'strands' of the Chipko movement had a predominantly male leadership.

6 Introduction to Jharkhand and Fieldwork Methodologies

Introduction

This Chapter is divided into two main parts. The first part provides background information on the geography, historical development, subsistence strategies, culture and politics of Jharkhand. The second part details the selection of the study villages and outlines the research methodologies used during the course of the fieldwork.

Background to the Fieldwork Area

The State of Jharkhand did not exist until November 15th 2000 when the central government allowed it to be created from the southern districts of Bihar. The Jharkhand region, however, has a longer history and is generally understood to consist of the Chota Nagpur plateau and parts of upland Santhal Parganas in South Bihar plus neighbouring Districts in Madhya Pradesh, West Bengal and Orissa (see figure 6.1).

Geographical Characteristics

Situated between twenty two and twenty three degrees north, Jharkhand has a more pleasant climate than the surrounding plains due to the relatively high elevation of the Chota Nagpur plateau (300-600 metres above sea level) and surrounding hills. Most of the State has abundant rainfall (1250-1650 mm per year) which falls mainly during the monsoon months (June to September) and drains quickly along a series of small river valleys that give the plateau an undulating appearance. Most of these valleys and the hills or outliers that rise above the level of the main plateau were, until recently, blanketed with dense stands of *sal*-dominated forest from which Jharkhand - literally 'land of forests' - got its name.[1] In addition to providing an important subsistence resource for the local human population, these forests also used to provide habitats for wild animals such as the tiger, leopard, bear, wolf, wild pig, nilgai (*Boselaphus tragocamelus*), sambar (*Cervus unicoler*) and antelope (Roy, 1915).[2]

Jharkhand's geology consists of a mixed metamorphic and sedimentary rock structure which contains rich deposits of iron ore, manganese ore, mica, copper, coal and limestone. The existence of such a geographically concentrated supply of minerals has provided a basis

Figure 6.1 The Jharkhand Region

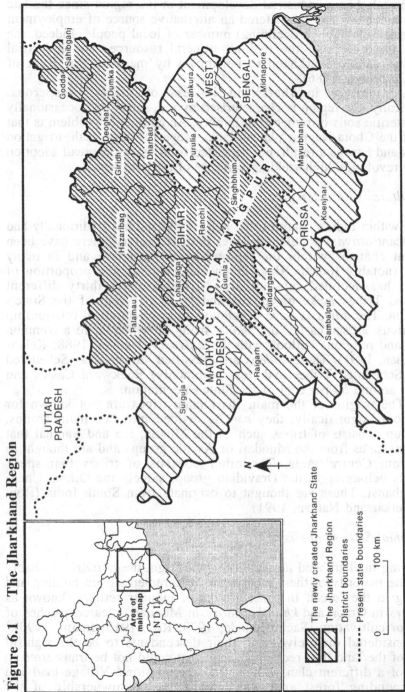

The newly created Jharkhand State

The Jharkhand Region

District boundaries

Present state boundaries

for large scale heavy industrial development in the region since the late nineteenth century and has offered an alternative source of employment for a significant (though declining) number of local people. Indeed, the national importance of the region's mineral resources and industrial output has caused Jharkhand to be known by many as the 'Ruhr of India' (Corbridge, 1986).

Agriculture in Jharkhand is less obviously prosperous. Pedologically, the region is characterised by red, sandy and occasionally loamy lateritic soils of rather poor quality. An additional problem is that much of the Chota Nagpur plateau lacks scope for large scale irrigation facilities and has, as a result, not benefited from the widespread adoption of green revolution high yielding varieties of rice and wheat.

Socio-cultural Characteristics

Situated within India's central tribal belt, Jharkhand has traditionally had a significant *adivasi* population. Over the last century, there have been important changes within the region's social structure and in many districts (notably Ranchi, Hazaribagh and Palamu), the proportion of *adivasis* has declined substantially. Nowadays over thirty different Scheduled Tribes make up approximately 30 per cent of the State's population. The *adivasis* have developed a close reciprocal relationship with various 'artisan castes' or *sadans,* many of which share a common cultural and religious outlook with the *adivasis* (Nathan, 1988; Kelkar and Nathan, 1991). These *sadans* are classified variously as Scheduled Castes, Scheduled Tribes and (less commonly) Backward Castes and make up around 50 per cent of the State's population.

The origins of the main Jharkhandi tribes are not known for certain, but linguistically, they can be divided into two main groups. One group consists of tribes such as the Munda, Ho and Santhal that speak languages from the Mundari or Asiatic groups and are thought to come from Central Asia. The other consists of tribes that speak languages belonging to the Dravidian group, namely the Oraon, Chero and Nagbansi. These are thought to originate from South India (Roy, 1915; Kelkar and Nathan, 1991).

Pre-colonial Settlement History

On arriving in Jharkhand around 2000 years ago, the *adivasis* probably gained the majority of their subsistence requirements from hunting and gathering in the forest. In time, groups of pioneer settlers (known as *bhuinhars* in Oraon and *khuntkattidars* in Mundari) cleared patches of jungle for cultivation (Sachidananda, 1979; Roy, 1915; Imam, 1990).[3] They considered themselves and their descendants to be the rightful owners of the land thus reclaimed: land that could not be transferred to settlers of a different clan without their permission.[4] Village land and the surrounding forest area was held under the ownership of the community as a whole. Inheritance of land is patrilineal and women were not (and are still not) allowed to own land except in temporary circumstances such as widowhood (Roy, 1915; Corbridge, 1986; Kelkar and Nathan, 1991).

Later *adivasi* settlers were either allowed to reclaim new (*rajhas*) land from the forest or were allocated (or sometimes given) *bhuinhari* or *khuntkattidari* land for cultivation. In either case, they were considered to be of slightly lower status than original settlers and tended, in most tribes, to be granted fewer rights within the community. Of even lower status were families from service and artisan castes which were allowed to cultivate little, if any *rajhas* land. Instead, a system evolved whereby they received a set amount of agricultural produce from each tribal family in exchange for their services (Sachidananda, 1979; Kelkar and Nathan, 1991). Nowadays, most *adivasi* villages have a number of *sadan* families plus a few Backward Caste or other non-tribal families living in them and are therefore more heterogeneous in terms of their socio-cultural, economic and political characteristics than Weiner's (1978) account of the 'single tribe village' suggests (Corbridge, 1986; Kelkar and Nathan, 1991; Devalle, 1992).

Mughal and British Colonialism

When tribal communities first settled in Jharkhand, they were their own landlords and paid tribute or rent to nobody. This situation changed in around the sixth or seventh century AD, when the local *adivasi* communities adopted the Chota Nagpur *Raja* as their leader and started to pay a fixed tribute to him (Roy, 1936). After 1585, the Chota Nagpur *Raja* was forced by the Mughal Emperor of Delhi to raise the level of tribute and sent in Hindu *jagirdars* to extract cash rents from the *adivasis*. This resulted in large scale land alienation from the *adivasis* as many *jagirdars* decided to take village land (*manjhihas*) for themselves as part of their share of the rent (Roy, 1915; 1931; 1932; 1936; Corbridge, 1991a).

When the British colonial administrators visited the area, they misunderstood the existing situation of rent collection and endorsed the position of the Chota Nagpur *Raja* by making him a *zamindar* or landlord. At the same time, they exacerbated the lot of the *adivasis*, particularly the 'subordinate lineages' as they bore the brunt of increased levels of tribute and the threat that the *jagirdars* would take their land away from them if they were unable to pay (Kelkar and Nathan, 1991). In addition, the British made alienable village lands that had traditionally been owned as common property and restricted the rights of local communities both to use forests and to reclaim forest land for cultivation (Roy, 1915; 1936). For some *adivasis*, the situation became so bad that they were forced to migrate for work in the tea plantations of Assam (Kelkar and Nathan, 1991).

Those that stayed became more and more unhappy about their worsening situation and started to agitate for change. The Kol rebellion of the early 1830s, and the Santhal *hul* (uprising) of 1855 both focused around the issues of land alienation, the exploitation of *adivasis* by Hindu money-lenders and merchants and the settlement of tribal lands by *'dikus'* or outsiders (Roy, 1946; Hoffman, 1961; Sinha et al, 1969; Sachidananda, 1979; Carrin-Bouez, 1990). These uprisings succeeded in challenging British rule in a major way and forced the British to recognise traditional systems of *adivasi* land ownership. In addition, a number of Christian missions in the area supported the *adivasis* in their desire for the restoration of *bhuinhari* and *khuntkattidari* lands and helped them to fight their cases through the courts (Roy, 1915; Corbridge, 1986; Kelkar and Nathan, 1991; Devalle, 1992).

Between 1869 and 1880, a major land survey was initiated by the colonial state to establish a record of rights detailing the location of *khuntkattidari*, *bhuinhari* and *manjhihas* lands. The result of this was the Chota Nagpur Landlord and Tenant Procedure Act which helped to restrict increases in rents but was unable to stop the alienation of tribal land. Again, discontent amongst the tribals grew and between 1890 and 1895, the Birsa Rebellion (led by Birsa Munda) gathered momentum in its quest for agrarian reform and an end to British rule (Roy, 1936; Sachidananda, 1979).

After crushing the Birsa uprising, the colonial state went on to recognise the *khuntkattidari* system of land ownership and forest use and to continue the process of land registration. The result was the Chota Nagpur Tenancy Act of 1908 which severely curtailed the alienation of *adivasi* land to non-tribals. The increased security of tenure associated with the act helped, in turn, to encourage land improvements and stimulate greater agricultural productivity (Roy, 1915; 1931; 1946; Corbridge, 1986; Kelkar and Nathan, 1991).

Unrestricted forest rights were fully retained only within the Munda *khuntkattidari* system. Most other forests were managed by *zamindars* either under a *katat* or a *rakhat* system. Under the *katat* system, villagers had unrestricted use of local forests upon payment of a fixed 'tribute' to the *zamindar*. Under the *rakhat* system, villagers' rights to forest produce were subject to certain restrictions such as the areas from which forest produce could be collected (Kelkar and Nathan, 1991).

The eventual success of the British in addressing some of the Jharkhandi tribes' major grievances was the result of *ad hoc* local governance rather than the implementation of a coherent tribal policy. Indeed, prior to the enactment of the Government of India Act 1870, British administration in many tribal areas was quite *laissez-faire* to the extent that the Paharia tribe of Chota Nagpur were virtually self-administered: an arrangement that had the added advantage of being very cheap (Corbridge, 1986). In other areas, local *adivasi* administrative structures such as the *Parha* system were maintained and village leaders were accepted as 'direct agents of the colonial government' (Kelkar and Nathan, 1991, p. 75).[5]

After 1870, an isolationist tribal policy (based on the concept of an easily distinguishable and homogeneous tribal community) was pursued by the British in an attempt to address the so-called backwardness of tribal people. The main aim of this policy was to establish special areas in which the 'savage yet noble' *adivasis* could be 'uplifted' and protected from the 'exploiting Hindus' (Singh, 1982; Tripathi, 1988; Corbridge, 1991a; Kelkar and Nathan, 1991; Devalle, 1992; Jewitt, 1995b). Despite the paternalistic rhetoric, however, financial inputs into these Wholly and Partially Excluded Zones (later Excluded and Partially Excluded Areas) were not particularly generous and little of what the British meant by tribal 'development' was achieved. What British isolationism did (albeit unwittingly) encourage many *adivasi* communities to do, however, was to develop a sense of difference based on their own 'tribalness'. This in turn engendered an unwillingness to tolerate further exploitation by outsiders (Singh, 1982a; Dasgupta, 1985; Corbridge, 1986; 1991a; Devalle, 1992).

The Post-colonial Period

Although after India's Independence, attempts were made to replace British isolationism with a more integrationist tribal policy, a number of isolationist measures such as the delineation of Scheduled Areas and the definition of Scheduled Tribes and Castes have been continued. A major change since 1947 has been the introduction of positive discrimination policies which provide college places and reserved jobs in the public sector for Scheduled Tribes and Castes (Sachidananda, 1979; Lal, 1983; von Furer-Haimendorf, 1989; Corbridge, 1991a).

While this assistance has helped to promote the establishment of a tribal elite, the quality of life for most *adivasis* and Scheduled Castes has not improved dramatically (Corbridge, 1986; 1991a; Nathan, 1988). The alienation of *adivasi* land still occurs (although it is much less common than before the enactment of the Chota Nagpur Tenancy Act) and exploitation of *adivasis* by non-tribal merchants, traders and moneylenders has become increasingly common (Devalle, 1992). This situation has not been helped by the fact that many Scheduled and 'Sub-Plan' areas contain a high proportion of non-tribals who naturally benefit from the financial assistance directed to these zones (Corbridge, 1986).

To compound matters further, India's tribal policy has been internally inconsistent; often being weakened by wider economic development programmes. The implementation of positive discrimination programmes, for example, actually helped to undermine assimilationist tribal policies by emphasising ethnicity and difference: the result of which has been seen in the form of Jharkhandi ethno-regionalism since the 1950s. Conflicts between tribal policy and wider economic development programmes, meanwhile, became widespread as north Bihar tried to exploit resource-rich Jharkhand as a kind of 'internal colony' (Devalle, 1992). Cases in point have included the replacement of *sal*-dominated forests with commercially valuable (but locally useless) teak monocultures and the acquisition of *adivasi* land for state-funded development schemes such as the building of irrigation dams, industrial plant, and mining projects (Munda, 1988; Kelkar and Nathan, 1991, Devalle, 1992; Alvarez, 1994; Karlsson, 2000).

Subsistence Strategies in Jharkhand

For most of Jharkhand's inhabitants, agriculture is still the main source of subsistence, although land holdings tend to be quite small (5 acres is a rough average) and the quality of the land poor. During the *kharif* (monsoon) season (June-November), most of the region's terraced valley bottom land (*don*) is planted with *don* (lowland) paddy (*dhan*). The drier, less fertile uplands (*tanr*) are planted primarily with *gora* (upland) paddy plus a variety of millets, pulses and oilseeds. As large scale irrigation is restricted by the region's hilly topography (only around 9 per cent of the southern plateau has irrigation - Kumar, 1998), winter (*rabi*) season cultivation is limited to land with good residual moisture or access to irrigation from streams, tanks and check dams. The main *rabi* crop grown is wheat although winter vegetables have become important as cash crops in recent years.

Significantly, only those rural families with good irrigation facilities or high quality land holdings can guarantee food self sufficiency in an average or poor year. As a result, most people supplement their food stocks with edible forest products including nuts, fruit, leaves, mushrooms, wild vegetables, tubers, game and liquor-making ingredients. Local forests also provide fuel wood, construction timber, medicinal plants, twigs for tooth cleaning, leaves for plate making and other items vital for everyday life.

Due to a decline in the quality and quantity of forests that are open to a growing rural population, however, there has been a reduction in the availability of many of these forest products. For most tribal communities, traditional subsistence strategies must now be supplemented with incomes from activities such as seasonal migration, agricultural labour, small scale business, cash cropping and NTFP sales, and (less so nowadays for *adivasis*) mining and quarrying (Sachidananda, 1979; Munda, 1988; Corbridge, 1986; Pathak, 1994).[6] For service and artisan castes with little or no land, the problem is even more acute as they have traditionally depended heavily upon forests both for raw materials and basic subsistence requirements (Kelkar and Nathan, 1991). As a result, a growing number of very poor families that can find no alternative source of employment are forced into destroying forest resources still further by selling fire wood for cash.

Jharkhandi Culture and Religion

Jharkhand is characterised by a strong sense of cultural identity which stems, in large part, from a common world view or outlook (Singh, 1982b; Kelkar and Nathan, 1991; Devalle, 1992). The region also has a *lingua franca*, known as Sadri or Nagpuri, which helps to unite the different tribes and to ease communication in a State characterised by a series of different languages and dialects.

The main religion for *adivasi* people in Jharkhand is a semi-Animist faith that centres around a supreme deity but also involves the worship of various village-based and nature spirits or deities (Roy, 1912; 1915; 1916; 1922; 1928; Chaudhuri, 1941; Sachidananda, 1979; Rosner, n.d.(a)).[7] Hinduism and Islam are the most important religions for non-Animists, although Jharkhand also has a significant population of both tribal and non-tribal Christians. Nevertheless, Hindus, Christians and Moslems often participate in the main socio-religious festivals of the region and Animists are certainly not averse to celebrating non-Animist festivals.

Jharkhand's *adivasis* are also quite liberal in their attitude towards gender relations and tribal women tend to have a significant economic role as well as enjoying a relatively high social status (Bhattacharaya, 1998; Sarin et al, 1998; Singh, 1998; Karlsson, 2000). Indeed, a tribal woman's 'economic value' is clearly indicated by the fact that when she marries, the boy's parents pay bride price (rather than receiving a dowry) to the girl's parents to compensate them for the loss of her labour. Couples tend to marry relatively late (usually late teens) compared to the mainstream Hindu population and love marriages (within the same tribe but not within the same clan) have a long tradition.

Some of the wealthier Backward Caste and caste Hindu households, by contrast, practice dowry-based arranged marriages and often opt for strategic marriage alliances with families from quite distant places. Women from better-off families often practice a loose form of *purdah* and try to minimise the work that they do outside the homestead. Gender divisions of labour tend to be especially pronounced in such households with men undertaking the lion's share of marketing, field-based agricultural work and forest-based gathering.

Although some gender and class-based divisions in task allocation are apparent in *adivasi* and Scheduled Caste households, the mobility of women from these communities is not restricted as a result of female seclusion and most play a major role both in agriculture and the gathering of forest produce. They also tend to have a greater say in household decision-making although village-level decision-making remains strongly male-dominated (Kelkar and Nathan, 1991; Sarin et al, 1998).

Jharkhand Politics

Since the 1950s, local politics within the Jharkhand region have been strongly influenced by a movement demanding the creation of a separate Jharkhand State within the Indian Union. At first, this movement was dominated by *adivasis* and called for a separate State for the region's tribal population. By the 1980s, the Jharkhand movement had broadened its scope to become more of a regional than an ethnic movement, but like many other tribal movements in India (Singh, 1982a), it has continued to emphasise the need for greater local control over Jharkhandi economic and cultural development.

The roots of the Jharkhand movement can be seen in the Santhal *hul* of 1855 and the Birsa rebellion of 1890 to 1895 which sought to restore some previous golden age of tribal rule by removing *dikus* from Jharkhand (Sinha, 1969; Corbridge, 1986, Devalle, 1986). At the same time, British isolationism followed by the post-colonial state's positive discrimination policies helped to stimulate the development of a politics of ethno-regionalism within Jharkhand as well as to encourage the region's Scheduled Tribes (and Castes) to press for greater access to state benefits (Sachidananda, 1979; Corbridge, 1986; Devalle, 1992).

In 1950, the Jharkhand Party was established and gained widespread support by capitalising on a long-standing anti-*diku* sentiment within the region (Sinha et al, 1969). Led by an Oxford-educated *adivasi* named Jaipal Singh, the Party campaigned for the establishment of a separate tribal State for the *adivasis* of Chota Nagpur, Santhal Parganas and neighbouring areas of West Bengal, Orissa and Madhya Pradesh (Sachidananda, 1979; Devalle, 1992).

This Jharkhand State, it was argued, would be independent economically (due to the region's rich natural resources), fairer politically (as *adivasis* would be ruled by other *adivasis* rather than by *dikus*) and united culturally and linguistically (Sinha et al, 1969; Munda, 1988). The latter argument was encouraged, to some extent, by the fact that the States Reorganisation Committee had chosen language as a basis for re-drawing India's State boundaries in 1956 (Corbridge, 1986).

Although the Jharkhand Party fared well in the 1952, 1957 and 1962 elections, splits within the movement's leadership coupled with a decrease in the proportion of *adivasis* to less than 35 per cent of the region's total population caused a decline in the Party's support base.[8] In response to this situation, the Jharkhand Mukti Morcha (JMM) party capitalised on increasing levels of economic differentiation within many tribal communities by changing the movement's emphasis away from an ethnic base towards a more regionalist (and class-based) alliance against external exploitation (Kelkar and Nathan, 1991; Devalle, 1992).

The JMM gained further support through its attempts to restore alienated land to its rightful owners and its willingness to challenge commercial forest exploitation. Indeed, during the late 1970s, the JMM actively encouraged protests over the cutting of local forests by contractors and condoned illegal tree felling (in state forests) by local people (Corbridge, 1986; 1991a; Devalle, 1992; Corbridge and Jewitt, 1997). In addition to its appeal for *adivasis*, this strategy almost certainly earned the support of poor and landless people (especially *sadans*), who were forced, as a result of their poverty, to depend even more heavily upon local forest resources for their subsistence. Its emphasis on gender equality also earned it the support of many female voters (Kelkar and Nathan, 1991).

Until the mid 1990s, however, the chances that a separate Jharkhand State would be created looked rather slim, due to the lack of co-ordination between the different Jharkhand parties and party factions and the central state's unwillingness to generate a 'domino effect' in other areas of ethno-regional tension (Munda, 1988). In August 1995, however, an important step forward was achieved when the Jharkhand Area Autonomous Council (JAAC) was set up to administer the Tribal Sub-Plan budget in south Bihar (interview with Ram Dayal Munda, 1995). During the last few elections, meanwhile, the Bhartiya Janata Party (BJP) had been trying to consolidate its electoral support base in the region by advocating the Jharkhand cause.

On coming to power at the centre in March 1998, the BJP reiterated its promise to create a separate Jharkhand State and on August 3rd, 2000, a Bill to this effect was moved in the Lok Sabha. On the stroke of midnight on Tuesday November 15th, 2000, Jharkhand became India's 28th State; the third (after Uttranchal and Chhattisgarh) to be created within a fortnight. The National Democratic Alliance formed the first government of Jharkhand under the leadership of the Dumka MP and former Union Minister of State, Babulal Murandi who was elected as Jharkhand's first Chief Minister. After Jharkhand's creation, a financial package of 50,000 lakh rupees was being considered by the central government to compensate the remaining state of Bihar for the loss of Jharkhand's mineral wealth.

The Fieldwork Villages

When I first went out into the field, my main aim was to undertake qualitative village level research into the micro-political ecology of local environmental (particularly forest) knowledges and management strategies and assess the extent to which they echoed radical populist representations of indigenous agro-ecological knowledges and community-based resource management systems. I was also keen to investigate the interactions between the technical and socio-cultural aspects of these knowledges and the extent to which (if at all) the Jharkhand movement had mobilised them as a means of resisting the imposition of commercially oriented forest management or state-funded development more generally.

Selection of the Research Villages

From the above brief description of some of Jharkhand's main characteristics, it is clear that the region is neither ethnically, socio-economically or politically as homogeneous as both British and post-colonial governments have tended to suggest. Quite obviously, this presented problems in terms of finding a fieldwork area that was both conducive to in-depth village level research and representative enough to enable my findings to be generalised.

I chose, therefore, to prioritise the need for in-depth research over that for representativeness and aimed to purposively select two villages that were reasonably 'characteristic' in terms of their size, ethnicity and subsistence base and in which it would be possible to live on a fairly long-term basis. In particular, I was looking for villages that had good forest cover, a predominantly *adivasi* population and were located in an area that was fairly underdeveloped but where Jharkhandi politics were important. Through contacts in the Hindustan Fertilizer Corporation,[9] my attention was drawn to Ambatoli; a fairly large village (1,547 hectares) situated in Bero Block, Ranchi District and fringed to the east by a large area (555 hectares) of Protected (but degraded) *sal* forest.[10] Its population according to the 1991 census was 1751 people, the majority (76.5 per cent) of whom were Scheduled Tribes (63.7 per cent Oraon, 17.3 per cent Mahli, 9.4 per cent Munda, 8.8 per cent Lohra and 0.8 per cent Chik Baraik according to my data).[11] The remaining 23.5 per cent of the population was made up of Scheduled Caste (17.5 per cent) and Backward Caste (6 per cent) families. In short, the village had a fairly representative ethnic mixture for the region that was almost perfect for the type of research that I wanted to do on agro-ecological knowledges, forest use and forest management by (and socio-cultural importance to) both *adivasis*, and non-*adivasis*.

Like most villages in this area, Ambatoli is not served by electricity or piped water and its inhabitants depend mainly on agriculture and forest produce for their subsistence needs. At 46 kilometers from Ranchi, the village is fairly remote (the nearest metalled road with any surface remaining on it was 7 kilometers away in 1993) and does not fall into the category of an 'atypical roadside village' (Chambers, 1983). Indeed, although there is a small shop and a so-called Post Office in one of the village's seven hamlets (*tolas*), almost all of the buildings in the village are built of mud and other locally available materials. The nearest permanent shops, doctor's surgery and chemist are 12 kilometers away in a small market town called Bero and while electricity is available there nowadays, in 1993, one needed to travel a further 15 kilometers to Itki (see figure 6.2).

For the purposes of my fieldwork, the only research criteria that seemed to be missing in Ambatoli were widespread support for and interest in the Jharkhand movement. When I inquired into the matter, it became apparent that although the movement had received much support at the village level between 1950 and the mid-1980s, dissatisfaction with its current lack of unity (plus doubt that the Centre would ever allow the creation of a separate Jharkhand State) had resulted in a significant loss of interest in the subject. Nevertheless, it was clear that even without an active interest in Jharkhandi politics, there were some very interesting linkages between so called technical and socio-cultural systems of agro-ecological, medicinal and silvicultural knowledge. In addition, forest issues were clearly extremely important to local people as well as being a major source of concern at the village level.

During the course of my fieldwork, however, I discovered that while little attention was paid to the day-to-day politics of the Jharkhand movement, many villagers (particularly the *adivasis*), possessed a strong sense of Jharkhandi identity. Although an associated pride about indigenous environmental knowledges was not particularly evident, local people had certainly been influenced by the Jharkhand movement's unwillingness to allow *dikus* to exploit the region's resources. Indeed, during my main period of fieldwork in 1993, the Jharkhand Mukti Morcha representative for the research area, Vishvanath Bhagat, was encouraging local people to prevent further forest decline by re-establishing traditional systems of forest management and protection.

As a result, I decided to give my research a dual focus. The first was a fairly general investigation into the juxtaposition of and interaction between technical and socio-cultural knowledges in the research area and the ways in which (if at all) the Jharkhand movement had influenced local perceptions of environmental issues. The second was a much more detailed and in-depth examination of the use of forests by different people and the problems caused by deforestation at the local level. In researching the latter, I sought to examine the socio-political factors influencing agro-ecological knowledge development and the extent to which village resource management systems resembled those described by India's forest intellectuals and more radical populist and eco-feminist writers such as Shiva.

This led me relatively quickly to my second research village, Jamtoli. Although quite close to Ambatoli (5 kilometers as the crow flies) and fairly similar in its ethnic composition (82 per cent Scheduled Tribe - most of whom are Oraon - and 18 per cent Backward Caste, according to the 1991 census), subsistence base and facilities, Jamtoli is rather different in size (857 hectares), population (1339 persons) and structure (it has only three *tolas*). Another very interesting and rather unusual feature of this village is its success in developing and maintaining community-based resource management systems for both irrigation water and forest land.

Unusually for this area, Jamtoli's 246.45 hectares (of which only 83.36 hectares remain) of *sal*-dominated forest is classified as a Reserved rather than a Protected forest. The reason for this is that Jamtoli was one of the few non-*zamindari* villages in the area during the British colonial period and its forest, which belonged directly to the crown, was an obvious target for reservation as it was sufficient in timber both for local use and commercial production.

Of particular interest for my research was the fact that unlike most Reserved forests, Jamtoli forest has actually been protected *from* the state (particularly forest contractors and corrupt forest officials who used to exploit it for their own profit) by means of a very effective system of autonomous, community-based forest protection. The forest is well known in the vicinity for its healthy appearance and because it is

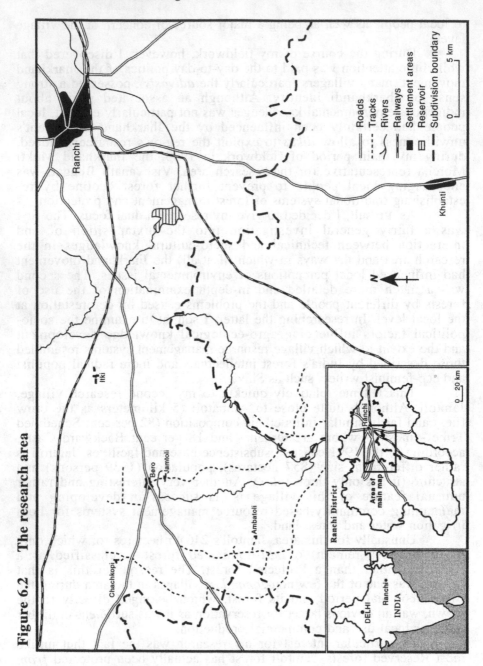

Figure 6.2   The research area

one of the few woodlands in Ranchi District that actually deserves to be called a forest.

When I pursued the issue of forest protection further, I found that although Ambatoli had had little success in sustaining village-wide autonomous forest management in the past, two of its hamlets were trying to imitate Jamtoli's initiative by protecting small areas of the village's rapidly degrading forest from further destruction by outsiders. Most villagers had also heard about the (then) new Bihar JFM programme and were interested in the possibility that JFM might unite the villagers and assist them in maintaining village-wide forest protection. In Jamtoli, meanwhile, local people were so enthusiastic about JFM that they had already started to re-arrange their autonomous forest management system along JFM lines. The combination of these events encouraged me to investigate the institutional ecologies assisting the establishment and maintenance of autonomous forest management as well as the practical value of populist discourse for helping to develop the new institutional arrangements needed for successful (and locally empowering) JFM.

Research Methodologies

Ambatoli Village Although conscious of criticism regarding researchers' unwillingness to undertake fieldwork during the monsoon (Chambers, 1983), I completed my main period of fieldwork in Ambatoli between January and June 1993 and did further research in the village between October and December, 1993. I also returned in April 1995, May 1996; August 1997 and March 1999 to update my research and to take note of the changes that had taken place since my main period of fieldwork.

Because I wanted my research to be village-based, I rented a room from an Oraon family in Ambatoli and stayed there (with the exception of weekly visits to the bank in Ranchi) between February and June 1993. During this time, I undertook a variety of research techniques starting with a questionnaire-cum-census survey of 100 households. Although my spoken Hindi is competent, I was concerned about missing the nuances of what people told me, so I employed a research assistant, Harish (who was working for the Hindustan Fertilizer Corporation as Ambatoli's village motivator) to help me with the administration of this survey and broaden my understanding of village agriculture more generally. The first part of the questionnaire survey sought information on household composition, socio-economic status, educational background, community affiliation, religion, primary and secondary employment, landholdings, agricultural techniques, cropping patterns and household survival strategies. The second part was devoted to eliciting information on the use and management of Ambatoli forest by right-holders from both Ambatoli and the neighbouring village of Patratoli. Most (but not all) of these interviews were held with men.

As social norms within rural India make it difficult for men to converse freely with women, I employed two *adivasi* women (Pyari Lakra and, later, Subani Kujur) to help me to obtain more detailed gender-related and socio-cultural information on agro-ecological and silvicultural knowledges. They proved invaluable in guiding me through the intricacies of Jharkhand village life and helping me to understand villagers' socio-cultural norms and development aspirations. With their help, I conducted a variety of fieldwork techniques such as direct and participant observation, group discussions and participatory appraisal techniques including social mapping, time lines, oral histories, wealth and preference ranking (Chambers, 1992) which were used to fill in data gaps and to cross-check the accuracy of the information obtained through the questionnaire survey (Chambers, 1983; Casley and Lury, 1987). On the basis of the initial questionnaire survey, I selected a purposive sub-sample of the main village population for more extensive discussions about intra-village (particularly gender-based) variations in the possession of (and control over) environmental knowledge. In particular, I was interested to discover whether villagers transferred agro-ecological knowledges, seeds or techniques from their natal to their marital villages (see Appendix II for a list of villagers interviewed on this topic). A number of separate interviews on forest use, knowledge and management were conducted in Ambatoli and Patratoli.

A second, slightly overlapping sub-sample was selected for more in-depth discussions about the socio-cultural aspects of village life (including agricultural and forest-related festivals), traditional and current forest management, forest usage patterns and forest lore more generally. In particular, I was interested to discover whether anomalies in agriculture, forest use and forest management strategies that showed up in the questionnaire survey data could be linked to differences in gender or socio-cultural, economic or educational background. I was also keen to learn about indigenous medicinal practices from local herbalists and to compare women's and men's use of, knowledge about and cultural interaction with village forests. Another area of interest for me was the extent to which silvicultural management strategies (such as coppicing and pollarding) vary between state-owned forests and individually-owned trees on private fields and homestead land.

This research was carried out mainly with the use of semi-structured discussions, direct observation and participant observation techniques and involved accompanying local people into the surrounding forests, observing their activities, asking questions and taking part (where appropriate) in their daily routines. Information on more sensitive or controversial issues such as gender-related differences in forest use and knowledges, the religious importance of forests, forest lore and the association of certain forest areas with evil spirits, was obtained from informal chats with groups of women (and, less often, men) who had come to know and trust me during the course of my fieldwork. Similar methods were also used to gain insight into villagers' attitudes towards the idea of a separate Jharkhand State and, more specifically, the extent to which the Jharkhand movement had helped to create an awareness of forest issues.

Villagers' opinions about both autonomous forest protection and the Bihar JFM programme were obtained primarily from semi-structured group discussions (some of them exclusively with women to encourage them to put forward their views) in each of Ambatoli's *tolas* as well as in Patratoli. More structured interviews were later carried out with those villagers who had played an important role in the past or were currently involved in the establishment of *tola*-based forest protection committees. Villagers who were hoping to be involved with the JFM programme when it was established in Ambatoli were also interviewed in detail about their thoughts on the scheme's potential benefits and problems and the way in which it would be administered.

More 'official' information on Ambatoli and its environs, local systems of agriculture and forest management, the Bihar JFM programme and forest policy more generally were obtained during formal, structured interviews; the main targets of which were the Forester in Bero and the Divisional Forest Officer (Ranchi East Division) in Ranchi. Secondary source data on agriculture and forest management in Ranchi District between 1939 and 1956 were obtained from census documents, Survey and Settlement reports and agricultural statistics dating from 1901 coupled with Annual Progress Reports of Forest Administration in the Province of Bihar. More general information on the linkages between local and scientific agricultural practices came mainly from interviews with researchers from the Birsa Agricultural University.

Following Peet and Watts, an attempts has been made in the empirical chapters that follow to provide 'a sort of political-ecological thick description' (Peet and Watts, 1996a, p. 38) of villager-environment interactions. In an effort to avoid orientalist misrepresentations of the 'others' that I was studying (Baudet, 1965; Said, 1978; Pagden, 1982; Inden, 1986; Pinney; 1988), I have let my respondents speak for themselves wherever my field notes permit accurate quotation.

Jamtoli Village Fieldwork in Jamtoli was carried out primarily between October and December, 1993, although I have made several return visits to assess the progress of the village's forest protection initiative. During my stay in the village, I rented a room from a man named Simon Oraon, who is the *Parha Raja* for twenty one villages in Bero Block and is also the person who initiated forest protection in Jamtoli in the early 1960s. Broadly speaking, the research methodologies that I used in Jamtoli were similar to those used in Ambatoli; the main difference being that my main priority was to conduct in-depth research on community-based resource (particularly forest) management and protection systems rather than to undertake a more broadly focused village survey. As a result, I completed only twenty one formal questionnaires and tried to make up the shortfall in basic census information with the use of participatory appraisal techniques and over twenty five group interviews.[12]

An additional difference with my research in Jamtoli was that I spent significantly more time with one 'key informant' (namely Simon Oraon) than was the case in Ambatoli. The reason for this was that I was living in his house and this gave me an ideal opportunity to question him informally about why he started autonomous forest protection in Jamtoli, how the forest is protected and managed on a day-to-day basis and why his system has succeeded where others have failed. This information was also supplemented with direct observation of villagers' daily routines, coupled with informal interviews and participant observation techniques when I was with people who felt comfortable with my presence.

As Simon is *Parha Raja*, I was also able have discussions with him and obtain very informed opinions about more general administrative, developmental, political and policy issues, including the Jharkhand movement and the Bihar JFM programme. As the programme was effectively running in Jamtoli during my main period of fieldwork, I was able gain more information about its objectives, benefits and administration than was the case in Ambatoli which did not establish JFM until 1998. I was also able to supplement (and cross-check) the information that Simon gave me during semi-structured group interviews in each of Jamtoli's three hamlets. In addition, I held in-depth interviews with several other members of Jamtoli's village forest protection and management committee (VFPMC) and used direct and participant observation techniques when I went into the forest with them on 'guard duty'.

In addition to his administrative work, Simon had carried out a large amount of developmental (particularly irrigation) work both within and outside Jamtoli and was, therefore, an excellent source of knowledge on how to improve local agriculture and manage common property resources. Indeed, Simon's indigenous agro-ecological and silvicultural knowledges were so much more extensive that those of anyone else that I met in Jharkhand, that cross-checking and triangulation proved to be very difficult. The view I took however, was that this information was valuable even though the verification of every fact was impossible. As I have no reason to believe that my 'key informant' in Jamtoli was telling lies, therefore, I have included, in Chapters Five, Six and Seven, a substantial amount of what he alone told me about Oraon history, local *Parha* administration, his own development work and, of course, village-based forest protection.

Notes:

[1] In addition to sal (*Shorea robusta*), other tree species including asan (*Terminalia tomentosa*), bar (*Ficus bengalensis*), bel (*Aegle marmelos*), bhelwa (*Semecarpus anacardium*), dumar (*Ficus glomerata*), dhaunta (*Anogeissus latifolia*), gamhar (*Gmelina arboria*), imli (*Tamarindus indica*), jamun (*Prunus cornuta*), karonda (*Carissa carandas*), karam (*Adina cordifolia*), karanj (*Pongamia pinnata*), kend

(*Diospyros melanoxylon*), *kusum* (*Schleichera oleosa*), *mahua* (*Madhuca indica*), mango (*Mangifera indica*), *parsa* (*Butea monosperma*), *pipal* (*Ficus religiosa*), *piyar* (*Buchanania latifolia*), *semar* (*Bombax malabricum*) and *sisam* (*Dalbergia sissoo*) used to be abundant in the forests of Jharkhand. During the 1950s, however, most of these forests were felled by *zamindars* and contract cutters and *sal*, as a result of its rapid regeneration through coppicing, became even more dominant than was the case previously.

2 Sarat Chandra Roy was an eminent Indian anthropologist who worked for many years with the tribal people of Chota Nagpur. Although Roy can be criticised nowadays for his 'kindly' paternalism and orientalism (Jewitt, 1995), his scholarly understanding of and empathy with *adivasi* life and culture in Chota Nagpur is probably second to none. Consequently, his work is drawn upon throughout the rest of this book.

3 The terms *bhuinhar* and *khuntkattidar* refer both to the original settlers of a village and to their male descendants of the same clan who are the only people entitled to inherit *bhuinhari* and *khuntkattidari* land.

4 According to Roy, *adivasi* clan members regard themselves as being 'of one blood' (Roy, 1915, p. 178) and there are strict social rules about inter- and intra-clan relationships, particularly marriage which is restricted to men and women from the same tribe but different clans (Roy, 1919; Hoffman, 1961). Most *adivasi* clans have animal or plant totems, the names of which they may take in preference to the general tribe name.

5 The *Parha* system is a traditional form of village administration run by village elders and headed by a *Parha* chief who is known to the Mundas and Oraons as a '*Parha Raja*' and to the Hos as a '*Manki*'. Each *Parha* consists of up to twenty two villages. The Santhals have a similar type of 'inter village council' known as a *Pargana* (Kelkar and Nathan, 1991).

6 Some Jharkhandi *adivasis*, particularly members of the Korwa and Birhor tribes, do not practice settled agriculture and rely very heavily on the gathering of forest produce and the hunting of game for their everyday subsistence (Kelkar and Nathan, 1991).

7 Jharkhandi Animism has merged, to some extent, with mainstream Hinduism from the plains (Kelkar and Nathan, 1991).

8 Corbridge (1986) shows that if British rather than current definitions of tribe are used (to remove the effects of the de-Scheduling of certain tribes), the *adivasi* population as a percentage of Chota Nagpur's total population would be significantly higher.

9 Ambatoli was selected as a project village under the Hindustan Fertilizer Corporation's Rainfed Farming Project. This project has since been taken over by Krishak Bharati Co-operative Ltd (KRIBHCO) as part of the DfID-sponsored Eastern India Rainfed Farming Project.

10 I am deeply indebted to my ex-partner, Stuart Corbridge, who travelled with me to India in January 1993, helped me in my search for a fieldwork village and accompanied me for a substantial part of my stay in Ambatoli. His moral support plus his knowledge and experience of the Jharkhand region were invaluable to me.

11 The district level census handbooks do not give detailed information on ethnic group at the village level, so I have supplied this information from my own census survey of 100 households (676 people).

12 The formal questionnaires gave information on only 177 people (out of a total of 1339), but group interviews and participatory rural appraisal techniques brought me into contact with around 300 different villagers.

7 Introduction to the Fieldwork Villages

Introduction

Although the main focus of the rest of the book is on the agro-ecological knowledges and resource management systems of villagers in Ambatoli and Jamtoli, it is first necessary to provide a more detailed introduction to the fieldwork villages and their inhabitants. To set the scene for the empirical work that follows, this Chapter firstly presents a brief account of Oraon settlement in the research area plus a more detailed physical and socio-cultural description of the two fieldwork villages. It then goes on to examine Ambatoli and Jamtoli from a micro-political ecology perspective. In particular, it examines the major inter- and intra-village and community differences in wealth, caste and social status and the effects that these have on local subsistence and household survival strategies.

Oraon Settlement, Culture and Administration

The Oraons are the most numerous tribe of Chota Nagpur and the fourth largest tribe in India. Although their origin is not certain, they are thought to have originated in South India, from where they migrated to the Son Valley in Shahabad District, Bihar (Roy, 1915). When they first arrived in Jharkhand in around the ninth century AD (Roy, 1912), they are thought to have subsisted from hunting and gathering. They later formed more permanent settlements and started to cultivate land reclaimed from the forest and to breed cattle. Forests and other public lands remained under community ownership whereas cultivated land (known as *bhuinhari* land after the *bhuinhars* or original settler clans and their descendants) was owned individually.

According to Roy, forcible land seizures by *jagirdars* and the sale of formerly protected jungle trees by *zamindars* stimulated many Mundas into moving to the south and east of the Chota Nagpur plateau and helped to pave the way for the in-migration of Oraons (Roy, 1936). An additional theory is that the Oraons (who claimed to have introduced plough agriculture into the Chota Nagpur plateau) managed, by virtue of their 'superior equipment for the struggle for existence ... better knowledge of agriculture, and their rapid multiplication due perhaps to better food secured through agriculture' (Roy, 1928, p. 27) to push the Mundas further to the south and east.[1] Nevertheless, a few Munda families were always welcomed in Oraon villages because they alone had the ability to propitiate the village deities that had been disturbed when the area was settled by Mundas (Roy, 1915). The Oraons also encouraged the settlement of a few *sadan* families to carry out basket-making, blacksmithing and cattle-minding activities (Roy, 1915).

Religion in the Research Area

The main religion for Oraons and Mundas in the research area is a semi-Animist faith known as *Sarna*. At the head of the *Sarna* pantheon is the supreme god *Dharmes* who is thought to be 'the guardian of all morality, punishing all offences against customary morality with sickness, death and other calamity' (Roy, 1928, p. 13). *Dharmes* is not worshipped as such, but his blessing is invoked before all major religious festivals and in times of dire emergency, a white cockerel or goat may be sacrificed to him (Roy, 1928). He is also 'remembered' on a daily basis by most Mundas and Oraons when they 'offer' him a few grains of rice and drops of water or liquor at meal times. Next in the hierarchy are village deities and spirits which are thought to control most day-to-day activities and to whom prayers and offerings are made periodically by the village priest (*Pahan*) to ensure that village harmony is maintained (Roy, 1928). The principal village deity is known as the 'old lady of the grove' (*Sarna Burhia* or *Jakra Burhia* in Sadri or *Chaala Pachcho* in the Oraon language, Kurukh) and is believed to control the other village deities and spirits. As *Sarna* worshippers have no temples as such, *Sarna Burhia* is worshipped in the village sacred grove (*Sarna* or *Jakra*) which is typically situated close to one or more *sal* trees and contains a cluster of rocks and stones that represent the main village deities. Worship usually consists of offerings, animal sacrifices and prayers for village health and prosperity. Third in the hierarchy are the family, clan or ancestral spirits which are believed to protect their living descendants from harm and are usually worshipped within the house by the head of household (Roy, 1922; 1928).

In addition to these predominantly beneficent deities and spirits, the *Sarna* pantheon contains groups of malevolent spirits or ghosts (*bhuts*) which live in trees or rocky places away from the sacred grove and work outside the control of *Sarna Burhia*. Their malevolence is usually attributed to the belief that they represent the disembodied spirits of people who died 'unnatural deaths' such as being murdered, killed by a wild animal, falling from a tree or (for women) dying during pregnancy or childbirth. To ensure that they do not cause harm to a village or its inhabitants, malevolent *bhuts* must be placated and propitiated with animal sacrifices (Roy, 1915; 1922; 1928). In extreme circumstances, the assistance of a 'ghost finder' or *bhagat* may be required and if all else fails, many *Sarna* worshippers resort to magic and witchcraft (Roy, 1915; 1928). In most cases, malevolent *bhuts* are propitiated every so many years with a specific sacrifice. The exact 'formula' of offerings or sacrifices varies according to the *bhut* in question and the nature of the request made of it. A buffalo is the biggest sacrifice that is usually offered and lesser spirits are propitiated with goats or chickens.

Not all Oraons in the research area are *Sarna* worshippers, and the *Sarna* faith is not itself discontinuous with all aspects of Hinduism (Kelkar and Nathan, 1991). German missionaries converted a large number of *adivasis* to Christianity from the middle of the nineteenth century onwards and a significant number of Oraons also joined Oraon revivalist movements during the late nineteenth and early twentieth centuries.[2] In addition, most villages have a small minority of Hindu Backward Caste villagers and Scheduled Caste *sadans* as well as a few semi-Hinduised tribal *sadan* families.

To all intents and purposes, the main religious leader in Oraon villages is the *Pahan* whose job it is to propitiate village gods and maintain equilibrium with the spirit world. In some villages the office of *Pahan* is hereditary, but in others, a new *Pahan* is 'chosen' every three years by *Sarna Burhia*. Usually, this involves the 'possession' of the retiring *Pahan*'s magic winnowing basket (*soop*) on which *Sarna Burhia* is believed to sit (Roy, 1928). In many villages, the *Pahan* is assisted in his duties by a *Pujar* and a *Soussar*.[3]

Village Administration

As the Oraons increased in number and their villages grew in size, secular headmen (*Mahtos*) were elected to deal with external affairs and village committees or '*Panches*' were established (Sachidananda, 1964). Elders from groups of seven, nine, twelve, twenty one or twenty two villages made up a *Parha*. This acted as an informal democratic socio-political institution which dealt with most forms of village-level administration, decision-making, conflict resolution and resource (notably forest) management (Roy, 1915).

Each *Parha* was presided over by a '*Parha Raja*', elected from within the *Parha*'s King or *Raja* village. The *Parha Raja* would hold regular *Parha* meetings near the *Raja* village's sacred grove and try to settle disputes that could not be resolved internally. Typical cases included intra-village conflicts, marriage disputes (where a pre-marital affair resulted in pregnancy and either the boy refused to marry the girl or the couple belonged to the same clan (*gotra*) and could not marry) and expulsion or re-admission of 'outcast' villagers to the tribe (Roy, 1928). The *Parha Raja* would also act, when required, as an intermediary between local people and representatives of the state. This was particularly important during the British period when literacy levels were much lower than today and many *adivasis* felt unable to represent themselves properly when conflicts over land ownership or forest use occurred.

In 1949, village-level administration changed somewhat with the introduction of the Gram Panchayat system of local governance throughout India (Sachidananda, 1964). Consisting of around 10,000 people (at the time of the 1950 census), each Panchayat has an elected leader (*Mukhia*) and a village committee (*Gram Samiti*) headed by a *Sarpanch*. The *Samiti*'s main responsibilities relate to village development and welfare (for which Block level funds are available) although the *Mukhia* may attempt to solve inter- and intra-village conflicts when required. Unlike the *Parha*, the Gram Panchayat system groups villages together on the grounds of administrative convenience rather than community-based relationships (Sarin, 1993). Unfortunately, the superimposition of this new local form of 'representative democracy' on top of the *Parha* system (plus the fact that the Gram Panchayat had direct access to local development funds) resulted in the breakdown of most *Parhas*.

For many Oraons, this change was greeted with dismay as they felt that they had lost the sense of being 'like one big *Parha* family that could solve all kinds of local problems' (discussion with Budhwa Oraon, 1993). Simon Oraon's *Parha* is one of the few that has survived and Simon himself tries hard to maintain the sense of community that many feel was lost after the introduction of the Gram Panchayat system. Interestingly, he still maintains that the area was ruled much better before India's Independence because 'the British didn't interfere with the *Parha* system and respected other *adivasi* customs and rules' (discussion with Simon Oraon, 1993). Elsewhere, this gap has been filled to some extent by the Jharkhand movement's emphasis on a wider sense of tribal identity and appeal as an alternative to non-*adivasi* governance and development (interviews with Tila and Simon Oraon, 1993).

The Fieldwork Villages

As mentioned in Chapter Six, the physical layouts of Ambatoli and Jamtoli are slightly different as Ambatoli is a larger, dispersed village made up of seven hamlets, whereas Jamtoli has only three hamlets. All hamlets (*tolas*) in both villages are surrounded by agricultural land which gives way to forest to the north and east in Ambatoli and the south, west and east in Jamtoli (see figures 7.1 and 7.2).

Layout and Physical Characteristics

Although there is no strict community-based spatial segregation in Ambatoli and Jamtoli, community groups often cluster together by choice. A number of hamlets also have small outliers consisting of a few houses occupied by a single community group: usually Mahlis or Lohras who despite their official status as Scheduled Tribes, are considered to be 'low caste' by the Oraons. Neither Oraons nor Mundas will take food or water from the artisan castes who try to cluster their houses where possible. Indeed, one hamlet in Ambatoli is inhabited almost entirely by Mahli families.

Figure 7.1   Ambatoli sketch map

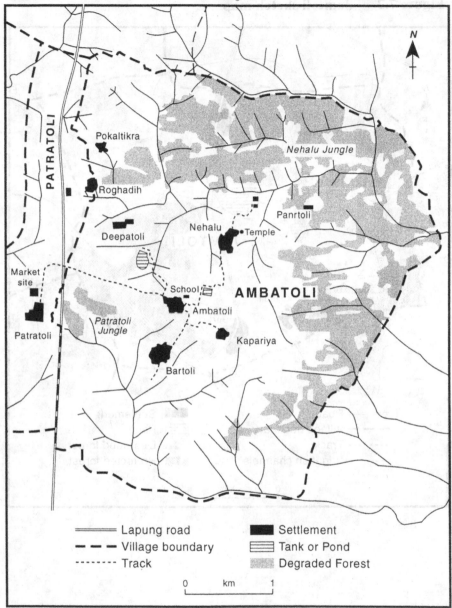

Legend:
- ═══ Lapung road
- ▬ ▬ Village boundary
- ········ Track
- ■ Settlement
- ▤ Tank or Pond
- ▒ Degraded Forest

0   km   1

Figure 7.2   Jamtoli sketch map

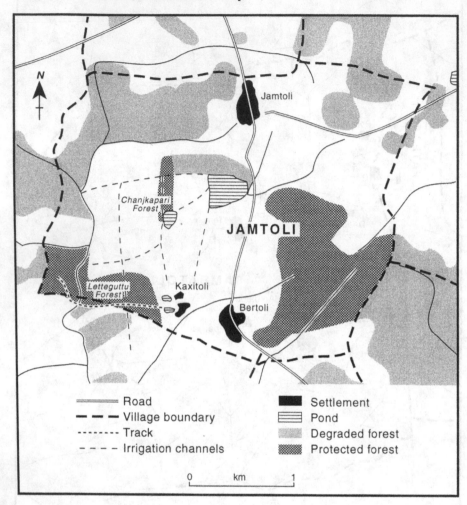

The houses in Ambatoli and Jamtoli are all built of mud and are usually constructed, extended and repaired by their owners, using local materials. The main roof support joists and beams are usually made of *sal* wood. Bamboo or other light wood is used as battens to support country tiles. Most villagers prefer this type of roof to thatch (which is usually made of locally available '*kher*' grass) as country tiles are more waterproof and can be made very easily at home.

Wealth and family numbers are the main influences on the size of individual houses, with many of the larger extended households having enough *bari zamin* (homestead land) to use as a vegetable garden. A few even have their own wells from which they can irrigate winter vegetables and draw their washing water without having to go to a communal well, pond, tank or stream. The vast majority of villagers fetch their drinking water from hand pumps (each hamlet has at least one) as they consider it to be cleaner than well or stream water. For bathing, however, most villagers prefer to use pond, stream or well water.

Socio-cultural and Religious Characteristics

As mentioned in Chapter Six, both Ambatoli and Jamtoli are mixed villages in terms of their ethnic composition and, like the mining villages in Singhbhum District examined by Corbridge (1986), do not conform well to Weiner's (1978) model of a 'single tribe' village. Jamtoli is perhaps as close as one might get, nowadays, to a 'typical' tribal village. My own census data indicates that 88.7 per cent of the villagers are classified as Scheduled Tribes and 11.3 per cent are classified as 'Other Backward Castes' (see table 7.1).

Table 7.1   Jamtoli's sample population by community group

Community group	Number of households	Number of individuals	% of village population
Scheduled Tribe	18	157	88.7
Backward Caste	3	20	11.3
Total	21	177	100

Source: Jewitt, 1996.

These figures are corroborated reasonably well by 1991 census data which shows a slightly lower proportion (82 per cent) of Scheduled Tribes and no Scheduled Castes. 'Other Backward Castes' are not listed separately in the census but can be assumed to make up the remaining 18 per cent. This category is not unproblematic, however, as the three Backward Caste households in my Jamtoli questionnaire survey all *think* of themselves as Scheduled Castes, even though they are not eligible for positive discrimination schemes. This situation raises some interesting issues regarding local versus official identities and the ways in which these have been contested in Jharkhand as part of a wider ethno-regional, political and economic struggle (Devalle, 1992; Raj, 1992). It is also an important reminder that although de-Scheduling by the central state has reduced the numbers of Scheduled Caste and Scheduled Tribe people in Jharkhand (Corbridge, 1986), it has not succeeded in removing their sense of *adivasi* or *harijan* identity.

Ambatoli is more heterogeneous in its ethnic composition than Jamtoli. My census data indicate a Scheduled Tribe population of 75.8 per cent, a Scheduled Caste population of 9.5 per cent and a Backward Caste population of 14.7 per cent (see table 7.2). These figures are fairly close to the 1991 census data which show a Scheduled Tribe population of 76.5 per cent, a Scheduled Caste population of 17.5 per cent and a Backward Caste population of 6 per cent. Again, much of the discrepancy in the Backward Caste figures may be accounted for by the Ambatoli villagers' own confusion over Backward Caste and Scheduled Caste classifications.[4]

Inter-community Relations The relationships between community groups and sub-groups in both Ambatoli and Jamtoli are quite cordial on the whole. Backward Castes, because they belong, nominally, to the mainstream caste Hindu population think of themselves, and are regarded by Scheduled Tribes and Scheduled Castes as (relatively) 'high caste'. The Sahus and Pradhans, in particular, are strict about not taking food or water from Scheduled Castes or Tribes. To some extent, however, the status differences between Scheduled Castes and Backward Castes are mitigated by their common feeling of superiority over the Scheduled Tribes and by the fact that they are all Hindus.

Table 7.2   Ambatoli's sample population by community group

Community group	Number of households	Number of individuals	% of village population
Scheduled Tribe	72	513	75.8
Scheduled Caste	12	64	9.5
Backward Caste	16	99	14.7
Total	100	676	100

Source: Jewitt, 1996.

In Ambatoli, religion is an important unifying factor as Hindus make up just under half (47.5 per cent) of the village sample population. This is due mainly to the fact that many Scheduled Tribe *sadans* profess themselves to be Hindus rather than '*Sarna* worshippers'. In Jamtoli, the Hindu population is small (11.3 per cent of the sample population) compared to the Animist population (62.2 per cent of the sample population), but the situation there is complicated by the fact that nearly 64 per cent of these Animists belong to the Tana Bhagat revivalist sect. In addition, the village has a significant proportion (26.5 per cent of the sample population) of Christian tribals.

Scheduled Tribes, Scheduled Castes and Backward Castes all tend to keep a certain amount of spatial and social difference between one another by living close to other members of their community, where possible, and taking food and water (in theory) only from people of their own community group. The Hindu Backward Caste villagers and many of the Scheduled Caste villagers consider the Scheduled Tribe *Sarna* worshippers to be 'unpure' for eating pork and sometimes beef and for (overtly) drinking alcohol. Casual and agricultural labour arrangements also tend to be made, in theory, within rather than between community groups.

In practice, many villagers (especially those who have spent time in the army or working in the brickfields) are not very strict about observing these rules because they have become accustomed to eating and working with people from other communities. At festival times, in particular (and more frequently in Ambatoli where most households regularly drink alcohol), villagers of all communities are commonly seen drinking and snacking together, if not actually sitting down to a meal. Cash-based agricultural and casual labour arrangements, meanwhile, are usually arranged on a fairly pragmatic basis, as those who need labourers to get their fields ploughed and harvested on time are not likely to make a fuss about which community their workers come from.

Intra-community Relations Relationships between community sub-groups, although usually cordial in both Ambatoli and Jamtoli, are often more complex than inter-community relations as they involve some quite subtle hierarchies. In both Ambatoli and Jamtoli, Oraons make up the majority (63.7 per cent and 82.2 per cent) of the Scheduled Tribe sample population and consider themselves on a par only with the Mundas in terms of social status. Oraons and Mundas do not intermarry, however, as both tribes have constraints on marrying inside the tribe and outside the clan and ideally, girls should also marry outside their natal villages. All of the Oraons and Mundas in Ambatoli are *Sarna* worshippers, whereas in Jamtoli, 46 per cent of the Oraon sample population are Tana Bhagats and 32 per cent are Christian.

In addition to a small Munda population, most predominantly Oraon villages have a number of *sadan* households (Sinha et al, 1969). A majority of these service castes are classified either as Scheduled Tribes (Mahlis, Lohras, Goraits, Chik Baraiks) or Hindu Scheduled Castes (Ghasis, Turis), although there are some that belong to the mainstream Hindu caste population (for example, Kumhars). Most of the Scheduled Tribe *sadans* are semi-Hinduised, even if they profess *Sarna* as their primary religion. Few, if any of them eat beef, although some will eat pork and (more frequently) chicken when they get the opportunity. The Kumhars and some of the Scheduled Castes, on the other hand, have secured the services of a Brahman priest for their *puja* (worship) and regard themselves as more 'pure' as chicken is the only meat that they will eat.

The reason for the presence of these artisan castes in Oraon villages is thought to be that Oraons developed such pride in their agricultural accomplishments that they came to view jobs like cattle minding (Ahirs), basket weaving (Mahlis, Turis), cloth weaving (Chik Baraiks), blacksmithing (Lohras), music and musical instrument making (Mahlis, Ghasis) and pottery (Kumhars), as being beneath them (Roy, 1915). To ensure that they didn't have to undertake such tasks, they developed reciprocal relationships with groups that had expertise in these occupations and encouraged one or two families to settle in their villages (Sinha et al, 1969).

In Ambatoli, Mahlis and Lohras make up the majority of the *sadan* sample population although another group of Scheduled Caste villagers known as Manjhis also undertake basket weaving work (see table 7.3). In Jamtoli, Lohras are the dominant service caste (11.3 per cent of the sample population). Mahlis and other *sadan* families are very few in number (and are not listed at all in my sample survey - see table 7.4) because Jamtoli was never a *zamindari* village and was not actually settled until the mid-nineteenth century. As a result, its development followed a slightly different pattern to many of the surrounding villages. In addition, the nearby Bero market has always been very central to life in Jamtoli and most villagers buy their baskets, pottery and other necessities from there.

Gender Relations   Gender relations in the research area are relatively egalitarian and *purdah* plus discrimination against women in the form of female infanticide and 'dowry deaths' are almost unheard of amongst *adivasis* (Harriss and Watson, 1987; Kelkar and Nathan, 1991; Henshall Momsen, 1991; Bhattacharaya, 1998; Murthy, 1998; Sarin et al, 1998). Although intra-community variations linked to age, social standing, education, marital status, etc. are apparent, most tribal and Scheduled Caste women have quite a high degree of autonomy and are heavily involved in marketing as well as in agriculture and the gathering of forest produce. In addition, the importance of sons is not as great as in many other societies (Harriss and Watson, 1987; Henshall Momsen, 1991) because although land inheritance patterns are patrilineal, families without a son can marry a daughter in the *ghar jamai* fashion. This enables them to 'adopt' a son-in-law into their family who will (hopefully) provide them with a grandson to inherit their property (Roy, 1915). In addition, *adivasi* women have well-defined land rights which consist of an entitlement to a share of any produce obtained from land

Table 7.3   Ambatoli's sample population by community sub-group

Community sub-group	Group	Number of households	Number of individuals	% of village population
Oraon	ST	44	327	48.4
Munda	ST	8	48	7.1
Mahli	ST	14	89	13.2
Lohra	ST	5	45	6.6
Chik Baraik	ST	1	4	0.6
Bokta	SC	5	25	3.7
Manjhi	SC	4	23	3.4
Swansi	SC	3	16	2.4
Pradhan	BC	5	39	5.8
Sahu	BC	9	49	7.2
Thakur	BC	1	6	0.9
Teli-Sahu	BC	1	5	0.7
Total		100	676	100

Source: Jewitt, 1996.

Table 7.4   Jamtoli's sample population by community sub-group

Community sub-group	Group	Number of households	Number of individuals	% of village population
Oraon	ST	16	129	72.9
Munda	ST	1	8	4.5
Lohra	ST	1	20	11.3
Ghop	BC	2	18	10.2
Pradhan	BC	1	2	1.1
Total		21	177	100

Source: Jewitt, 1996.

belonging to their family and a responsibility to manage land on behalf of male relatives if they should be widowed (Roy, 1928; Kelkar and Nathan, 1991).

In some areas, however, these traditions are gradually being eroded by the steady influx of 'plains Hindu' practices such as the giving of dowries rather than bride price (Sarin et al, 1998). Although Kelkar and Nathan's claim that bride price has almost disappeared amongst Oraons is not apparent in the research area, there is evidence of a declining female sex ratio throughout Jharkhand. In Bero Block, for example, census data shows that the female to male sex ratio declined from 1153:1000 to 936:1000 between 1921 and 1991 and in Ranchi District as a whole, it fell from 1022:1000 in 1881 to 927:1000 in 1991 (Census of India, 1901; 1921; 1991).

In Ambatoli and Jamtoli the sex ratio declined from 1049:1000 and 1105:1000 to 994:1000 and 943:1000 respectively between 1971 and 1991 (see figure 7.3).[5] Although it is possible that seasonal, predominantly male out-migration may be partly responsible for this increasing 'maleness' at the village level, the steady decline in the sex ratio for Ranchi District as a whole after 1901 seems to indicate that other factors are at work (Census of India, 1971; 1991).

The Distribution of Land and Wealth

Although intra-community variations in wealth are significant in both villages, for the sake of simplicity, the level of analysis here is kept to that of individual communities. Interestingly, perceived differences in social status between the Scheduled Tribes, Scheduled Castes and Backward Castes are not always reflected in landholdings and wealth indicators. In both villages, the Scheduled Tribe sample populations hold the majority of the total land (72.4 per cent in Ambatoli and 90.4 per cent in Jamtoli), although the Backward Caste sample households hold a significantly larger proportion (22.8 per cent) in Ambatoli than in Jamtoli (9.6 per cent). Ambatoli's Scheduled Caste sample households own a very low proportion of land (4.8 per cent) relative to the size of the community (14.7 per cent of the sample population).

Amongst the Scheduled Tribe sample populations, the Oraons have the largest proportion of village land (53.8 per cent in Ambatoli and 82.4 per cent in Jamtoli), followed by the Mundas (11.8 per cent and 4.6 per cent). The Mahlis and the Lohras, being primarily artisan castes rather than agriculturists hold significantly smaller amounts of land in both villages (Ambatoli: Mahlis 4.4 per cent; Lohras: 2.2 per cent; Jamtoli: Lohras 3.6 per cent). Amongst the Backward Caste sample groups, the Pradhans and Sahus in Ambatoli (10.4 per cent and 10.2 per cent) and the Ghops in Jamtoli (9.2 per cent) hold the greatest amounts of village land in percentage terms. None of Ambatoli's Scheduled Caste communities hold more than 2.6 per cent of the village land.

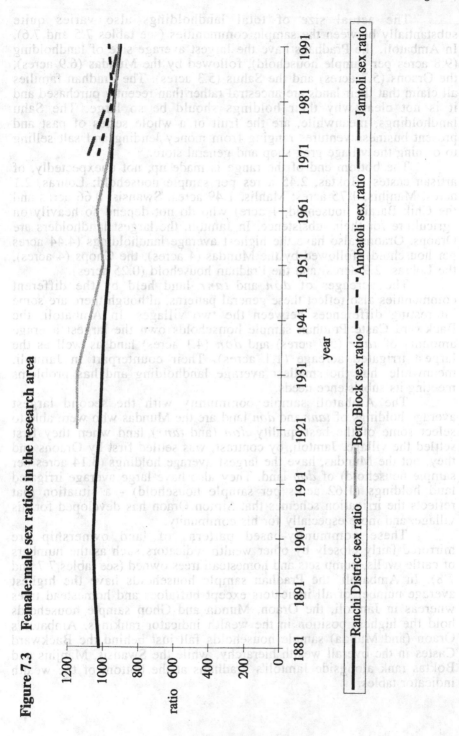

Figure 7.3 Female:male sex ratios in the research area

The actual size of total landholdings also varies quite substantially between the sample communities (see tables 7.5 and 7.6). In Ambatoli, the Pradhans have the largest average size of landholding (9.8 acres per sample household), followed by the Mundas (6.9 acres), the Oraons (5.7 acres) and the Sahus (5.3 acres). The Pradhan families all claim that their lands are ancestral rather than recently purchased and it is not clear why their holdings should be so large. The Sahu landholdings, meanwhile, are the fruit of a whole series of past and present business ventures ranging from money lending and salt selling to owning the village grog shop and general store.

The bottom end of the range is made up, not unexpectedly, of artisan castes (Boktas, 2.45 acres per sample household; Lohras, 2.1 acres, Manjhis, 1.75 acres; Mahlis, 1.49 acres; Swansis, 0.66 acres and the Chik Baraik household, 1 acre) who do not depend so heavily on agriculture for their subsistence. In Jamtoli, the largest landholders are Oraons. Oraons also have the highest average landholdings (4.44 acres per household), followed by the Mundas (4 acres), the Ghops (4 acres), the Lohras (2.9 acres) and the Pradhan household (0.25 acres).

The acreages of *don* and *tanr* land held by the different communities also reflect these general patterns, although there are some interesting differences between the two villages. In Ambatoli, the Backward Caste Pradhan sample households own the largest average amounts of *tanr* (5.7 acres) and *don* (4.1 acres) land as well as the largest irrigated acreage (1.1 acres). Their counterpart in Jamtoli, meanwhile, has the smallest average landholding and has problems meeting its subsistence needs.

The Ambatoli sample community with the second largest average holdings of *tanr* and *don* land are the Mundas who were able to select some of the best quality *don* (and *tanr*) land when they first settled the village. Jamtoli, by contrast, was settled first by Oraons and they, not the Mundas, have the largest average holdings (3.14 acres per sample household) of *don* land. They also have large average irrigated land holdings (1.62 acres per sample household) - a situation that reflects the irrigation schemes that Simon Oraon has developed for his village, and more especially for his community.

These community-based patterns of land ownership are mirrored fairly closely by other wealth indicators such as the numbers of cattle, wells, pump sets and homestead trees owned (see tables 7.7 and 7.8). In Ambatoli, the Pradhan sample households have the highest average number of all indicators except buffaloes and homestead trees whereas in Jamtoli, the Oraon, Munda and Ghop sample households hold the highest position in the wealth indicator rankings. Ambatoli's Oraon (and Munda) sample households fall just behind the Backward Castes in the overall wealth hierarchy, while the Swansis, Manjhis and Boktas rank alongside Jamtoli's Pradhans at the bottom of the wealth indicator tables.

Table 7.5   **Ambatoli land holdings (average acres per household)**

Community	Total land	*Tanr* land	*Don* land	Irrigated land
Oraon	5.73	2.98	2.74	0.66
Munda	6.93	3.62	3.31	0.23
Mahli	1.49	0.93	0.56	0.32
Lohra	2.10	1.20	0.90	0.30
Chik Baraik	1.00	0.50	0.50	0.50
Bokta	2.45	1.21	1.24	0.26
Manjhi	1.75	0.87	0.87	0.25
Swansi	0.66	0.66	0.00	0.00
Pradhan	9.80	5.70	4.10	1.10
Sahu	5.30	2.84	2.46	0.45
Thakur	5.00	3.00	2.00	0.50
Teli-Sahu	5.00	2.50	2.50	0.00
Average	4.68	2.51	2.17	0.50

Source: Jewitt, 1996.

Table 7.6   **Jamtoli land holdings (average acres per household)**

Community	Total land	*Tanr* land	*Don* land	Irrigated land
Oraon	4.44	1.29	3.14	1.62
Munda	4.00	2.00	2.00	0.25
Lohra	2.90	0.70	2.20	0.70
Ghop	4.00	1.50	2.50	0.75
Pradhan	0.25	0.12	0.13	0.00
Average	4.11	1.27	2.84	1.36

Source: Jewitt, 1996.

Table 7.7   Ambatoli wealth indicators (average numbers per household)

Community	Cattle	Buffaloes	Wells	Pumpsets	Homestead Trees
Oraon	3.68	0.18	0.77	0.40	13.90
Munda	3.12	0.00	0.38	0.25	20.50
Mahli	1.14	0.14	0.14	0.14	6.57
Lohra	0.80	0.00	0.40	0.20	6.80
Chik Baraik	2.00	0.00	1.00	0.00	3.00
Boktas	1.40	0.00	0.60	0.20	10.00
Manjhi	0.50	0.00	0.50	0.50	15.25
Swansi	3.00	0.00	0.00	0.00	7.60
Pradhan	6.00	0.00	1.00	0.80	20.40
Sahu	3.88	0.00	0.55	0.44	11.20
Thakur	5.00	0.00	1.00	0.00	7.00
Teli	4.00	0.00	0.00	0.00	1.00
Average	3.01	0.10	0.58	0.34	12.50

Source: Jewitt, 1996.

Table 7.8   Jamtoli wealth indicators (average numbers per household)

Community	Cattle	Buffaloes	Wells	Pumpsets	Homestead Trees
Oraon	4.44	0.44	1.56	0.37	91.3
Munda	4.00	0.00	1.00	0.00	218.0
Lohra	5.00	0.00	1.00	1.00	0.00
Ghop	0.00	2.00	1.50	0.50	67.5
Pradhan	0.00	0.00	0.00	0.00	14.0
Average	3.80	0.52	1.43	0.38	87.0

Source: Jewitt, 1996.

Subsistence Strategies in the Research Area

With the conversion of much forest land for cultivation during the last century, agriculture has long replaced hunting and gathering as the main form of subsistence for most people in the research area. But for many, a history of land alienation and fragmentation means that household food stocks must be supplemented in other ways. For poorer villagers who have neither the qualifications or skills to get jobs in government service or business nor the land to make money from cash cropping, the most common ways of doing this are by undertaking casual labour or seasonal migration.

Casual labour is usually undertaken on a temporary basis within the village and is carried out whenever work is both desired and available. In 1993, male daily labourers were usually paid 15 rupees per day for their own labour plus that of a bullock team (with associated agricultural implements) and 12 rupees per day for their labour alone.[6] Women were usually paid 8-10 rupees per day. The main hiring periods are in May to July (land preparation and *kharif* sowing) and September to November (*kharif* harvest and *rabi* sowing). In addition, many casual labourers are hired between March and June for help with house construction and extension activities. Seasonal migration, meanwhile, has been quite widespread since the late 1970s when bad harvests coupled with widespread forest loss put existing household survival strategies under a lot of strain. The most popular destinations are the brickfields of Uttar Pradesh and West Bengal. Indeed, this type of work has become particularly widespread amongst unmarried youths who are attracted by the opportunities that migration offers to see big cities like Calcutta and to spend some of their earnings (usually around 5000 rupees) on Western consumer goods.

Nevertheless, local forests are still very important for meeting subsistence needs (many villagers regularly collect fruit, nuts, fuel wood, timber, thatch, leaves for plates and bowls, mushrooms, medicine and even game) and for magico-religious purposes (Kelkar and Nathan, 1991). Degraded forests also provide a potential source of agricultural land for villagers whose fields lie close to the forest boundary. Although this is supposed to be illegal (Pathak, 1994), villagers have sometimes bribed the forest guard to turn a blind eye.

The Sale of Produce from Individually-owned Trees

In agreement with Chambers and Leach's (1990) and Nesmith's (1991) research which shows the importance of individually-owned trees as a source of 'insurance' for resource poor households, villagers in Ambatoli and Jamtoli do occasionally cut and sell large trees when they are faced with short-term cash deficits. A family from Ambatoli, for example, made 3000 rupees from the sale of a mature *jamun* (*Prunus cornuta*) tree in May, 1993 and the going rate in April 1995 for 180x10x10 cm posts for house construction was 50-55 rupees.

More commonly, however, local people see homestead trees as a regular rather than an emergency source of income as they provide an annual supply of fruit, seeds and leaves. Villagers who plant bamboo on their homestead land also gain an income from selling it to Mahli and Manjhi basket-weavers for around 30 rupees per mature 'stalk'. As in many other areas of the developing world (Chambers and Longhurst, 1986), the income generating capacity of homestead trees and tree products are particularly important during the pre-harvest period when food stocks are low.

In Ambatoli, 89 per cent of the sample households own homestead trees and in 1993, thirty one sample families sold homestead tree produce, earning an average of 129 rupees. Interestingly, the sale of homestead tree products in Ambatoli was heavily concentrated amongst the wealthier sample villagers who have a larger than average number of trees per household (21 compared to the village sample average of 12.5): a situation that reflects these villagers' entrepreneurial skills rather than their need for cash to meet subsistence requirements.

In Jamtoli, incomes from homestead tree products (mainly fruit, seeds and leaves) are higher than in Ambatoli as over 95 per cent of the village's households own homestead trees and in 1993, 57 per cent of the sample families sold homestead tree products, earning an average of 154 rupees. Unlike in Ambatoli, the sample households that sold these products in 1993 owned an average of only 53 trees (compared to the sample average of 87 trees) per household: a situation that reflects the importance of household tree product sales to Jamtoli's food- and cash-deficit households. Interestingly, Simon Oraon's family which owns over 700 trees uses them for home consumption only, but often allow other villagers to help themselves to fruit, free of charge, if they need to meet short term requirements for cash.

An even more profitable cash earning strategy in Jamtoli (which is practised by only a few families in Ambatoli) is *lac* (shellac) cultivation. The annual returns vary between 6000 and 10,000 rupees and all that is needed is a twig that has already been colonised by *lac* insects plus a suitable host tree (*Ziziphus mauritiana*, *Schleichera oleosa*, *Butea monosperma* and *Ficus retusa* are particularly favoured) into which the twig can be put. At the end of the season, the *lac* encrusted bark is stripped off and taken to the *lac* agent at Bero market who buys it for 50 rupees per kilogram. Almost all villagers who have suitable trees try to cultivate *lac* and their incomes depend mainly on the number of suitable host trees that they own.

Conclusion

To summarise then, the ease with which local people can satisfy their subsistence requirements (plus a small cash income) varies significantly between and within Ambatoli and Jamtoli. Although Jamtoli's land holdings and other forms of wealth are quite heavily concentrated within the Oraon community, most villagers cultivate enough food to last them an entire year and are also able to obtain fuel wood and timber from Jamtoli's protected forest area. In Ambatoli, by contrast, economic inequalities are more pronounced, fewer villagers regularly produce enough food to last them all year and fuel wood is more difficult to obtain. As a result, daily labour and migration provide very important sources of income for the land-poor *sadans* as well as for a number of the poorer Oraon households.

As later Chapters will show, Jamtoli's villagers have much in common with the empowered and knowledgeable forest communities celebrated by forest intellectuals such as Guha. Indeed, villagers' 'unity', ethnic homogeneity, willingness to co-operate and ability to maintain their *Parha* has been both drawn upon and strengthened by Simon Oraon in his efforts to promote greater agricultural productivity and community-based forest management. Ambatoli, like many other villages in the research area fits the populist model less well, however. Instead of being prepared and ready to re-establish village-based resource management, it's ethnic heterogeneity, problems in sustaining intra-village unity and marked inequalities in land and other wealth indicators have impeded the development of village-wide autonomous forest protection. Unfortunately, populist critiques and alternative development models often underestimate such difficulties and therefore offer little practical help in overcoming them. They also sometimes overemphasise the disadvantages of modern farming methods as well as the state's ability to impose modernising development discourse upon unwilling local recipients.

The extent to which subsistence strategies in Ambatoli and Jamtoli have been affected by the adoption of modern agricultural technologies and forestry practices will be addressed more fully in Chapters Eight to Twelve along with the implications of this for the displacement (or otherwise) of traditional agro-ecological and silvicultural knowledges. Both the agricultural section (Chapters Eight and Nine) and the forest section (Chapters Ten to Twelve) contain a separate Chapter on gender which examines the relevance of populist, eco-feminist and WED discourse to agricultural and forest management in Jharkhand.

Notes:

[1] Roy (1915) suggests that, *Kurukh*, the name that the Oraons use for themselves may be linked to the Sanskrit 'Krs' (to plough).

[2] One of the most popular and long-standing of these is the Tana Bhagat movement which was established to rid the Oraon country of *bhuts* which were believed to be at the root of Oraon misfortune. Like a number of other tribal reform movements in India, the Tana Bhagat movement had a strong socio-economic aspect to it in that it aimed to try to raise Oraons to the economic and educational levels reached by many Christian converts (Roy, 1981; Singh, 1982a; Hardiman, 1987; Devalle, 1992). To purge themselves and their villages, followers of the Tana Bhagat movement ceased to worship village *bhuts* and gave up animal

sacrifice, meat and alcohol (Roy, 1928; 1981). During food shortages, however, many followers tired of the restrictions on meat-eating and alcohol consumption and reverted back to the *Sarna* faith, although many have retained their 'Bhagat' (or 'Bhaktain' for married women) titles.

3 To make their duties less onerous, the *Pahan*, *Pujar* and *Soussar* are each allocated an additional amount of land on which they are expected to cultivate grain for ceremonial use during their terms of office. The usual amounts given are 2 acres for the *Pahan*, 1 acre for the *Pujar* and 0.5 acres for the *Soussar* (Roy, 1928).

4 In particular, five 'Pradhan' families identified themselves as Scheduled Castes and were thought by some villagers to be a subset of the Scheduled Caste Bokta community. Official Government of Bihar listings identify Pradhans as a Backward Caste community, however, and it is this classification that I have used here.

5 1971 is the first year for which village level data census data is available for Ambatoli and Jamtoli.

6 During my main period of fieldwork in 1993, there were around 45 rupees to the UK pound and one kilogram of rice cost between three and seven rupees.

8 Technical and Socio-cultural Systems of Agro-ecological Knowledge in Jharkhand

Introduction

Within agriculture as well as forestry, recent challenges to top-down development thinking have helped to highlight the importance of local participation in the development process. The transfer of Western agricultural technology into tropical ecosystems has attracted particular attention as a result of the socio-environmental problems that have accompanied the intensive irrigated cultivation of modern seed varieties (Johnston and Kilby, 1975; Byres, 1981; 1983; Agarwal, 1986a; Bates, 1988; Shiva, 1991a; Alvarez, 1994; Swaminathan, 1996; Greenland, 1997; Krishnakumar, 1999). Indeed, a number of radical populist, eco-feminist and anti-development theorists have become concerned that Western technologies and ideas are displacing more economically, culturally and environmentally appropriate indigenous alternatives (Gupta, 1992; 1996; Yapa, 1996). As a result, they argue for a radical development alternative based on indigenous technology, community-based organisation and environmental (often women's) knowledge.

One of the problems with such supply side populist ideas is that they sometimes have little in common with the aspirations of the people being 'developed'. Moreover, they often fail to acknowledge that such people may be capable of articulating what they want (as well as what they don't want) from development and of making coherent demands in national as well as local political fora. A growing literature on peasant resistance and new social movements testifies to the fact that local people can sometimes organise themselves very successfully to achieve specific development goals and are becoming increasingly skilled in making their voices heard (and causes known) in key political arenas (Ghai and Vivian, 1992b; Vivian, 1992; Blauert and Guidi, 1992; Schuurman, 1993; Alvarez, 1994; Omvedt, 1994; Escobar, 1995; Peet and Watts, 1996a; 1996b; Kaufman, 1997b). Indeed, this type of grassroots action could offer a means of mixing tradition and modernity to achieve a more locally appropriate and bespoke form of development. Where participation is directed from above rather than from below, however, simplistic and romanticised assumptions about the socio-political congruity of stakeholders' or community interests can promote the development of inappropriate development and resource management interventions.

Bearing in mind the different viewpoints advocated by proponents of top-down, bottom-up, radical populist and autochthonous development, this Chapter will go on to examine the interaction of traditional and modern technology within the context of rice-based peasant farming systems in Jharkhand. Particular emphasis is placed on the extent to which populist concern about the displacement of traditional farming methods by modern agricultural practices (and the impacts of this on local livelihoods and ecologies) are echoed by local people in one of the many predominantly rainfed areas 'left out' by the green revolution (Lipton and Longhurst, 1989; Agrawal and Sivaramakrishnan, 2000b). An attempt is also made to investigate the micro-political ecology of local agricultural knowledge possession, paying special attention to linkages between the technical and socio-cultural aspects of indigenous knowledges.

Using census, Survey and Settlement and agricultural statistics dating from 1901 coupled with detailed ethnographic work published in 1915, the following section charts the main agricultural changes that have occurred in Ranchi District over the last century. It is argued that although population increase and the subdivision of land holdings have stimulated agricultural intensification, this has not been accompanied by either the type or scale of problems experienced in the more intensively farmed wheat-based green revolution areas. Instead, modern agriculture has brought significant benefits (notably increased food security and employment) and there has been little evidence to date of either technology displacement or biodiversity loss.

Before moving on to investigate more recent agricultural change at the village level, an attempt is made in the second section of the Chapter to address the traditional bias towards the technical rather than the socio-cultural aspects of indigenous agricultural knowledge. The complexity of local understandings of natural phenomena is captured well by S.C. Roy's observation that Oraons do not 'attribute all the phenomena of Nature and all the good and ills of life to spiritual agencies - to personal spirits, ghosts and deities or to mysterious impersonal forces and powers' (Roy, 1928, p. 4). Instead, they recognise the influence of natural causes and accept that human agency has the power to determine the outcome of certain events.

Nevertheless, all of the major *Sarna* religious festivals are closely connected with agriculture (Roy, 1915) and the correct religious observances made in association with the different agricultural operations are often just as meaningful as the techniques practised. Moreover, it is important to recognise that to most villagers, little distinction exists between so-called technical or socio-cultural knowledges: both are equally real and valid means of understanding and trying to control events that go on around them. In view of this, the section emphasises the value of emic research into the socio-cultural aspects of agricultural knowledge as a means of explaining cultivation practices in the research area and providing a basis for a more informed and appropriate type of development (Bebbington, 1992; Salas, 1994; Agrawal, 1995).

In the third and fourth sections of the Chapter, the research emphasis shifts towards village-level agricultural change and draws upon empirical data collected from Ambatoli and Jamtoli between 1993 and 1999. The main concern here is to examine the extent to which modern farming methods are displacing or de-legitimising traditional farming techniques and cropping patterns in the study villages. It is argued that although pollution from agro-chemicals and declining water table levels will require careful monitoring, agricultural intensification seems to have brought relatively few environmental problems to date. It is also suggested that although several traditional technologies have been somewhat marginalised by modern agricultural practices, there is at present quite a healthy and harmonious mixture of both old and new techniques. Moreover, the widespread adoption of modern agricultural practices in one of the villages has been associated with substantially better food security, reduced levels of seasonal migration and greater village-wide co-operation.

Agricultural Change in Ranchi District

Ranchi District makes quite a good case study for examining the displacement of traditional agricultural technology as it is typical of a rather 'backward' rainfed area and has a long history of detailed ethnographic investigation that gives insight into agricultural change over time (Roy, 1912; 1915; 1928). Another interesting factor is that the District's dominant tribal group, the Oraons, view themselves primarily as cultivators and have great pride in their agricultural prowess. In contrast to some of the more remote forest-based Jharkhand tribes, Oraons have long preferred to clear forests for agriculture rather than hunt and gather in them. Also, unlike tribes such as the Mundas which own all village land as a corporate body, the Oraons have traditionally maintained their forests and other public lands under community ownership but allowed individual family ownership of cultivated land: a situation that seems to have both enabled and favoured agricultural development (Roy, 1915). In theory, therefore, one would expect to find the most dramatic agricultural changes, socio-environmental impacts (and displacement of traditional technology) in Oraon-dominated areas.

Traditional Agriculture in Ranchi District

When S.C. Roy was undertaking his ethnographic work on the *adivasis* of Chota Nagpur during the early part of the last century (Roy 1912; 1915; 1928), agriculture formed a major socio-religious focus for the Oraons as well as being their prime economic activity. Nevertheless, cultivated food crops were usually supplemented with game, fruit, flowers, leaves, leafy vegetables, mushrooms and tubers from nearby forests. The flower of the *mahua* tree (*Madhuca indica*), which can either be eaten or distilled to make country liquor, provided a particularly important relief mechanism in times of famine, sometimes forming the equivalent of two months food supply (Hallett, 1917).

The equipment used for cultivation was fairly simple. Cattle drawn implements including wooden single furrow ploughs (with iron ploughshares), harrows and levellers were all made locally. Clods were broken with a wooden mallet and crops were harvested with a sickle and threshed by cattle walking over them. Winnowing was undertaken with a bamboo winnowing basket (*soop*) and husking was carried out with the use of a foot-operated wooden rice pounder known as a *dhenki*. The main form of transport used was a bullock cart made of wood and bamboo. Being expensive to build, wells were not particularly common and were used primarily for drinking and bathing water. The main systems used for drawing water were either a bucket on a rope or a *latha* which consisted of a long wooden lever pivoted between two posts and weighted at the end furthest from the well. A rope holding a metal pail was attached to the other end and could be drawn down to collect and lift water out of the well.

As irrigation was limited to those Oraon households with wells on their uplands or fields close to streams and natural depressions, cultivation was restricted almost exclusively to the *kharif* (monsoon) season (June-November). Rice was the principal food crop and, depending on the rainfall, took between 84.4 per cent and 61.2 per cent of Ranchi District's net cultivated area between 1901 and 1917 (Government of India, 1911a; 1911b; Reid, 1912; Hallett, 1917). The best paddy has traditionally been grown on *don* land which can be divided into four categories with different harvesting times and productivity levels (see table 8.1).

Table 8.1   Productivity levels of different land types in Ranchi District

Land type	Yield (maunds[1] per acre)
don 1	19
don 2	19
don 3	15
don 4	9
tanr 1	8
tanr 2	4
tanr 3	2

Source: Reid, 1912.

The lowest *don* land (*don* 1) is very fertile, continually wet and has traditionally been double cropped with winter as well as summer rice. There is only a very small amount (760 acres) of this type of *don* in Ranchi district (Reid, 1912). In most areas, the best quality rice land is *don* 2 which is found in fertile and water-retentive shallow valley bottoms. This land has traditionally been planted with fine varieties of *aghani* rice which are cut in the month of *Aghan* (December). *Don* 3 is slightly more elevated and less fertile and has traditionally been planted with *khartika* rice which is harvested in *Khartik* (November). *Don* 4 is situated in the upper part of the shallow valleys and is less fertile and more freely draining. It has traditionally been planted with varieties of coarse *bhadoi* rice which are harvested in *Bhado* (October).

At the time when S.C. Roy was writing, *don* lands were cultivated annually, manured every second or third year and ploughed three or four times every year before broadcasting or transplanting. In order to alleviate the mono-cropped nature of *don* paddy cultivation, farmers used to plant different varieties of paddy in their fields each year and alternate transplanting with broadcasting (Sinha, 1976).

The drier and poorer quality uplands (*tanr*) have traditionally been planted with '*gora*' (upland) paddy in rotation with other crops (Reid, 1912). Like *don* land, *tanr* has also been classified according to its quality and productivity. *Tanr* 1 is the best quality upland and consists of fertile land lying in or near the village which is manured regularly and cultivated every year. There is relatively little (20 square miles) of this type of *tanr* in Ranchi District and it often corresponds to what villagers regard as their homestead or *bari* land. *Tanr* 2 is further away from the village but is relatively level and stone-free with a reasonable depth of soil. *Tanr* 3 is usually quite far from the village and consists of either very stony or sloping land with thin soil and low moisture retaining capacity. Most upland crops are broadcast at the beginning of the monsoon in June.

In more remote areas, *tanr* land was often left fallow once every two or three years although Reid was of the opinion that as 'cultivation becomes more intense and the lands are more liberally manured and better cultivated, this class of *tanr* will, no doubt, greatly improve' (Reid, 1912, p. 4). In Bero block, however, cultivation was fairly intensive even in 1912 and the 'average uplands' were 'cultivated approximately every year' (Reid, 1912, p. 62).

Most *tanr* has traditionally been cultivated in rotation to maximise the use of manure and to conserve soil fertility. At the beginning of the cycle, the fields were manured and a millet called *marua* (*Eleusine corocana*) was sown; sometimes intercropped with small amounts of *bodi* (*Vigna catiang*). In the second year, benefiting from the manure of the previous year, *gora* paddy was sown. Often on better quality *tanr* that retained some residual moisture, *arhar* (*Cajanus indicus*) was inter-cropped with the paddy and harvested with the *rabi* (winter season) crops the following March. In the third year, *urad* (*Phaseolus roxburghii*) was sown and was sometimes followed on the better *tanr* by a winter (*rabi*) food crop of gram, lentil, *kurthi* (*Dolichos biflorus*) or an oilseed such as linseed, mustard or *surguja* (*Guizota abyssinica*). In the last year, the coarse millet *gondli* (*Panicum milaire*) was planted as it yields even on the poorest land. Being harvested in August, *gondli* was often followed by a late-*kharif* crop of *surguja* or *kurthi* (Reid, 1912). By the time the *surguja* was harvested, the soil had become friable and ready for ploughing and manuring for the next crop cycle (Reid, 1912; Hallett, 1917; Sinha, 1976).

Those Oraons who could afford the time and labour to dig wells on their uplands often grew irrigated *rabi* crops such as mustard, chillies, turmeric, onions, aubergines and potatoes or made nurseries for paddy seedlings which would later be transplanted onto *don* land. Many of the better-off Oraons also owned a plot of *bari* or homestead land which they could supervise closely and cultivate intensively. Such land was commonly enclosed and planted with kitchen vegetables such as *jhinghi* (*Luffa acutangula*), okra, beans, maize, potatoes and aubergines (Roy, 1915).

Apart from the cultivation of hardy crops such as *bodi*, *arhar*, lentil, *kurthi* and oilseeds on the more moisture-retaining *tanr* or garden vegetables on intensively managed *bari* land, *rabi* crops were not important until relatively recently.[2] The main limitations were the water-logging of the lowlands, the dryness of the uplands and the difficulty of digging wells in an area with hard base rock and a low water table (Sinha, 1976). As a result, only 0.2 per cent of Ranchi District's gross cultivated area was irrigated in 1911 (Government of India, 1911a; 1911b). Significantly, however, the District Gazetteers report that famines were strongly linked to areas with a high proportion of *tanr* lands relative to *don* as 'water drains away from the uplands (*tanr*) and even the higher low lands (*don*) with great rapidity; and unless the crops growing on these lands receive a plentiful and continuous supply of rainfall, they wither and die very quickly' (Reid, 1912, p. 5).[3] Indeed, according to Hallett (1917) famines 'due to deficiency of rainfall may be said to be the only natural calamity from which the inhabitants of Ranchi can suffer ... When famine or scarcity has occurred, it has been due almost invariably to the failure of the rains in the latter part of August and in September and October' (Hallett, 1917, p. 134).

In areas where tanks had been constructed for irrigation, however, Hallett found the impact of drought to be much less severe. Perhaps thinking of the implications of this for agricultural taxation as well as famine alleviation, he bemoaned the fact that well irrigation was 'an expensive labour which the aboriginal cultivator will not make' (Hallett 1917, p. 116). The District's Survey and Settlement Officer, Reid, agreed with this analysis arguing that given the prevalence of rivers and streams in the district, much 'could be done in this respect by the judicious expenditure of capital' (Reid, 1912, p. 5). He also stressed that without irrigation, '*rabi* crops are of little importance and can never be relied on as affording even a partial substitute for the main crop of the year' (Reid, 1912, p. 5). Foreseeing the land fragmentation that rapid population increase would bring, Hallett predicted that farmers would be 'compelled to learn what are to them new methods of cultivation, to improve their uplands by the use of manure, and to make good crops a certainty in their lowlands by means of irrigation' (Hallett, 1917, p. 114).

Agricultural Intensification and Modern Farming Practices

Hallett was, of course, right about land sub-division and the role of population increase in forcing agricultural change. Over time, land holdings became fragmented and household survival strategies became more tenuous. The average farm size of 8.72 acres recorded in the 1927-35 Revisional Survey and Settlement Report fell to the current average of around 5 acres per household. In addition, forest decline reduced the availability of many forest fruits, nuts, leaves, herbs, mushrooms, wild vegetables, tubers, liquor-making ingredients (notably *mahua* flowers) and game. Bad harvests in the late 1970s were a significant factor in forcing many households to diversify their economic activities. Many poorer farmers took on additional work as agricultural or migrant labourers while the better-educated sought to move into off-farm employment such as business or government service.

Most families that retained farming as their primary economic activity were forced to consider how to increase the productivity of their holdings. Farmers with land close to degraded forests were sometimes able to come to an 'arrangement' with the forest guard to extend their area of cultivation. For other owner-cultivators, intensification was the obvious option. Between 1930 and 1960, the net sown area increased by around 7 per cent whereas the number of working animals increased by about 22 per cent percent and ploughs by 40 per cent (Bose and Ghosh, 1976).

Nowadays, therefore, very little agricultural land remains uncultivated during the *kharif* season. *Don* lands are still planted almost exclusively with winter paddy but are manured at least every other year in March or April. The broadcasting of *don* paddy has largely given way to transplanting with paddy seedlings usually being sown on upland seedling beds in June and transplanted (usually by women), onto wet, bunded *don* fields in late July or August.

More changes have occurred on the uplands, many of which are now far more intensively farmed than before. The upland fallows common in remote areas a century ago have all but disappeared as villagers use annual applications of manure (and increasingly chemical fertiliser) to enable these lands to be cultivated at least once in a year. In areas like Bero Block where annual upland cultivation has a long history, a shift to yearly manuring has caused a decline of the traditional four year *marua*, *gora* paddy, *urad* and *gondli* rotation with inter-crops of *bodi* and *arhar*. Reflecting wider Indian food preferences that favour rice and wheat and regard coarse millets as 'poor people's crops' that cannot be served to guests without attracting disdain (Shiva, 1988; 1991a; 1991b; Dixon, 1990), *gora* paddy has increased at the expense of *gondli*. Far from being displaced, however, *gondli* is still frequently grown on poor land that other crops will not tolerate.

Marua, which is used to make both *roti* and rice beer, is grown more widely than before, however; possibly reflecting the reduced availability of *mahua* flower liquor as a result of forest decline. To maximise *marua* yields, some farmers make seedlings beds in June and transplant the young plants onto *tanr* land in July. Most, however, still broadcast their *marua* and *gora* paddy in June, often inter-cropping it with small amounts of *arhar*, *bodi* and *urad* as they believe this to be less 'exhausting' for the soil. Most of these crops are grown on all three types of *tanr* land, although the poorest quality *tanr* 3 land is usually reserved for hardy crops like *gondli*, *arhar*, *kurthi* and gram.

As in colonial times, the *marua* and *gora* paddy crops are still often followed, on fields with good residual moisture, with late-*kharif* or early *rabi* season crops such as gram and maize or oilseeds such as mustard, linseed and *surguja*. In addition, pumpkins and gourds are now sometimes grown on the lowest-lying *tanr* (and occasionally the more elevated *don* land) after the paddy harvest.

By far the most important source of intensification, however, has been the spread of irrigation which has made *kharif* crops less vulnerable in dry years, increased the importance of *rabi* season *tanr* cultivation and opened up opportunities for using MVs along with chemical fertilisers and pesticides. While the environmental implications of this may become more pronounced in future, they have to date been limited and drastically outweighed by the greater freedom from famine and food security that irrigated cropping offers in the event of a failure in the *kharif* crop.

As villagers made efforts to dam streams, create irrigation tanks and dig wells on their uplands, the first winter crops that they grew consisted mainly of *rabi* wheat, pulses (peas, *arhar*, *masoor* and *khesari*) and oilseeds. By 1966, vegetables were grown on around 32,000 acres in Ranchi District and in 1969-70, 2.68 per cent of the gross cultivated area was irrigated (Kumar, 1970; Tripathy, 1989). The most common vegetables grown at that time were potatoes, aubergines, okra, tomatoes, peas, cauliflowers, beans and carrots. Wheat cultivation increased significantly during the green revolution period when MV seeds became widely available. As awareness about cash crop markets increased, villagers also experimented with commercially valuable vegetables such as potatoes and peas alongside kitchen vegetables for their own use. In time, most of the good quality irrigated *tanr* and *bari* land came to be planted with wheat, potatoes, sweet potatoes, tomatoes, peas, aubergines, beans, okra, maize, gourds, *jhingi* (*Luffa acutangula*), pumpkins, radishes, onions, chilli and garlic. The best fields are now usually planted with improved varieties of wheat and vegetables which are well supervised and heavily fertilised with farm yard manure, compost and, increasingly, chemical fertiliser. On more remote *tanr* land where irrigation is available or moisture availability good, villagers often plant sweet potatoes and hardier local potato varieties.

Benefiting from Bero Block's higher than average irrigated acreage (12.56 per cent of the 1981 net cropped area compared to 4.53 per cent for Ranchi District as a whole - see table 8.2),[4] local farmers responded to the demand by both local and distant urban centres (including Calcutta) for crops such as winter peas, tomatoes, potatoes and cauliflowers. Initially, the dependence of these crops on timely, regular watering and (to a lesser extent) chemical fertiliser and pesticide meant that villagers with favourable land and capital assets were often in a better position to cultivate them than others: a situation that has tended to accompany the spread of modern agricultural technology throughout the developing world (Byres, 1981; 1983; Shiva, 1988; 1991a; Conway and Barbier, 1990; Dixon, 1990). Nevertheless, rural Jharkhand's relatively low levels of economic inequality (compared to many other regions), restrictions on the sale of land to non-tribals and limited scope for large-scale irrigation have largely enabled it to avoid the scale of agrarian class differentiation experienced in more advanced agricultural regions (Johnston and Kilby, 1975; Byres, 1981; 1983; Brass, 1990). In addition, opportunities for resource poor villagers to take up irrigated *rabi* cropping have been increased by the availability of Block-level loans for pumpsets (costing Rs 18,500) and irrigation wells. To maximise the use of these new irrigation facilities some farmers have started to grow summer vegetables such as okra, beans, gourds and pumpkins as well as new paddy varieties on their most productive *bari* land.

Table 8.2   Changes in the proportion of irrigated land in Ranchi District and Bero Block

Year	Ranchi District (%)	Bero Block (%)
1900-11	0.2	no data available
1916-17	0.19	no data available
1969-70	2.68	no data available
1980-81	4.53	12.56

Sources: Government of India, 1911b; Hallett, 1917; Kumar, 1970; Government of India, 1981; 1991.

The recent increase in dry season vegetable growing owes much to the fact that it can be both profitable and high yielding. It can also be carried out by resource poor farmers as it requires little land and is more labour than capital intensive. A study of Bero Block by the Xavier Institute of Social Services (XISS) between 1979 and 1984 (XISS, 1984), for example, found that potatoes had the highest average productivity (36.88 maunds per acre) of all crops. The category 'other vegetables' also had a high average yield (17.6 maunds per acre) and profitability compared to more traditional *rabi* crops such as wheat (14.59 maunds per acre). It seems no wonder, therefore, that irrigated *rabi* cultivation focusing on high yielding and commercially valuable vegetables like potatoes has been such a widespread intensification strategy in this area. By reducing pressure to reclaim declining areas of waste and forest land for cultivation, the increase in double cropping has also had significant environmental benefits.

From the point of view of rural livelihoods, meanwhile, intensification has been significant in helping to increase agricultural productivity and food security in the face of uncertain monsoon rainfall, rapidly increasing population, fragmented land-holdings and declining opportunities for forest-based subsistence. The option of supplemental irrigation for the staple *kharif* paddy crop has been a major factor in reducing the risk of crop failure in dry years. It also benefits both owner-cultivators and agricultural labourers by providing the option of growing labour intensive *rabi* crops to compensate for paddy shortfalls. As a result, it has helped resource poor farmers to meet at least some of their subsistence needs while landless villagers have benefited from more year-round agricultural employment. In Bero Block, for example, the proportion of villagers (as a percentage of the total working population) classified as 'cultivators' changed little between 1981 and 1991 (from 82.9 per cent to 82.4 per cent) indicating that intensification has not been accompanied by increased landlessness during this period (see table 8.3). The percentage of agricultural labourers declined very slightly between 1981 and 1991 (from 10.9 per cent to 9.9 per cent of the total working population) but the actual numbers of people involved increased significantly (by 22.7 per cent) and in line with the percentage increase in population.

Table 8.3   **Changes in the number and proportion of cultivators and agricultural workers as a percentage of all workers in Bero Block between 1981 and 1991**

	All workers	Culti- vators	% of cultivators	Agri- cultural labourers	% of agricultural labourers	% of cultivators and agricultural labourers
Bero 1981	22955	19037	82.93	2500	10.89	93.82
Bero 1991	29912	24638	82.37	2956	9.88	92.25

Sources: Government of India, 1981; 1991.

Tradition within Modernity

Yet in spite of these changes, agriculture in Chota Nagpur still bears many similarities to how it was 100 years ago with rice as the major crop, dominating almost all of the *don* and much of the *tanr* land during the monsoon season. Indeed, *kharif* season cultivation in areas like Bero Block that have no recent tradition of *tanr* fallows remains remarkably similar to the descriptions given by Reid (1912), Hallett (1915) and Roy (1915). Even average rice yields have changed little since the onset of the green revolution. The XISS study carried out between 1979 and 1984, for example, reported average *don* paddy yields of 12.33 maunds per acre and *gora* paddy yields of 8.47 maunds per acre (XISS, 1984). These correspond closely and not particularly favourably with the working average of '14 maunds of *dhan* per acre *don*, and 7 maunds per acre *tanr* (net cropped)' (Reid, 1912, p. 115) noted in the Settlement and Survey report of 1902-1910: a situation that highlights the importance of irrigated *rabi* cropping to the District's food security.

Also, and in significant contrast to the main wheat-based green revolution areas like Punjab, there has been little change in the agricultural equipment used in the region. Far from making a shift to labour displacing machinery, most farmers still use cattle-driven single furrow wooden ploughs and harrows, arguing, as they did in colonial times (Reid, 1912; Kumar, 1970) that their cattle are not strong enough to manage 'improved' implements. Tractors are rarely feasible due to the small size of the plots and the limited capital of most farmers, so bullock carts are still the primary means of transporting bulky goods. Sickles are still used for harvesting, cattle for threshing, *soops* for winnowing and *dhenkis* for husking. Although kerosene driven pumpsets are now commonly used for irrigating *rabi* crops, many wells still have the traditional *latha* system in place.

Overall, therefore, secondary source data on Ranchi District and Bero Block displays relatively little evidence to support populist concerns that traditional agro-ecological knowledge is being displaced by modern agricultural methods or that local labourers are losing work as a result of mechanisation. Nor is there any significant evidence to date that agricultural intensification in the area has been associated with either increased landlessness or biodiversity loss. Indeed, all of the crops grown a century ago are still grown today and many of the inter-cropping systems remain the same. More localised agricultural change, along with its ecological impacts, will be examined in detail in the third part of the Chapter. Before moving on to this village-level data, however, it will be helpful to provide a more comprehensive picture of agriculture in the research area by outlining some of the main socio-cultural knowledge systems that both influence and are influenced by the technical agro-ecological practices discussed above.

Socio-cultural Knowledge Systems

Traditionally, agriculture has played a central role in most of the region's major *Sarna* and Hindu religious festivals and many close interactions remain between agriculture, religion and magic in Jharkhandi culture. Particularly since the 1980s, however, the linkages between technical and socio-cultural knowledge systems have taken on an important political ecology dimension as part of a wider Jharkhandi self-determination strategy.

The Jharkhand Movement

As mentioned in Chapters Two and Six, the Jharkhand *andolan* was a prime example of a powerful and long-standing political movement that sought to counteract the power of the central and Bihar State governments and seize control of Jharkhand's political, economic and cultural resources. Rather than rejecting development as a Western 'disease' (Rahnema, 1988), however, the movement's campaign hinged on a desire for greater control over the process as well as over the region's resources. Indigenous socio-cultural knowledges were seen as central to the future of the region and local environmental management strategies were frequently asserted as a form of resistance against 'official' development (particularly industrialisation, large scale dam building and commercial forestry).

For many years, the lack of unity within the Jharkhand movement was a major factor preventing it from providing the basis for autochthonous development. Indeed, local interest in the Jharkhand movement declined substantially after the mid-1980s when, in spite of many economic blockades and endless discussions with the Indian central government (Gupta, 1992a; 1992b), there were no signs that a separate Jharkhand State (or even Union Territory) would be granted. But although many people stopped actively supporting the movement, most *adivasis* maintained a strong sense that they wanted to be ruled by their 'own people' rather than by outsiders. This, they believed, would bring a big improvement in self esteem, if not in actual food consumption. There was also a widespread feeling (especially amongst Scheduled Tribes and Scheduled Castes who stood to gain from 'reserved jobs') that a separate Jharkhand State would enable local people to benefit from better access to education and government service employment.

The election of the local Jharkhand Mukti Morcha (JMM) candidate, Vishvanath Bhagat, as a Member of the Legislative Assembly in March 1995 reflects these beliefs as well as a feeling amongst his *adivasi* supporters that an *adivasi* leader could take care of their problems in the way that a *diku* could not. In response, Vishvanath has attempted to strengthen local village culture and festivals, particularly through a revival of the *Parha* system. Indeed, one of the main reasons why Vishvanath first became involved in politics was that he was appalled by the way the local authorities, particularly the police, refused to respect *adivasi* customs and caused trouble for local people when they practised traditional cultural activities (discussion with Vishvanath Bhagat, 1993).

Adivasi *Religious Festivals*

The first major *adivasi* festival of the agricultural year is *Sarhul* which takes place in March or April and must be celebrated before villagers plough their land. According to Roy, the festival celebrates the ceremonial 'marriage' of the *Pahan* and his wife which symbolises the marriage of Mother Earth with the Sun (Roy, 1915; 1922; 1928). In Ambatoli and Jamtoli, this marriage is usually represented by the 'marriage' (entwining) of two paddy seedlings. The seed paddy from the previous harvest is then blessed by the *Pahan* and put aside for sowing time (Roy, 1915).

The second most important religious festival for maintaining agricultural prosperity is the Hariari festival in June-July when sacrifices are offered to *Dharmes*, *Sarna Burhia* and other village deities to secure plentiful rain and a good harvest (Roy, 1928). The transplanting of *don* paddy seedlings usually follows soon after this festival and the first seedlings must be planted by the *Pahan*. Also in June or July, the older villagers celebrate *Burhia Karam* during which they pray for good monsoon rains. Of central importance to this festival is the *karam* (*Adina cordifolia*) tree which is associated by the villagers with good rainfall (Roy, 1928).

The third major agricultural festival is *Kajlota/Karam* which takes place in August, just before the upland paddy is harvested. On *Kajlota* day, the *Pahan* prays to the village deities for the protection of the ripe paddy from theft or damage by evil forces. Later on, a group of male villagers fetch *kend* (*Diospyros melanoxylon*) *sal* or *sinduar* (*Vitex negundo*) twigs from the jungle for use in the *Danda Katta* (cutting of the evil teeth) ceremony the following (*Karam*) day. The purposes of this ceremony are to banish the 'evil eye' and 'evil mouth' from the village and to promote general prosperity (Roy, 1928). After it has been performed, the *kend*, *sal* or *sinduar* twigs collected the previous day are hung with a *bhelwa* leaf containing rice used in the ceremony. This is planted on *don* fields to protect them from the 'evil eye' (Roy, 1928). After lunch, three branches of the *karam* tree are set up in the village *akhara* (dancing place) as '*Karam Raja*' (Karam King) and the villagers pray to them for good rains, good crops and general village prosperity. The following day, three young men or women take the three *karam* branches to a nearby pond or stream to be 'drowned': an action that is believed, through sympathetic magic, to stimulate good rainfall (Roy, 1928).

Hindu Festivals

The major Hindu festivals that are celebrated in the village are *Holi* in March and *Diwali* in November. *Holi* is regarded by most villagers as a kind of 'new season' festival to welcome the new leaves on the trees and a new agricultural year. In Ambatoli and Jamtoli, *Holi* is celebrated both by Hindus and non-Hindus, with the village *Pahan* often sacrificing chickens and giving thanks to *Sarna Burhia* for past harvests and praying for good ones in future. *Diwali* is also celebrated by non-Hindus, although the *Pahan* does not usually have any direct role. Most villagers regard it as a kind of post-harvest festival, although it is the village cattle and buffaloes that are given *puja* (worship) and generally made the centre of attention.

In addition to these major festivals, there are a number of occasions when Hindu and non-Hindu villagers undertake *puja* either to give thanks or to pray for something that has affected them personally. During the course of my fieldwork, the only occasion that such *puja* was directly related to agriculture was when the Oraon family that I was living with had their newly constructed well 'blessed' in a semi-Hinduised fashion by the village *Pahan*.

Malevolent Bhuts, Dains and other Magico-religious Beliefs

In addition to these festivals, villagers often attempt to ensure the well being of their families, crops, livestock and other possessions by periodically sacrificing animals to propitiate the malevolent *bhuts* that live in and around the village. The most common malevolent acts that *bhuts* are believed to be responsible for include causing the monsoon rains to fail, damaging crops, stealing paddy to give to *bhut*-worshipping villagers and causing illness amongst non-worshipping villagers or their animals.

The most powerful village *bhuts* are generally given the biggest sacrifices: usually a goat, cow or buffalo. The most powerful sacrifice of all, however, is a human sacrifice which is believed to prevent almost any kind of calamity and to pacify the most malevolent *bhuts*. Indeed, several villagers informed me that the only way to guarantee good monsoon rains is to propitiate village *bhuts* with human sacrifices. The blood from a human sacrifice is also believed to guarantee bumper harvests when mixed in with the paddy seedlings to be planted in the fields. Indeed, the power associated with human sacrifices is such that most villagers are terrified of becoming the victim of an *otanga*: a man who prowls around deserted fields and jungle areas between March and November (the main agricultural season) looking for an opportunity to catch someone to use (or sell) as a human sacrifice. Many villagers also live in fear of local opportunistic 'head cutters' who might be tempted to murder somebody in their own or a nearby village in an attempt to obtain better crops or more material wealth.[5]

In many cases, malevolent *bhuts* are assumed to be acting at the behest of a witch or *dain* (who can be male, but is more usually female - Roy, 1928; Kelkar and Nathan, 1991), rather than independently. Although inciting villagers to commit murder is not thought to be particularly common, most of the deeds attributed to witchcraft do show an underlying motive of '*hisingar*' or jealousy of other people's good fortune.[6] Typical examples include the use of the 'evil eye' by a *dain* to stimulate the *bhuts* under her control to cause damage to crops that are growing in a field or to a tree that is fruiting particularly well. Most *dains* are thought to be active only between June and October when they can do most damage to villagers' standing crops or (dwindling) paddy stores. Indeed, many villagers in Jamtoli believe in a kind of 'Robin Hood' *dain* who sends out *bhuts* to steal paddy from the rich to give to the poor.

Witches are also thought to be able to cause harm to a villager's family or possessions. Many villagers are convinced that if a *dain* 'looks' at their children or livestock, they can cause them to fall ill. Detecting this damage is very difficult, however, as *dains* are thought to be able to disguise themselves or their *bhuts* as animals. One woman from Ambatoli told me that a *dain* (in the form of a cat) had licked the foot of one of her water buffaloes and the following morning the buffalo had died (discussion with Goindi Orain, 1993).

To prevent a *dain* from causing damage through the use of the 'evil eye', villagers in Ambatoli and Jamtoli employ two main strategies. The first is the use of a broken pot to 'break' the gaze of the evil eye through sympathetic magic (Roy, 1928) and the second is the use of brightly patterned pots, tiles or other items to distract the evil eye away from the object of interest (Roy, 1915; 1928). Both methods are commonly used to protect field crops, vegetables or fruit and close to harvest time, most paddy fields and vegetable patches sport either a broken or white patterned pot on a stick to ward off the evil eye. Even along roofs where pumpkins are trained during the monsoon, a white patterned tile can often be seen 'guarding' the growing crop.

If these precautions do not work or if a *dain* uses strategies other than the 'evil eye' to cause damage, villagers usually take more direct action. As a first step, most try to remedy the situation by offering animal sacrifices to the *bhut* that is working at the behest of the *dain*. If this doesn't work, they usually send for a *bhagat* (ghost finder) to cast spells and identify the *dain* responsible for the trouble. The *bhagat* then confronts the *dain* and orders her to control or appease the *bhut* that is responsible for the problem (Roy, 1928). In many cases, compensation is demanded by the injured party and if the *dain* agrees to this, she is usually absolved by the *bhagat*. If the *dain* refuses, however, she is identified publicly and sent away from the village.

Given that witch identification (as well as witchcraft itself) is often based on *hisingar* and can act as an important means of limiting upward mobility within the local area, accusations of witchcraft can have very serious implications for the *dain* and her family. It can also have serious gender implications as it is usually women rather than men who are declared as witches in Jharkhand. Paul Olaf Boddings, who worked in Santhal Parganas went as far as to suggest that witch-finders are 'unmitigated scoundrels' who can 'denounce anyone they like' (quoted in Kelkar and Nathan, 1991, p. 95): a situation that helps to reinforce men's power over women and provides a means of restricting women's 'non-conformism or deviance from the rules that are being established' (Kelkar and Nathan, 1991, p. 99).

Throughout the world, investigations into the political ecology behind witch hunts have often revealed wider struggles over wealth and property rights (Kittredge, 1972; Thomas, 1973; Boyer and Nissenbaum, 1974). In Jharkhand where widows are allowed to have temporary control over land, accusations of witchcraft often accompany attempts by 'male agnates to remove the threat to their property rights posed by widows' life interest in land' (Kelkar and Nathan, 1991, p. 93). A recent accusation of witchcraft in Ambatoli provides support for this theory. In 1990, Silamuni Orain's mother-in-law, a widow, was accused and found 'guilty' of witchcraft when a villager went to a *bhagat* complaining of stomach pain. She was forced to leave the village and her son had to mortgage out over half of her land to pay the compensation required by her 'victim'. In the mean time, Silamuni and her husband must work as daily labourers to buy the food that they would ordinarily have been able to grow themselves, and to try to pay the remainder of the compensation required to get their land back.

In 1993, Silamuni's mother-in-law came back to Ambatoli emboldened by the accusations of witchcraft and emphasising her own power as a *dain*. As a result, many villagers are even more wary of her than before and try to keep their children and cattle away from her 'evil eye'. According to the theory that deviant behaviour by women is greeted with accusations of witchcraft (Kelkar and Nathan, 1991), however, it is likely that the finger will again be pointed at Silamuni's mother-in-law with the result that her family will lose the rest of their land.

Agricultural Change in Ambatoli and Jamtoli

Turning now to agricultural change in the fieldwork villages, a more detailed investigation will be made into the environmental problems associated with modern farming methods and the extent to which they are displacing or de-legitimising traditional farming techniques and cropping patterns. An attempt will be made to investigate whether local people see modern agriculture as the key to greater livelihood security and cultural stability (as do the central Andean peasants studied by Bebbington, 1990a; 1992; 1996), or whether they see traditional knowledges and local agricultural expertise as central to a more populist development alternative. An additional area for investigation is the extent to which Jharkhandi agricultural institutions have tried to challenge the 'one-way transfer of technology from the industrial world to the Third World' (Altieri et al, 1987, p. 196) by attempting to unite local and scientific agricultural knowledges.

Ambatoli

The 1981 census lists Ambatoli as having 1543.1 acres of unirrigated cultivable land and 10.16 acres of irrigated land. Agricultural fields are small and 1993 data from my 100 sample households indicate an average land holding of 4.68 acres made up of 2.17 acres of *don* land and 2.51 acres of *tanr*. Only a few acres of *don* land is irrigated from the village's small streams and ponds, but the total amount of *tanr* land irrigated in 1998 was around 45 acres: a significant increase since 1981. Nevertheless, only 41 per cent of Ambatoli's sample households regularly grow enough food to meet their annual subsistence requirements and most have to find additional sources of income to make ends meet. 61 per cent of the village sample population undertakes some form of casual labour and 10.8 per cent undertakes regular seasonal migration.

In the 1991 census, 706 members (80.95 per cent) of the village's working population were classified as 'cultivators' and 116 (13.23 per cent) were listed as 'agricultural labourers' giving a total of 822 villagers (93.73 per cent of the village's working population) who depended on agriculture as their primary economic activity.[7] As was mentioned in Chapter Seven, however, land and wealth is distributed quite unequally both within and between communities.

Ambatoli's agricultural practices should, in theory, be more 'modern' because it was adopted by the Hindustan Fertilizer Corporation (HFC) as a 'Rainfed Farming Project' village (now taken over as part of the Krishak Bharati Co-operative's Eastern India Rainfed Farming Project) in the late 1980s and has had a full time 'village motivator' living there for much of the time since then. The aim of the Rainfed Farming Project is to undertake a 'sustainable cropping system approach' to 'develop risk management strategies for resource poor farmers ... based on their total participation' (Hindustan Fertilizer Corporation, 1990, p. 1). At the cutting edge of the project are the village motivators who identify the main agricultural constraints faced by local people and introduce new crop varieties, inter-cropping techniques and alternative income generating strategies.

Significantly, Ambatoli's adoption of HFC-distributed MVs of upland *gora* and lowland *don* paddy corresponds well with the classic model of green revolution technology diffusion: the wealthier farmers adopting first as they can afford the risks and inputs associated with the new technology, and the poorer farmer adopting later when the advantages of the new seeds have been demonstrated (Chambers Pacey and Thrupp, 1989; Lipton and Longhurst, 1989; Conway and Barbier, 1990; Yapa, 1996). In 1993, MV paddy varieties were planted by only twelve of Ambatoli's 100 sample households. All of these were relatively

well-off with favourable landholdings; many of which had potential for
supplemental irrigation.[8] The total area that they devoted to these
varieties was 20.4 acres (4.51 per cent of the net cropped area) of which
14 acres were planted with *don* paddy and 6.4 acres with *gora* paddy.
The remaining 278 acres (61.4 per cent of the net cropped area) of *don*
and *gora* paddy plus a further 80 acres of non-paddy *tanr* crops
including *marua*, *arhar*, *bodi* and *gondli* were all planted with
traditional varieties. As a result of farmers' preference for maintaining
traditional varieties and growing small quantities of MVs alongside
them, the number of crops and crop varieties grown today is greater
than 100 years ago (see table 8.4): a situation that contrasts with
populist fears about biodiversity loss in more developed agricultural
regions (Shiva, 1988; 1991a; 1991b; Gupta, 1992; 1996).

**Table 8.4   Crops commonly grown in Ranchi District in 1915
compared to 1993**

Year	Cereals	Millets	Pulses	Oilseeds	Vegetables	Spices
1915	*Don* paddy	*Marua*	Arhar	Linseed	Aubergine	Chilli
	Gora	*Gondli*	Bodi	*Surguja*	Bean	Turmeric
	paddy	*Bajra*	Horsegram	Mustard	Potato	Garlic
	Wheat	*Jowar*	Kurthi		Onion	
	Barley		Urad		Pumpkin	
	Summer		Gram		Bottle gourd	
	paddy				Okra	
	Maize				Radish	
					Jhingi	
1993	*Don* paddy	*Marua*	Arhar	Linseed	Aubergine	Chilli
	Gora	*Gondli*	Bodi	*Surguja*	Bean	Turmeric
	paddy	*Bajra*	Horsegram	Mustard	Potato	Garlic
	Wheat	*Jowar*	Kurthi		Onion	Ginger
	Barley		Urad		Pumpkin	
	HYV *don*		Gram		Bottle gourd	
	paddy				Okra	
	HYV *gora*				Radish	
	paddy				*Jhingi*	
	HYV wheat				Bitter gourd	
	Summer				Cauliflower	
	paddy				Tomato	
	Maize				Pea	
					Spinach	
					Sweet potato	

Sources: Roy, 1915; Jewitt, 1996.

The twelve families who grew MV paddy plus a handful of the wealthier *adivasis* were also the only sample villagers who applied chemical fertilisers (in addition to farmyard manure) to their *don* paddy and chemical pesticides to their *don* and *gora* paddy. Nevertheless, the average MV *don* paddy yield of 21 maunds per acre (ranging from 8 to 40 maunds, depending on rainfall, land quality and access to supplemental irrigation) was not significantly greater than the average yield of 20.4 maunds per acre (ranging from 5 to 40 maunds) for traditional *don* paddy grown on good land but without chemical fertiliser. MV *gora* paddy yields ranged from 12 to 15 maunds per acre while local varieties yielded between 5 and 15 maunds per acre.[9] The variation reflects both rainfall and the quality of the land, with MV *gora* paddy tending to be planted only on the better quality *tanr*. Inter-cropping seemed to make little difference to *gora* paddy yields although farmers felt that it improved soil quality and allowed a degree of risk minimisation on medium quality *tanr*. Typical yields for a 1 acre plot of *gora* paddy or *marua* with inter-crops of *bodi*, *arhar* and *urad* were 7.5 maunds of *gora* paddy or *marua*, 0.5 maunds of *bodi* or *arhar* and 0.25 maunds of *urad*.

Due to the limited residual moisture available in 1992-3, *kharif* and late-*kharif* season vegetable cultivation was quite restricted with only twelve sample villagers, all with relatively good land, planting part of their *tanr* land to crops such as sweet potato, maize, potato, pumpkin, pea, bean, okra, tomato, gourd and aubergine. Only three sample villagers, all with over five acres of land, grew late-*kharif* oilseeds on their *tanr* land and were aiming to make a good profit from them at the local market in Bero. In wetter years, however, the cultivation of these crops is more widespread.

Rabi season vegetable cultivation was practised by 34 per cent of the village sample population in 1993 and, because of its dependence on irrigation,[10] was dominated by villagers with access to wells and pumpsets or with landholdings close to the village's two ponds. Interestingly, the increased availability of pumpsets in the village (around a third of Ambatoli's sample households owned one in 1993) has not brought about a decline in the use of *lathas*. Indeed, most villagers use their pumpsets only during the dry season, whereas *lathas* are used to extract bathing and drinking water throughout the year. They also act as substitutes for pumpsets when mechanical problems arise or kerosene shortages occur.

The amount of land planted with *rabi* season crops is usually determined by the success of the previous paddy harvest. In poor years, villagers grow more vegetables which are used both for home consumption and for cash sales. In 1993, 45 acres of irrigated *rabi* crops were planted by the sample villagers, most of whom aimed to grow at least a few kilograms of vegetables as cash crops. Most villagers planted potatoes, peas and cauliflowers in September or October, to enable them to be sold as early vegetables for the Calcutta market. Potatoes were particularly popular as they yielded well (up to 80 maunds per acre on good irrigable land) and often fetched around 4800 rupees per acre.

Rabi wheat was cultivated by only 14 per cent of Ambatoli's sample population in 1993. It is usually sown a little later, often on good quality *tanr*, to reduce the amount of fertiliser required. A few households also grow oilseeds (linseed, *surguja* and mustard) as *rabi* season cash crops to supplement household budgets during the lean (monsoon) season. Other *rabi* vegetables and spices such as onions, ginger, maize, garlic and potatoes are usually planted in December, although some farmers inter-crop early and late *rabi* crops to maximise cropping intensities and profits. Typical examples include 'early and short' with 'late but tall' combinations such as wheat, tomatoes, aubergine or onions planted with maize or *surguja*. Others further increase *rabi* profits by growing summer vegetables (harvested in July) and reducing the amount of *kharif* season *gora* paddy that they grow in favour of late-planted *urad* with *bodi* and/or *arhar* combinations.

As irrigated *rabi* and summer crops do not benefit from the nutrients carried in the monsoon rainfall and runoff, it is common practice to apply chemical fertilisers. In 1993, 30 of the 34 rabi cultivators used them on at least one of their fields, but often applied them in less than optimal doses (heavily supplemented with manure) to minimise capital expenditure. Indeed, some villagers with good quality *tanr* 1 land frequently get good yields using manure and household compost only. Others were concerned that their soils would become exhausted by (or their crops dependent on) chemical fertilisers and tried to avoid this by maximising their use of manure and household compost supplemented with leaves and flowers from forest and homestead trees. The same villagers who used chemical fertilisers also applied small quantities of chemical pesticides (but not herbicides) to their most valuable *rabi* crops.[11] On the rest of their crops, they relied on ash dusted on the leaves of affected plants to deter insects.

Primarily as a result of the small amounts of agro-chemicals applied plus the use of organic alternatives, water pollution from pesticides and chemical fertilisers is not (yet, at least) perceived to be a problem in the village. A number of villagers with wells on elevated land, however, are concerned about falling water table levels as a result of increased levels of irrigation. Although this is not a problem for villagers with better quality, less hilly *tanr* land, it may, given the environmental problems associated with intensive irrigated cultivation elsewhere (Greenland, 1997), be something that requires close monitoring in future. From a socio-economic perspective, meanwhile, the influx of more modern farming practices seems to have been fairly benign to date. Intensification, especially irrigated *rabi* cropping, has enabled villagers to cope with population increase, forest decline and land fragmentation without a significant increase in landlessness or unemployment. Although the percentage of cultivators (as a proportion of the total working population) declined from 88 per cent to 80.95 per cent between 1981 and 1991, the proportion of agricultural labourers increased from 3.4 per cent to 13.23 per cent (see table 8.5).[12] Most local people see land fragmentation and not land sales as the main cause of this shift; a situation which suggests that cultivators losing out from this have gained from the employment that intensification has offered. Indeed, the percentage of villagers involved in agriculture (cultivators plus agricultural labourers) increased over this period from 92 per cent to 93.73 per cent of the total working population.

Table 8.5 **Changes in the number and proportion of cultivators and agricultural workers as a percentage of all workers in Ambatoli between 1981 and 1991**

	All workers	Culti-vators	% of culti-vators	Agri-cultural labourers	% of agricultural labourers	% of cultivators and agricultural labourers
Ambatoli 1981	525	465	88.00	18	3.40	92.00
Ambatoli 1991	877	706	80.95	116	13.23	93.73

Sources: Government of India, 1981; 1991.

Jamtoli

The 1981 census lists Jamtoli as having 1416.56 acres of cultivable land of which, at that time, 57.43 acres were irrigated. More recent data from my field research indicates an irrigated area of over 200 acres and average landholdings of 4.11 acres (of which 1.36 acres are irrigated) made up of 2.84 acres *don* and 1.27 acres *tanr*. According to the 1991 census, 478 members (68.29 per cent) of the village's working population were listed as 'cultivators' and 205 (29.29 per cent) were classified as 'agricultural labourers', indicating agriculture as the main economic activity for a total of 683 villagers (97.57 per cent of the village's working population). As mentioned in Chapter Seven, land is concentrated more heavily amongst the Oraons who first settled in the village. Nevertheless, income differentials (and poverty levels) have been substantially reduced by a very successful irrigation scheme initiated by Simon Oraon.

The Adoption of Modern Farming Practices in Jamtoli

In contrast to Ambatoli, Jamtoli is not a Rainfed Farming Project village yet it is characterised by a significantly higher degree of modern farming coupled with active and ongoing experimentation with new seeds and farming techniques. Although its topography is slightly more favourable than Ambatoli's, the difference in cultivation reflects, in large part, the agricultural expertise, ability to innovate and, for want of a better expression, 'charismatic leadership' of Simon Oraon.

Although Simon had to withdraw from primary school to help on the family farm, he was nevertheless able to conceptualise, plan and organise the finance for and the construction of five check dams which irrigate a substantial portion of the village's *don* and *tanr* land. He developed an interest in irrigation as a boy when he dammed a small stream on his father's land and succeeded in growing *rabi* crops. He developed a more active interest in agricultural development when he became *Parha Raja* over 40 years ago. Taking advantage of a stream running through the village from a nearby hill, he built a shallow (two and a half metres deep) dam and experimented with *rabi* season vegetables and summer paddy. His success with these crops encouraged him to expand Jamtoli's irrigable potential, so he gathered together over 100 local people to work as daily labourers on the construction of a bigger (seven metres deep) dam. To extend the area irrigated by this dam, three kilometers of irrigation channel were constructed, a branch of which cuts through Kaxitoli's main protected forest area (Letteguttu Forest) and irrigates some of the land that lies to the south. To obtain the irrigation water from the channels, most villagers simply have to break down the walls of smaller feeder canals to let water into their fields. Those who own land adjacent either to the check dams or to the major irrigation channels usually obtain water directly from these sources using either a *latha kunta* or a kerosene driven pump set.

As a direct result of the irrigation provided by this dam, Jamtoli villagers were able to survive the very poor monsoons in 1977-78. Simon also helped a number of neighbouring villages through this drought by lending them pump sets and obtaining Block funding for well construction, pumpsets, fertilisers, manure and seeds. As a result, they were able to cultivate 250 acres of *rabi* season wheat and vegetables.

The problems caused by the 1977-78 droughts enabled Simon to persuade villagers from Jamtoli and Kaxitoli hamlets to give up 6 acres of their land for the construction of an even bigger dam. Most of the total cost of the dam was met from Block Development funds in Bero. A large part of this money went to the local villagers who worked fairly solidly between 1978 and 1980 as daily labourers on the construction of the dam.

This dam has three main outlets, two of which release water directly into paddy fields between Jamtoli and Kaxitoli. The third feeds into an irrigation channel that cuts through Kaxitoli's smaller protected forest (Chanjkapari forest) and branches off to supply water both to the land between Chanjkapari and Letteguttu forest and to the second main check dam. In total, the new dam enabled over 150 acres of land to be irrigated.

But Simon did not stop there. In 1990, he started the construction of an irrigation-cum-fish pond close to Kaxitoli hamlet. This pond is used mainly to irrigate *garmi* paddy, although many villagers also use it as a place for watering and washing animals. The edges of this pond are planted with papaya and banana trees to help stabilise the banks and to provide fruit for local villagers.

Indeed, this smaller, multi-purpose pond was such a success that in 1993, the Block Development Officer authorised funding for the construction of a second pond to provide fish, bathing water and irrigation for *rabi* season vegetables. The construction of this pond brought the total amount of irrigated land in the village to over 200 acres, with many of the irrigation channels still supplying water to *garmi* paddy in early June.

Simon's next plan is to install a pipe to conduct water directly from this pond to nearby fields and to line the open irrigation channels with concrete to prevent seepage loss. He is also continually asking for Block loans to build wells in Jamtoli and other villages within his *Parha* so that they can grow *rabi* crops. To date, he estimates that he has helped to obtain funding for around 1800 wells in the Bero area.

As a result of Simon's efforts in Jamtoli, food security has increased dramatically as villagers can provide supplemental irrigation for their paddy when rainfall is poor and maximise cropping intensities by growing *kharif*, *rabi* and even summer crops. Indeed, higher paddy yields plus the profits made from (and employment generated by) winter vegetables has almost halted the need for seasonal migration.[13] Nowadays, over 75 per cent of Jamtoli's households grow enough food to last them throughout the year: a factor that has helped to significantly increase the village's sense of unity as well as promoting a culture of co-operation for further development work and agricultural experimentation.

Interestingly, most Jamtoli villagers cultivate local rather than high yielding varieties of *don* and *gora* paddy although around 60 per cent regularly use chemical fertiliser on their *don* paddy (to maximise the yield of their staple crop) and 28 per cent use chemical fertiliser on their *gora* paddy.[14] On fields close to check dams, canals and wells, average paddy yields are higher than elsewhere with irrigable *don* regularly yielding up to 50 maunds per acre and irrigable *tanr* yielding over 15 maunds per acre. Yields on non-irrigable *don* usually range from 15 to 30 maunds per acre while those for non-irrigable *tanr* range from 5 to 12 maunds per acre.

Another advantage of irrigation in Jamtoli is that it has made the cultivation of late-*kharif tanr* crops significantly more widespread than in Ambatoli, with around 35 per cent of the village population cultivating MV vegetables (mainly maize, tomatoes and pumpkins) and 24 per cent cultivating oilseeds (mustard and *surguja*). Some farmers also grow late-*kharif* pumpkins on their *don* land but the fields must be elevated enough not to be waterlogged but within reach of irrigation if the need arises.

In the *rabi* season, the major factor influencing villagers' adoption of wheat and vegetable cultivation has been not wealth, as in Ambatoli, but the proximity of their land to a source of irrigation. By 1993, Simon's assistance in obtaining Block-funded well and pumpset loans for villagers with no irrigable land had made *rabi* season cultivation possible for over 75 per cent of Jamtoli's households.[15] The most popular early vegetables now grown on *don* land are peas, potatoes, tomatoes, cabbages, aubergines and pumpkins which are planted after the paddy is cut and harvested between January and March. Wheat, onions and garlic are usually planted in November and harvested in February or March. Some farmers then plant a second crop of peas, potatoes, tomatoes or even summer paddy on their lowest *don* land which is harvested in late May or June. Such triple cropping is not as cash intensive as it sounds, however, as *don* land (being more fertile) requires less fertiliser than *tanr* and rarely needs irrigating until January or February.

The most popular *rabi* season *tanr* crops are potatoes, peas, wheat and tomatoes, although many villagers also cultivate garlic, onions, aubergines, cabbages, *bodi*, pumpkins, mustard, maize and cauliflowers. Less common crops include coriander, ginger, okra, chilli and sugar cane, but villagers are constantly trying out new seeds and searching for new niches in the cash crop market. As in Ambatoli, *rabi* crops in Jamtoli are usually irrigated either two or three times per week. The use of chemical fertilisers is more widespread than in Ambatoli with almost all *rabi* cultivators using at least some to supplement their applications of manure and compost. The other main inputs are also applied rather more intensively in Jamtoli with around half of the sample households using pesticide and a fifth using herbicide.

The Reasons behind Jamtoli's Adoption of Modern Agriculture

In spite of the fact that Jamtoli is not a Rainfed Farming Project village, agriculture there is more intensive and less differentiated by wealth than is the case in Ambatoli. The primary reasons for this are Simon Oraon's emphasis on village development and modern agriculture, the success of his assistance to the most resource poor farmers and the interest that most other villagers take of his own agricultural experiments and innovations. Simon's contacts with the local Block Development Officer and other agricultural practitioners also helps the village as a whole both to learn about different agricultural techniques and to get hold of new seeds and agro-chemicals. Although most villagers are keen to practice traditional farming techniques where they meet local requirements, they also see modern farming methods as a key means of meeting household food requirements and achieving increased livelihood security. Following Simon's lead, a number of villagers have become very interested in experimenting with MV vegetable seeds and several have attended Block organised farming courses. The proximity of Bero market also helps as villagers are aware of any changes to the prices and demand structures of both agricultural inputs and crops.

Another important factor is the manner in which Jamtoli's success has fed upon itself. A prime example is Simon Oraon's flow irrigation system which has given the village significantly greater food security as well as generating substantial local employment (both during its construction and afterwards) by allowing more intensive agriculture. Although between 1981 and 1991 there was a decline in the percentage of cultivators in Jamtoli (as a proportion of the total working population) from 84.1 per cent to 68.29 per cent, most villagers attributed this primarily to the recent expansion of nearby Bero which brought landless and non-agricultural families into the village. Indeed the increase in the percentage of agricultural labourers (as a proportion of the total working population) from 12.14 per cent to 29.29 per cent between 1981 and 1991 indicates the importance of intensification in softening the impact of population increase, land fragmentation and the loss of agricultural land to irrigation tanks (see table 8.6). Moreover, the overall proportion of villagers employed primarily in agriculture (cultivators plus agricultural labourers) increased from 96.24 per cent to 97.57 per cent of the total working population.

Table 8.6		Changes in the number and proportion of cultivators and agricultural workers as a percentage of all workers in Jamtoli between 1981 and 1991			
	All workers	Culti-vators	% of cultivat-ors	Agri-cultural labourers	% of agricultural labourers	% of cultivators and agricultural labourers
Jamtoli 1981	346	291	84.10	42	12.14	96.24
Jamtoli 1991	700	478	68.29	205	29.29	97.57

Sources: Government of India, 1981; 1991.

A major socio-cultural benefit of Jamtoli's modern farming practices that is greatly envied in neighbouring villages is its role in reducing the need for seasonal migration. In addition to strengthening village unity, the year-round presence of most villagers has provided a strong foundation for co-operative development work including a very successful forest protection system (Jewitt, 1995a; 2000a). Simon has also used his influence with Block Development Officers to obtain government owned common land for cultivation by the poorest villagers in Jamtoli: a situation that has significantly improved levels of support for his work as well as the capacity to pay off outstanding loans. Indeed, Jamtoli's forest protection and flow irrigation systems have increased levels of outside interest in the village to the extent that villagers have started to compete with one another to impress the state agricultural officials, extension workers and NGOs who come to see Jamtoli's modern farming methods and triple-cropping systems.

Are 'Modern' Farming Practices Displacing 'Traditional' Agro-ecological Knowledges?

Although it cannot be denied that modern agriculture, especially irrigated cultivation has been significant in increasing food security and profits, it is nevertheless important to assess these benefits in the light of potential environmental problems and the risk of a displacement of traditional technology. As noted earlier, the last 100 years have brought little change in the actual farm equipment used in the region except for the introduction of kerosene pumpsets. Instead of displacing labour or traditional technology, however, these pumpsets have created employment by allowing winter cropping and are largely used in conjunction with traditional *lathas*. Unlike the large agricultural machinery found in areas like Punjab, the relative cheapness (coupled with the availability of Block loans) and land-augmenting nature of pumpsets makes them accessible even to resource poor farmers who can form co-operative groups to buy and use them.

It is important to acknowledge, however, that the introduction of green revolution seeds and agrochemicals coupled with more intensive farming and a widespread adoption of irrigated *rabi* cropping has brought quite dramatic change; the implications of which may not be fully apparent for many years. Chemical fertiliser use, for example, has increased significantly over the last 20 years and has brought a decline in the quantity and range of organic nutrients used by farmers. Formerly, cattle manure and household compost were used in conjunction with other nutrients including ash, *parsa* (*Butea monosperma*) flowers and leaves and *karanj* (*Pongamia glabra*) flowers, leaves and seed cake. Now they are often supplemented only with chemical fertiliser. Likewise, pesticide use has increased in conjunction with *rabi* vegetable cultivation and the traditional practice of using ash on vegetable foliage to repel or kill insects has become less widespread. Herbicide use is as yet relatively uncommon, although the labour-saving implications of its spread could be serious.

Similarly, the environmental problems associated with intensive irrigated cultivation coupled with large applications of agro-chemicals are well known (Greenland, 1997; Pingali et al, 1997) and could provide grounds for future concern. To date, however, the research area seems to be coping quite well environmentally. Being a predominantly rainfed area, irrigated well- and tank-based *rabi* cropping is too small scale to cause the types of waterlogging, salinisation and soil toxicity that have been associated with major irrigation schemes (Shiva, 1988; 1991a). Nor are yield declines resulting from a degradation of the paddy ecosystem likely to be a problem as the dry winter season prevents continuous wetland cropping (Greenland, 1997). Groundwater pollution, meanwhile, has been limited by the relatively small amounts of agro-chemicals currently used although this and falling water table levels may both become more widespread in future.

MV seeds are another aspect of modern farming that have attracted much criticism in the populist and anti-developmentist literature for their environmental damage and limited overall benefits (Shiva, 1988; 1991a; 1991b; Apffel Marglin and Marglin, 1990; Yapa, 1996). In the study villages, however, most farmers continue to use traditional varieties of paddy alongside MVs during the *kharif* season because they taste better and often yield as well on poor land with fewer inputs. As in other parts of the developing world (Shiva, 1988; 1991a; Lipton and Longhurst, 1989; Dixon, 1990) they are also liked for their ability to produce good quality, long straw that can be used as cattle fodder, thatch, bedding and for making items such as grain storage baskets. In contrast to populist fears about biodiversity loss resulting from MV cropping (IDS Workshop, 1989b; Conway and Barbier, 1990; Shiva, 1991a; 1991b; Gupta, 1992; World Resources Institute, 1994), therefore, evidence from the study area suggests that MVs can help to increase overall varietal diversity in rainfed environments during the *kharif* season where risk minimisation is essential for food security.

In the *rabi* season, meanwhile, irrigation reduces the risk of losses caused by uncertain rainfall and makes it more economic for villagers to invest in MV wheat and vegetables accompanied by modern agrochemicals. A profit-oriented emphasis on irrigated *rabi* vegetable cultivation has also played a major role in enabling resource poor farmers to buy MV seeds and associated inputs for intensive cultivation. Even landless farmers sometimes cultivate winter vegetables as cash crops on their homestead land. And as the Jamtoli case study has shown, irrigated *rabi* cultivation has enabled many villagers to meet their food requirements without having to migrate. It has also allowed them to eat a more varied and balanced diet which, although perhaps signalling a change in traditional food preferences (Shiva, 1988; 1991a; Dixon, 1990), may simultaneously represent a shift to better nutrition and healthier eating habits.

Populist concerns that the displacement of traditional technology may reduce options to secure future autochthonous development (Appadurai, 1990; Gupta, 1992), meanwhile, seem to have little relevance to the study area. There appears to be relatively little evidence that the introduction of modern farming practices into the study villages has created significant landlessness, unemployment, or a loss of traditional farming practices and crop varieties. Although some traditional practices including the use of ash as a pesticide and a wide variety of composting materials as fertiliser have been displaced somewhat by the increased use of agro-chemicals, they have certainly not disappeared. In contrast to Gupta's (1992) concern that it is usually the poorest farmers who suffer most from a loss of traditional technology or increased prices of modern inputs, evidence from Ambatoli and Jamtoli suggests that the smallest landholders are the most likely to use a judicious mixture of agro-chemicals with organic manures and pesticides. Almost everybody that I spoke to in both Jamtoli and Ambatoli could name at least four or five people who still use these technologies and most people said that they would re-adopt them if agro-chemicals increased greatly in price. Indeed, during a series of in-depth interviews in both Jamtoli and Ambatoli, nobody that I spoke to could think of any agro-ecological techniques used by their parents that had been lost completely during their life times.

Whilst this is not particularly surprising given the continuity of agriculture in the research area over the last 100 years, it is important to avoid the assumption that indigenous knowledge systems are static because they contain so called traditional knowledges. Farmers in Ambatoli and Jamtoli have long conducted experiments to test the suitability of new seeds and agricultural techniques in a manner that is not dissimilar to (although less refined than) the work of agricultural research stations. Indeed, farmers are constantly adjusting their agricultural practices because this is the normal way in which indigenous agro-ecological knowledges develop and adapt to change (Boserup, 1970; Bebbington, 1990a). The main difference in recent years is that local peoples' subsistence strategies have been squeezed by population increase and land fragmentation, so they have seized opportunities to integrate more modern practices into their agro-ecological knowledge bases. Rather than focusing on the displacement (or otherwise) of traditional technology, therefore, it is perhaps more pertinent to ask whether current rates of change in farming practices are so unlike anything that has occurred in the past that a future co-existence of traditional and modern farming techniques and knowledges will be impossible.

Certainly in the short term, this seems unlikely because the current mixture of agro-ecological knowledges is a very positive and practical one for an area where agriculture itself is far from predictable. Contrary to Nandy and Visvanathan's claim that the adherents of Western *episteme* have no respect for 'competing systems' (Marglin, 1990a, p. 25), it is also a mixture that receives a substantial amount of support from KRIBHCO's Rainfed Farming Project as well as from the main agricultural university in Chota Nagpur, the Birsa Agricultural University (BAU) in Ranchi.

In many ways, researchers from the BAU seem to have taken heed of bottom-up emphases on learning from (not talking at) local farmers. In particular, they try to work closely with local people on their own farms and aim to improve 'farmers' technologies rather than to import completely new technologies' (Birsa Agricultural University, 1992, p. 145). They have also taken issue with the idea that local and scientific knowledges are incompatible (Banuri and Apffel Marglin,

1993a; Scoones and Thompson, 1994b; Drinkwater, 1994a; 1994b; Agrawal, 1995) and have been very successful in their efforts to build on farmers' experiments and to combine local and modern farming techniques in a cost-effective and ecologically sensitive manner. Moreover, they have tried to show that this can be done in ways that, far from disturbing or de-legitimising local knowledges, integrate well with wider development aspirations: not least the need to adjust to changing economic circumstances and to meet household subsistence requirements.

One example of this is the BAU's effort to build upon and expand the ways in which local farmers use a combination of organic nutrients and chemical fertilisers to cut down cultivation costs (interview with Professor Mohsin, 1993). Another is the trials undertaken to test the effectiveness of locally available *neem* (*Azadirachta indica*) oil as a foliar insect repellent spray, *neem* leaves to reduce tuber moth damage during potato storage and *kusum* (*Schleichera oleosa*) cake as a nutrient-rich cattle feed. The university also conducts frequent experiments on pest repellent and nitrogen fixing inter-crops, zero tillage, optimum crop sowing times and standard varietal trials in different farming environments (Birsa Agricultural University, 1992; 1995).

So in addition to taking steps to minimise the environmental impacts of intensive agriculture, the BAU also seeks to encourage the use of traditional technologies that have environmental or financial benefits for local farmers. An important example of this is the efforts of BAU researchers to raise farmers' awareness about the problems of mono-cropping and the advantages of maintaining varietal diversity. Indeed, a significant underlying principle that the BAU shares with many local people is that if water pollution becomes a problem or if the agricultural input price rises envisaged by Appadurai (1990) do occur, farmers would resort to using a predominance of labour intensive irrigation methods coupled with organic fertilisers and pest repellent techniques.

Conclusion

In this particular region, therefore, it seems that much has been learned from the more advanced agricultural areas about the potential of intensive agriculture as well as its socio-economic and environmental impacts. Populist critiques of the transfer of technology model and development assistance have been tremendously valuable in both publicising such problems and helping to stimulate a shift towards more participatory approaches which involve working with farmers and building upon indigenous agro-ecological knowledge. They have also highlighted the dangers of biodiversity loss in a world that is changing rapidly both physically and culturally as well as raising awareness of the psychological impacts that modern technology can have on indigenous social and belief systems.

Research stimulated by the latter has helped to highlight the importance of 'emic' investigations (Bebbington, 1990a) as a means of promoting more culturally acceptable and locally appropriate forms of development. It has also been important in raising awareness of the interlinkages between socio-cultural and technical knowledges which in turn has potential for providing insights into why particular forms of environmental management occur. As the second section of this Chapter illustrates, magico-religious practices are central to the success of agriculture in the research area. A good harvest is not just the result of the possession of well developed agro-ecological knowledges coupled with careful land preparation, hardy or high yielding seeds and the correct use of organic or chemical fertilisers and pesticides. It also reflects a more general state of harmony between male and female villagers, village deities and malevolent *bhuts* and *dains*. Their 'invisible world' and the restrictions that it places on villagers' use of space affect environmental management more generally, as certain areas are left fallow for religious ceremonies and others are haunted by malevolent *bhuts* that will exact their revenge if disturbed.

Nor are modern seeds and cropping techniques assumed to be immune to the evil eye. MVs are guarded, like local varieties, by painted or broken pots in the pre-harvest period and care is taken to protect kerosene driven pump sets from the powers of *dains* or malevolent *bhuts*. Indeed, following the logic behind *hisingar*, one would expect modern agriculture and development more generally to increase the importance of understanding the socio-cultural aspects of agro-ecological knowledges.

In spite of such mechanisms to reduce upward mobility and the potential of modern agriculture for increased class differentiation (Byres, 1981; 1983; Brass, 1990; 1994), however, most villagers in Jamtoli and Ambatoli have no hesitation in choosing a modern rather than a populist development pathway. Indeed, unlike many green revolution critics who highlight the economic and environmental disadvantages of intensive irrigated (especially wheat-based) farming systems and downplay the fact that there has never before been an 'increase in food production that was remotely comparable in scale, speed and spread' (Lipton with Longhurst, 1989, p. 1) most Jharkhandis are well aware of the benefits of modern agriculture. They also recognise that without the intensification that has taken place in the region, many people alive now would be dead. Faced with rapid population increase, fragmented land holdings and forest decline, a shift to double cropping, particularly irrigated *rabi* MV cultivation, literally represented a life-saving option for some and a substantial increase in annual food security (not to mention better nutrition) for many more.

Local farmers are not therefore throwing away perfectly good indigenous technology because they are told to adopt modern farming methods by state authorities that cannot resist the influence of multinational agro-chemical, seed and biotechnology corporations.[16] Rather, they adopt new techniques because they believe that it is worthwhile for them to do so. Similarly, if MVs really are so 'resource wasteful', vulnerable to pests and diseases and 'compared to the cropping systems they displace ... are not 'high yielding' or 'improved' at all' (Shiva, 1988, p. 122), surely local people would stop growing them. After all, peasant farmers in India have to be adaptable to survive. In Ambatoli and Jamtoli, most farmers are very skilled in maintaining a judicious mix of modern and traditional seeds and technologies to minimise risk, environmental damage and unnecessary capital expenditure. In recognition of this, the BAU has been keen to promote a participatory and socio-culturally sensitive approach to development which tries to take the best from indigenous and modern agricultural practices.

Given local farmers' clearly articulated desire for increased food security (and profits if possible), therefore, it seems prudent to be wary of the radical populist and anti-developmentist assumption that local people should and could 'go back' to simpler existences based on traditional agricultural practices (Esteva, 1987; Shiva, 1988; 1991a; 1991b; Apffel Marglin and Marglin, 1990; Sachs, 1992; Alvarez, 1994; Escobar, 1995). Although traditional societies may appear 'affluent' (Sahlins, 1972) in terms of their 'satisfaction of basic and vital needs' (Shiva, 1988, p. 12), other studies suggest that they often encounter difficulties in meeting their requirements for carbohydrates and water (Kelkar and Nathan, 1991), not to mention basic health care and education.

Many supply side populist views also ignore the fact that local people are not silent and helpless 'others', but are capable of choosing to what extent they want to access the things that development has to offer. As Bebbington's (1990a; 1990b; 1992; 1996) work in Ecuador illustrates, local people are often keen to modernise their indigenous farming techniques when this offers them greater food security, cultural cohesiveness and a way out of undesirable household survival strategies like migration. In many ways, Jamtoli's approach has been similar as intensive irrigated farming has enabled villagers to give up seasonal migration and has helped to increase food security and wider village unity to a much greater degree than in Ambatoli where subsistence is more precarious and seasonal migration common. As a means of achieving local self determination, Simon Oraon's strategy has been very effective as it combines indigenous and modern agro-ecological knowledges in a way that helps to strengthen local culture and meets villagers' evolving socio-economic and political aspirations. Yet it is a strategy that has been based on working *with* the state rather than resisting state-imposed development. Indeed, it has been the evident success of Simon's clever combination of indigenous and modern farming methods that has enabled him to obtain state funding for a locally appropriate form of development that has improved villagers' livelihoods in both material and socio-cultural terms.

Obviously, Simon Oraon is a rather exceptional character who has considerable local support and political clout. His success should not, therefore, be taken as an illustration of what other local leaders could be expected to achieve. Nevertheless, his ability to work closely with state officials does perhaps suggest that state-financed and locally-desired development are not necessarily as mutually opposed as many anti-developmentists seem to suggest. The skill of local people in selecting and adopting the aspects of modern agriculture that bring real benefits also shows that given the willingness of the state to provide appropriate support, some villagers can be quite capable of controlling and directing their own development.

Notes:

[1] 1 maund is roughly equal to 40 kilograms.

[2] In the Ranchi District Gazetteer of 1917, however, it is noted that market gardening was practised near to Ranchi and Lohardaga towns (Hallett, 1917).

[3] Ranchi District suffered severe droughts associated with food scarcity in 1820, 1823, 1827, 1837, 1866, 1873, 1920-1, 1927-8, 1945-6, 1948-9, 1952-3 and 1957-9 and declared famines in 1896-7, 1899-1900, 1907-8 (Reid, 1912; Hallett, 1917; Kumar, 1970).

[4] It has not been possible to update this information from the 1991 census as the Ranchi District level handbook is not yet available and landuse data is not contained on the 1991 cer.sus disks.

[5] While I was doing my fieldwork in Ambatoli in 1993, there were three 'sacrifice' murders and two attempted murders within 4 kilometers of where I was staying. Only one of these cases, a man from Bartoli who had been 'told' by a village *bhut* to sacrifice his own son, was reported to the police because he was regarded as being mentally ill. This man is now in gaol. The other cases, although not condoned by local people, were dealt with internally by the local community. Once when I was at a neighbouring village market, a youth who had tried to cut the throat of an elderly relative was dragged to the market place and publicly beaten.

[6] *Hisingar* is a common concept in Hindu as well as *Sarna* culture.

[7] Most of the remaining 6.27 per cent were artisans (bamboo weavers, blacksmiths etc.). The rest were either businessfolk or had waged jobs outside the village.

[8] The best *don* fields are irrigated by small streams and villagers with *tanr* fields close to wells and ponds can provide supplemental irrigation to their *kharif* paddy crops if the monsoon rains are poor.

[9] The highest yields typically come from fields close to ponds, streams or other sources of supplemental or 'emergency' irrigation.

[10] I came across only four families who owned irrigated land and did not grow vegetables at all.

[11] 42 per cent of Ambatoli's sample households used pesticide on either their *kharif*, *rabi* or summer crops.

[12] Interestingly, the proportion of female workers, cultivators and agricultural labourers all increased significantly in both Ambatoli and Jamtoli between 1981 and 1991. Although some of this change may reflect an increase in female-headed households (resulting from male seasonal migration), in-depth interviews on

agricultural change suggest that it has more to do with how women's work was classified in the two censuses.

13 Only 2.25 per cent of the sample population undertook seasonal migration in 1993 and most of these villagers migrated out of curiosity (as a means of earning money to buy consumer goods or visit major cities like Calcutta) rather than necessity.

14 The only person experimenting with IR36 on his *don* land was Simon Oraon.

15 At this time, Jamtoli's sample villagers had an average of 0.38 pump sets, 1.43 wells and 1.36 acres of irrigated land per household compared to 0.34 pumpsets, 0.58 wells and 0.5 acres of irrigated land for Ambatoli's sample households.

16 See, for example, Shiva's (1988; 1991a) accounts of the detrimental impacts of the green revolution.

9 Gender and Agriculture in the Research Area

Introduction

Picking up on the gender-environment discussions set out in Chapters Three and Five, this Chapter examines eco-feminist/WED ideas about women's 'subsistence perspective' and highly developed agro-ecological knowledges in light of empirical data from the research area. As *adivasi* society is characterised by a fairly high degree of gender equality, Jharkhand provides an excellent opportunity for investigating the extent to which the State's gender divisions of labour reflect those portrayed in the wider WED literature. Also, given the detailed ethnographic and gender-related research that has already been carried out in Jharkhand (Roy, 1915; 1928; Kelkar and Nathan, 1991), the region is ideal for testing Vandana Shiva's hypothesis that women, not men, are the main 'selectors and custodians of seed' (Shiva, 1992, p. 211).

In particular, an attempt is made to explore the political ecology of gender-environment relations with respect to agricultural participation, decision-making and knowledge articulation. In response to frequent calls for more empirical research on the 'social relations which limit [women's] power to form environmental knowledges' (Jackson, 1994, p. 133), an investigation is also made into some of the main gender-, religion-, class- and caste-related variations in the development, vocalisation and transfer of agricultural knowledge. Relatedly, issues regarding the communication of knowledge are also investigated: notably the extent to which local agro-ecological knowledges are transferred by women when they move, upon marriage, from their parents' to their husbands' villages.[1]

More specifically, the first section of the Chapter emphasises the micro-political ecologies, socio-economic factors and actor oriented issues that influence inter- and intra-gender task allocation in both *adivasi* and non-*adivasi* communities.[2] The second section examines inter-community and intra-household gender bargaining strategies and the effects that these can have on women's knowledge development and participation in household decision-making. The third section investigates gender-variations in agro-ecological knowledge possession and agricultural decision-making, paying particular attention to the difficulties that many women have in developing their agricultural expertise. This theme is carried through to the fourth section which examines women's transfers of agricultural techniques and seeds from their parents' villages to their husbands' homes.

Gender Divisions of Agricultural Labour

Although Scheduled Tribe men and women are fairly equal in social status, Jharkhandi society, like many societies elsewhere (Boserup, 1970; Blaikie, 1985; Chambers, 1983; Henshall Momsen, 1991; Sontheimer, 1991; Jahan, 1992), is characterised by well-defined divisions of agricultural labour. Some tasks are fairly loosely defined as 'male' or 'female' for practical reasons, but cause no problems if they happen to be carried out by the opposite sex (Roy, 1915; 1928; Kelkar and Nathan, 1991). Others are very strictly gendered and are enforced by taboos which, because they usually apply only to women, help to 'legitimise male control over the production process' (Kelkar and Nathan, 1991, p. 62) and maintain the subordinate position of women.

One of the most widespread taboos that affects women's participation in agriculture is the prohibition on ploughing which is universal to the Jharkhand tribes (Roy, 1912; 1915; 1928; Kelkar and Nathan, 1991). In Ambatoli and Jamtoli, most *adivasi* women believe that if they tried to plough (or, some believe, even touch a plough) a whole series of calamities would occur including the monsoon failing and everybody in the village falling ill. Local perceptions of why this taboo exists seem to relate more to the lesbian connotations behind a woman 'penetrating Mother Earth' than to gender-based struggles for control over agricultural production. Indeed, the sexual implications were quite clear in women's laughing comments that it would be 'unnatural' for women to plough 'Mother Earth'; only men could do that without 'damaging' her. Nevertheless, one woman did come forward with the more cynical interpretation that 'women would no longer be men's property if they could plough' (discussion with Sila Lohrain, 1993).

If a woman did attempt to plough, local custom dictates that a meeting should be held to inform everybody of the incident and to decide upon the punishment. The woman in question would usually be fined and sent away and if she refused to go, her hair and ears would be cut off and she would be chased from the village. Similar taboos and punishments are also associated with thatching or roofing work, although the origins of this seem to lie as much in the male dominated nature of religious practices (clan spirits are believed to live in roofs and must be propitiated by males) as in a desire for control over property (Kelkar and Nathan, 1991).

Other activities that are taboo to women include driving a bullock cart and carrying items on a shoulder pole. According to Kelkar and Nathan, the method by which *adivasis* carry items is so sex specific that one way of inquiring about the sex of a new baby is 'to ask whether the child will carry on the head or the shoulder' (Kelkar and Nathan, 1991, p. 58). The allocation of most other agricultural tasks is only loosely gendered but often reflects intra-household bargaining power as well as more general social structures that favour men and make it easier for them to get access to items of transport (usually bullock carts or cycles). Land preparation work involving the use of bullocks is carried out primarily by men whereas manual field preparation (using a hoe or a large wooden mallet to break up clods of earth) is often done by women.

Although transplanting and weeding are considered to be 'female tasks' throughout the Indian subcontinent (Harriss and Watson, 1987; Jahan, 1992) they are not carried out exclusively by women in tribal areas. In the research area, women do the lion's share of the transplanting work as they see it as the female equivalent to ploughing in terms of field labour input. If transplanting is behind schedule, however, it is quite acceptable for men to help out. Weeding work in Ambatoli and Jamtoli is more female-dominated than in other areas of Jharkhand (Kelkar and Nathan, 1991), although again, it is quite acceptable for men to help out when required. Harvesting work, by contrast, is more akin to the wider Jharkhand model described by Kelkar and Nathan (1991), with men and women participating fairly equally in work gangs to ensure that the paddy harvest is completed on time. The main exception is that when the harvest is transported home, women carry it on their heads while men bring it on a shoulder pole or in a bullock cart. Even when women are not working in the fields themselves, many carry out food to their menfolk.

Threshing, which is usually carried out on rocky ground where grain can be left to dry in the sun, is usually carried out by both men and women. In contrast to Kelkar and Nathan's (1991) claim that threshing with animals is carried out only by men, it is not uncommon to see children (and sometimes women) in the research area walking cattle over paddy stalks to separate them from the grain (see plate 9.1). Winnowing is fairly equally divided between men and women, although they tend to use different tools. Women usually sieve the grain through a metal colander whereas men tend to use a winnowing basket which relies on wind to carry the chaff away as it falls. Husking work, on the other hand, is a predominantly female task as it is carried out at home and can easily be combined with other household work. In terms of the amounts of time spent undertaking agricultural tasks, therefore, men tend to do more field-based work than women who have a variety of other tasks to complete including forest-based work, housework and child care.

Indeed, this situation echoes traditional WID emphases on female double work burdens (Boserup, 1970), as women are expected to carry out their household chores in addition to doing agricultural work. Many of the poorest women have a triple work burden as they undertake periodic wage labour to make ends meet. At the other end of the spectrum, however, women from better-off households often choose to 'buy themselves out of' the most tiresome tasks by hiring casual labourers to undertake this work. Indeed, this tendency echoes wider Indian trends for women to withdraw from market, field- and forest-based labour as their socio-economic status improves (Agarwal, 1997a).

One aspect of gender divisions of labour in Jharkhand that is not examined by Kelkar and Nathan is the extent to which they vary

Plate 9.1   A woman using cattle to thresh crops

between *adivasis* and the non-*adivasi* communities that now make up around 70 per cent of the region's population. As my own fieldwork was conducted in *adivasi*-dominated villages, it is only possible to outline the most obvious non-tribal deviations in task allocation from the general *adivasi* patterns. Nevertheless, even amongst villagers who follow similar subsistence strategies, some interesting community- and class-related differences in gender divisions of agricultural labour can be identified.

The most significant exceptions were found amongst Ambatoli's better-off Hindu Backward Caste households. Perceiving themselves to be at the top of the village socio-economic hierarchy, they often opt for strategic marriage alliances and expect their brides (many of whom marry in their mid-teens or before) to practice a loose form of *purdah*. This usually means avoiding field-based agricultural work or forest-based gathering, but the reality is significantly affected by economic factors: notably the extent to which the families in question can afford to buy in agricultural labour to replace that of female household members. In the wealthiest Backward Caste families, therefore, women's agricultural work consists mainly of vegetable cultivation and other activities that can be carried out close to or actually within the homestead. Poorer Backward Caste households, by contrast, find *purdah* impractical as they don't have the means to replace women's labour with that of paid workers. As a result, their gender divisions of labour are fairly similar to those of the Scheduled Tribes. The same is also true of most Scheduled Caste *sadan* households.

Interestingly, many Oraon women hold the view that 'Sahu and other Backward Caste women only do housework while Scheduled Tribe women do field-based agricultural work as well as forest-based gathering' (discussion with Goindi Orain, 1993). As a result, field-based agricultural work tends to be strongly male dominated in the wealthiest Backward Caste families as men are unable to delegate these tasks to their womenfolk to the same extent as *adivasi* and Scheduled Caste men.

Intra-household Bargaining Power and the Re-negotiation of Work Allocation

In addition to these economic and community-based influences on gender divisions of labour, it is important to recognise the fluidity and frequent re-negotiation of work allocation in accordance with intra-household power dynamics (Kandiyoti, 1988; Hobley, 1996; Jackson, 1993a; Cornwall, 1998; Gujit and Kaul Shah, 1998a). In joint households where married sons share a home with their parents, intra-gender bargaining over favoured tasks can be quite heated. Factors such as marital status, age, number of children and individual personalities can have a very important influence on the household pecking order. Where economic and labour resources permit, mothers-in-law and their daughters often 'bag' the most favoured jobs, allocating the more tiresome and repetitious jobs to the young married women.

In addition, young married women often feel intimidated by their mothers-in-law and find themselves with significantly more tasks to complete than they had in their parents' place. Indeed, out of eighteen Ambatoli women with whom I conducted in-depth interviews, eleven said that they were made to change their cooking and housework methods by their mothers-in-law. Many felt both miserable and at sea when they first moved to Ambatoli. A typical comment was 'I felt unhappy and uncomfortable here for the first two or three years ... Although my mother taught me how to cook and do housework I had to relearn everything when I came here. Sometimes my mother-in-law scolded me and ordered me about' (discussion with Rahil Orain, 1993).

Insecurities were particularly pronounced amongst women who married young to men they hardly knew. Although this system is increasing amongst *adivasis*, it is still more common for them to have 'love' or 'semi-arranged' marriages where the couple are allowed to get to know one another and back out, if necessary, before the wedding ceremony. The tendency for *adivasi* girls to marry boys just a few years older than themselves (and whom they have often known since childhood) also helps to encourage a more equal relationship. Backward Caste and (to a lesser extent) Scheduled Caste women, by contrast, tend to marry at a younger age (often to significantly older husbands), with a dowry. They also tend to suffer greater dislocation as they usually move further from their natal places than *adivasi* women. Niramala Sahu, a twenty six year old Backward Caste woman from Ichack, (over 100 kilometers away) in Hazaribagh District was unable to speak the local dialect when she arrived in Ambatoli. She told me: 'I was very upset when I first married and cried because I missed my family. My husband's family tried to reassure me, but I couldn't understand the local language very well. I wanted to run back home, but what could I do? My family paid a big dowry to get a good husband for me' (discussion with Niramala Sahu, 1993).

In many cases, women's bargaining power within the household reflects wider perceptions of 'economic value' and 'deservedness' (Boserup, 1970; Agarwal, 2001); often being higher amongst educated and wage-earning women. It also tends to increase with age; especially after having children (particularly sons). Women also tend to gain greater autonomy when their household structure changes from joint to nuclear. Jitan Manjhi, a forty five year old Scheduled Caste *sadan* (basket weaver) woman commented: 'My mother-in-law used to dictate what I should cook, but I got more independence when we became separate households. I much prefer to be in charge of my own household' (discussion with Jitan Manjhi, 1993).

Gender Divisions in the Possession and Control over Agro-ecological Knowledges

Before investigating men's and women's agro-ecological knowledges in detail, it is important to emphasise that gender is only one of many factors influencing knowledge possession. As the previous sections indicate, the capacity of individuals to avoid, delegate or buy themselves out of the most burdensome tasks is strongly influenced by wider socio-economic and political ecology factors as well as inter- and intra-gender power relations. These, when coupled with the frequent re-negotiation of task allocation, have knock-on effects for local people's ability to accumulate and articulate agro-ecological knowledge. Just as importantly, inequalities in land ownership or control over environmental resources act as a major influence on the extent to which local 'situated agents' (Bebbington, 1994; 1996) can acquire and develop their environmental knowledge.

In an attempt to illustrate this, the rest of this section highlights how environmental knowledge possession is influenced by class, caste, socio-cultural norms, age and individual circumstances as well as by gender. It also attempts to intervene in the ongoing debate between eco-feminists and their critics over gender variations in agro-ecological knowledge. Vandana Shiva, for example, argues that women have better-developed environmental knowledges than men and act as major agricultural innovators as well as the main 'selectors and preservers of seed' (Shiva, 1992, p. 210). In contrast to this view, Kelkar and Nathan (1991) maintain that even in hill and tribal areas like Jharkhand where gender relations are significantly more egalitarian than in plains Hindu society, men have better-developed agricultural knowledges as well as the primary role in environmental decision-making. Even more importantly, Agarwal argues that India's patrilineal inheritance patterns encourage men to develop a long-term interest in the condition of their land whereas women usually have only indirect access to (and very limited control over) it (Agarwal, 1994a; 1994b).

Data from the research area largely support Kelkar and Nathan's and Agarwal's arguments as it is men, as the main landowners, who make the vast majority of agriculture-related decisions including what crops and inter-crops to plant, how much to plant, when to plant them, where to plant them and what inputs to use. It also tends to be men who conduct agricultural experiments, have dealings with (usually male) agricultural extension officers and Block Development Officers and learn about new seeds, agricultural techniques, marketing strategies and so on. Even in share-cropping households, most agricultural decisions are made by men from either the owner's or share-cropper's family and amongst landless villagers, it tends to be men, as heads of household, who determine which family members are to seek paid agricultural or other labour.

Most women, therefore, depend heavily upon men to decide whether and to what degree they actually contribute to field-based agriculture. The extent to which Ambatoli's Backward Caste women can develop their agro-ecological knowledge, for example, depends largely on whether their husbands can afford to replace their labour with that of hired workers. Most Oraon and *sadan* women, meanwhile, are usually quite knowledgeable about the tasks that they are responsible for undertaking but still regard agriculture as their husbands' primary responsibility.

Another significant factor that restricts women's ability to obtain locally specific agro-ecological knowledge is the region's predominantly patrilocal residence system. This helps to ensure that men have much greater familiarity with the specific ecology and pedology of their farm land as well as a lifetime's experience of growing crops on it. Most married women, by contrast, have to start from scratch when they move from their natal villages: a situation that can present a very steep learning curve for women coming from town-based, non-agricultural households or areas where agricultural practices are significantly different. In addition, few newly married women feel that their suggestions would be welcomed by their husbands and parents-in-law. The three recently married women that I interviewed in Ambatoli were afraid to make any comments about agriculture in case their husbands or parents-in-law felt insulted.

The problems that many women face in accumulating local environmental knowledge are often compounded by the difficulties they have in both vocalising the knowledge that they do possess and contributing to household decision-making more generally. Out of my eighteen in-depth interviewees, only six made regular contributions to agricultural decision-making; usually concerning sowing times, the acreage of particular crops that they should cultivate and the possibility of trying new crops. Reflecting Bina Agarwal's findings that women's muteness varies according to community and region, I found significant differences within Ambatoli in the extent to which women felt confident about articulating their views. Generally speaking, *adivasi* and Scheduled Caste women were more assertive than the better-off Backward Caste women in their contribution to agricultural decision-making; mainly because they undertake more field-based agricultural work and therefore have the practical experience on which to base their suggestions.

Although the desire to observe a loose form of *purdah* is undoubtedly a limiting factor on the autonomy of the wealthier Backward Caste women, there are important issues other than socio-economic ones that influence women's role in agricultural decision-making. Particularly significant are age, intra-household bargaining power, specific circumstances and individual personalities. Generally speaking, most recently married women are anxious to make a good impression and avoid questioning decisions made by their husbands or in-laws. Almost all of my interviewees admitted to feeling unsure of themselves in their new surroundings and being hesitant about making agricultural suggestions. One woman said: 'When I first came to

Ambatoli, I was sixteen and I was very shy about talking to people. I was frightened of my mother-in-law and tried hard not to annoy her or my husband. Even though I had cultivated vegetables and worked in the fields at my parents' place, I didn't feel able to make any suggestions about agriculture here. My husband's family have been farming this land for years. What could they learn from me?' (discussion with Neera Lohrain, 1993).

Despite a general perception amongst Ambatoli women that their role in household decision-making increases with age (and especially after having children), many women have remained hesitant about making agricultural suggestions. Twelve of my eighteen in-depth interviewees said that they played no role in agricultural decision-making and ten of these had been married for over five years and had children. One of these women simply said that she preferred to leave the agricultural decisions to her husband and son while another two blamed their unwillingness to make suggestions on a lack of good quality land. Jatri Orain, for example, commented: 'I would like to grow vegetables but I don't want to suggest it as we don't have any good irrigated land' (discussion with Jatri Orain, 1993). Out of the four women who grew up in towns, meanwhile, three still felt that they knew much less about agriculture than their husbands.[3] The fourth, an *adivasi* woman called Sila Lohrain who grew up in a suburb of Ranchi commented: 'When I first came to the village ... I didn't know anything about agricultural work ... I learned most things about crop cultivation from my mother-in-law and I learned the rest by watching local women' (discussion with Sila Lohrain, 1993).

Particularly significant, however, is the reluctance to participate in agricultural decision-making amongst older women who have been active in field-based agriculture for many years. Amongst my interviewees, there was a definite perception that agriculture was not a domain in which women should interfere as the land belongs to their husbands. One of my most self-confident respondents commented: 'In the past, I have tried to make suggestions to my husband about sowing times and what crops to grow, but he doesn't listen to my views. He says that I don't do proper work, only housework, so why should he listen to me? He tells me that he would take more notice of what I say if I were earning a wage' (discussion with Sila Lohrain, 1993). Similarly, a very assertive and articulate Oraon woman who had cultivated sugar cane and maize in her parents' village said: 'I don't feel able to suggest growing these crops in Ambatoli because my husband would only laugh at me or be annoyed with me for interfering' (discussion with Hiramuni Orain, 1993). In a similar vein, two tribal *sadan* women of the Mahli (basket weaving) community who asked their husbands about growing vegetables were greeted with the response 'I don't know how to grow vegetables, I only know how to do basket work' (interviews with Pairo and Sanicaria Mahli, 1993).

To some degree, the impasse faced by these four women can be accounted for by the fact that they are all married to men for whom farming is a secondary occupation (Hiramuni's husband runs a flour mill and the others' husbands are blacksmiths or basket makers) and have little real interest in agriculture. Also, and in spite of the fact that these women are quite capable of managing the family's agricultural land (indeed, Sila, Hiramuni and Pairo, were all very self-confident and assertive women), their husbands would not give the necessary financial backing for this. Consequently, these women have ploughed most of their agricultural expertise into their homestead vegetable gardens over which (because they live in nuclear households) they have complete authority.[4]

Transfers of Agro-ecological Knowledge

Putting eco-feminist/WED ideas about women's special link with nature into the spatial context of patrilocal residence patterns, one would expect women to be responsible for major transfers of agro-ecological knowledge when they move to their husbands' villages after marriage. The best known theory of knowledge diffusion over space is the 'distance-decay' model (Abler, Adams and Gould, 1971; Blaikie, 1975; Haggett, 1983) which predicts that proximity is the major control on the likelihood of innovations spreading. In the case of agricultural innovations and knowledge, however, it seems likely that distance would act as a stimulant rather than a hindrance to diffusion as it is usually associated with changes to environmental conditions and agricultural practices. One might therefore hypothesise that women who come from the most distant villages would generate larger knowledge transfers than women who marry more locally; often within the same agro-ecological zone. One would expect men's knowledge transfers, by contrast, to be fewer as they tend to have less exposure to different areas and knowledge bases. Notable exceptions to this are men who are married in the '*ghar jamai*' fashion and go to live with their in-laws (Roy, 1915).

Generally speaking, data from Ambatoli supports the idea that distance is associated both with environmental change and different farming practices. Ambatoli has a lower than average proportion of irrigated land compared to Bero Block as a whole (12 per cent in 1981) and is less strongly market-oriented than villages situated close to Bero town. Most villages in the neighbouring Blocks of Sisai, Chanho, Bandhra and Kara have fairly similar edaphic conditions and agricultural practices, although Sisai and Kara have less irrigable potential than Bero but a strong market orientation. Bandhra and Chanho, by contrast are flatter and fairly well irrigated.

There is also evidence to suggest that a linkage exists between community group and marriage distance (see figure 9.1). Out of my sample of eighteen in-depth interviewees, the three Backward Caste women moved 79 kilometers from their natal villages, the two Scheduled Caste women moved an average of 38 kilometers and the

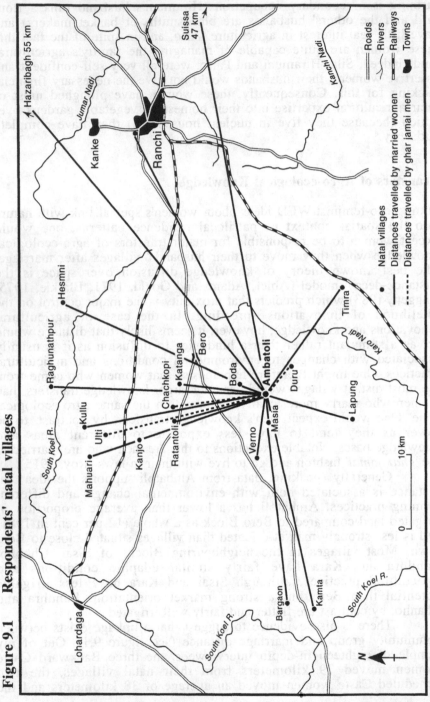

Figure 9.1   Respondents' natal villages

thirteen Scheduled Tribe women moved an average of just under 21 kilometers.

Despite this linkage, however, there is little data to indicate either that women from distant villages transfer more agro-ecological knowledge than those who grew up nearby, or that women make more knowledge transfers than men. Out of my eighteen interviewees, only five transferred seeds or agricultural techniques and all of these women grew up within 24 kilometers of Ambatoli, in villages with broadly similar agricultural practices. Indeed, women's class, community, education, age, input into the household budget, husbands' occupation and individual personalities appeared to be far more important than distance and environmental change as factors influencing agro-ecological knowledge transfers. Significantly, however, both of Ambatoli's *ghar jamai* men made important knowledge transfers and although one of these men moved only a short distance from his natal place, the fact that both had travelled quite widely seemed to be important in influencing their transfers of agricultural techniques.

In many respects, the reasons given by my thirteen 'non-transferring' interviewees for not bringing seeds or agricultural techniques from their natal villages reflect the patterns of agricultural knowledge possession and decision-making noted above. Two of the thirteen grew up in towns with parents who were not involved in agriculture and two felt that their land was too poor for experimenting with new crops. Four said that there was no point in them bringing seeds as their husbands had little interest in agricultural experimentation and would laugh at their ideas.

Despite the fact that the two better-off Backward Caste women grew up in quite different agro-ecological zones, both claimed that they didn't bring seeds from their natal villages as 'the crops grown there are exactly the same as those grown in Ambatoli' (interviews with Niramala and Rina Sahu, 1993). Their knowledge of different agricultural practices, meanwhile, was limited by the fact that in an attempt to maintain a loose form of *purdah*, their parents had discouraged them from contributing to field-based agricultural work.

Interestingly, the remaining three 'non-transferring' women were all *adivasis* who had contributed to agriculture in their natal places but claimed that the practices there were exactly the same as in Ambatoli. In Gandri Orain's case, this was not surprising as her natal place, Boda, is situated just five kilometers from Ambatoli. Neera Lohrain and Laki Mahli, by contrast, both come from villages 32 kilometers away with better irrigable potential than Ambatoli and a wider range of crops. Consequently, one would have expected some 'distance-increased-diffusion' to occur, even if only to the women's well watered and manured homestead gardens.

Nevertheless, it is important to recognise the transfers of seeds and techniques made by five of my in-depth interviewees and to acknowledge that this does lend some support to eco-feminist discourses about a special women-environment link. The most significant seed transfer was made by Silamuni Orain from Kanjo village, Bandhra Block (24 kilometers away as the crow flies) who claims to be the first person to bring *gongra* (squash gourd) and *jhingi* (*Luffa acutangula*) seeds to Ambatoli. Silamuni used to spend a lot of time in her mother's homestead garden and when she visits her parents, she often brings seeds from there to her garden in Ambatoli. She has also given a number of her mother's seeds to villagers in Ambatoli who were interested in growing them. Her next plan is to be the first person to plant sugar cane in Ambatoli and she intends to ask her parents' advice on how to cultivate it as they have had a lot of success with it. At the moment, however, there is no suitable land on which she and her husband can grow this crop.

Another impressive, although slightly conflicting seed transfer claim is that of Nauri Orain (from Ratantoli village, 14 kilometers from Ambatoli) who argues that she was the first person to grow *jhingi* and *gongra* with seeds from her mother's garden. She also believes herself to be the first person in Ambatoli to grow *lokee* (bottle gourd) and, like Silamuni, she tried to encourage its cultivation by giving spare seeds to her neighbours.

A third seed transfer claim was made by Sattan Lohrain (from Katanga village, 13 kilometers away) who brought French beans and cucumbers to Ambatoli from her mother's village. As a girl, Sattan did a lot of work in her mother's vegetable garden where she used to grow French beans, ladies finger, potatoes, *lokee* and *gongra* for sale at Bero market. When Sattan first grew these vegetables in Ambatoli, her neighbours were very curious and asked her for seeds and advice on how to grow them.

The only other interviewee to bring seeds from her natal place was Etwari Orain who grew up in Masia village, just five kilometers away and does not therefore conform very well with the idea of 'distance-increased-diffusion'. Etwari brought *jhingi, bodi*, sweet potato, cauliflower and cabbage seeds from her mother's vegetable garden but does not claim to be the first person to grow these vegetables. Nevertheless, they were new to her husband's family, as were a number of wild vegetables like '*chimti sag*', '*chera sag*', '*suga sag*', '*matak sag*' and '*kattay sag*'[5] that she sometimes collects in Masia but which have been largely displaced in Ambatoli by cultivated vegetables.

Interestingly, and perhaps in keeping with the importance of female-dominated homestead gardens as destinations for these four women's seed transfers, the only agricultural technique transferred by a woman was *marua* (*Eleusine corocana*) transplanting; transplanting being a predominantly female task in the research area. The person responsible for transferring this technique was Fagni Pradhan, a Backward Caste woman from Timra village, 18 kilometers away. Because her parents were too poor to keep her in *purdah*, Fagni learned a lot about agriculture as a child, including how to increase *marua* yields by transplanting rather than broadcasting. Fagni, whose claim is backed up by her husband, says that she was the first person to bring *marua* transplanting to Nehalu hamlet and has encouraged its spread by teaching other villagers how to do it.

Despite the absence of support for a 'distance-increased-diffusion' of women's agricultural knowledge, it is possible to make some interesting observations about the influence of class, community and gender on the movement of seeds and farming techniques over space and between people. Out of the five women who transferred seeds or techniques, four were *adivasis*, and all came from villages within 24 kilometers of Ambatoli. Even with the limited sample of eighteen women, this suggests that factors other than marriage distance are important. The failure of the two Sahu women and the success of four Scheduled Tribe women in transferring agricultural knowledge can be explained partly by the different degrees of gender equality and female autonomy within these communities. The relatively high 'economic value' of *adivasi* women and the lack of restrictions on their mobility enables them to obtain agricultural knowledge from quite an early age. The two Sahu women, by contrast, grew up in fairly well-off families which could afford to keep their women out of field-based agricultural work: a situation that has made it difficult for them to gain agricultural experience or to feel confident about making agricultural suggestions to their husbands or in-laws. Nevertheless, these factors fail to account for why four *adivasi* women made knowledge transfers and the other nine didn't and why the only agricultural technique transfer was made by a Backward Caste woman. Here, an actor oriented perspective can be useful for investigating the individual circumstances and personality related factors working within wider socio-economic and political structures.

Indeed, one significant factor that characterises all five of the 'transferring women' is their self-confidence and autonomy both within and outside the household. In addition, they all come from farming families and were heavily involved in both field-based and homestead garden cultivation before they married. Their husbands, meanwhile, all take an interest in agricultural innovation and give their wives free rein to experiment in their homestead gardens. Indeed, this interest in agricultural experimentation is something that the five transferring women obviously have in common but is lacking amongst two of the three *adivasi* women (plus the two Sahu women) who claimed that agriculture in their natal villages is the same as in Ambatoli. But before concluding that natal to marital place knowledge transfers are most likely to be made by women who are interested in agriculture (as well as having the autonomy to act on this interest), it is important to look at the agro-ecological knowledge transferred by Ambatoli's *ghar jamai* men.

As mentioned earlier, both of Ambatoli's *ghar jamai* men were responsible for bringing new agricultural techniques to Ambatoli. In addition to this, a discussion with the widow of a third *ghar jamai* man revealed that he regularly used to transfer paddy and vegetable seeds from his natal village of Chachkopi, thirteen kilometers away. Interestingly and reflecting men's greater responsibility for field-based agriculture, the *ghar jamai* knowledge transfers mostly consisted of agricultural techniques and major crop seeds. Tila Oraon who is a retired serviceman from the nearby village of Duru did not have a *ghar jamai* marriage although he was widely regarded as a *ghar jamai* man. His wife was born in Ambatoli and he bought land there with his army savings and chose to settle in the village when he retired. During his Army career, he travelled widely and saw many different agricultural practices. This encouraged him to experiment on his own land and he is now regarded as one of the village's expert farmers. Although his travelling supports the idea of 'distance-increased-diffusion' his main knowledge transfer claim is that his mother showed him how to transplant *marua* and he taught his wife this technique over thirty years ago, well before Fagni Pradhan arrived in the village.[6]

Interestingly, Ambatoli's other *ghar jamai* man, Etwa Mahli, also claims to have brought *marua* transplanting to the village, but as he settled more recently than both Tila Oraon and Fagni Pradhan, it seems likely that he introduced the technique only to his own hamlet, Deepatoli. Etwa originates from Ulthi village which is around 30 kilometers away in Bandhra block, but worked in Ranchi for a while before marrying and moving to Ambatoli about thirteen years ago. Like Tila, Etwa did not marry in the *ghar jamai* fashion, but moved to Ambatoli at his wife's and in-laws' request when his wife was ill and pregnant with their first child. Also like Tila, he supports the idea of 'distance-increased-diffusion' quite well as when he was working in Ranchi, he became curious about the many different crops and paddy varieties sold in the market and now tries to bring new varieties back to experiment with whenever he goes to town. He especially likes to bring good quality, long-grained *don* paddy which his in-laws are very fond of eating on special occasions.

Amongst the *ghar jamai* men, therefore, there does appear to be evidence for 'distance-increased-diffusion' of knowledge although this reflects their experience of travelling as much as of growing up in different villages. As far as Ambatoli's women are concerned, however, community-based, economic and cultural factors seem to be more important than distance in influencing knowledge transfers. Nevertheless, substantially more data would need to be collected on the diffusion of agricultural knowledge from villages with different ecological conditions to test this hypothesis further.

Overall, natal to marital village transfers of seeds and agricultural techniques do seem to be a significant factor in the diffusion of agro-ecological knowledge, although they tend to reflect underlying structural inequalities as well as actor oriented factors. The women's transfers, in particular, reflected the personalities and experiences of the people concerned and the extent to which they could overcome wider community-related, cultural and socio-economic inequalities to develop agro-ecological knowledge and put it into practice. The *ghar jamai* men, by contrast, showed rather more consistency in their knowledge transfers: something that should not really surprise us given the male-dominated nature of agricultural decision-making and men's closer links to the main sources of agricultural innovation; notably seed sellers, extension workers and Block Development Officers. It is also usually men rather than women who get opportunities to attend agricultural training courses. So, although women's agro-ecological knowledges transfers can be very significant, there seems to be little doubt that men are Ambatoli's primary seed selectors and 'custodians' (Shiva, 1992) as they have better opportunities for developing and articulating agricultural expertise.

Conclusion

In a number of ways, therefore, empirical data from Ambatoli highlights the shortcomings of eco-feminist/WED discourse in terms of its failure to challenge (or even recognise) the social constraints that prevent women from developing and utilising their environmental knowledges. It also indicates that discourses which romanticise women's special link with nature and invisibilise men's agricultural expertise (Leach, Joekes and Green, 1995) require urgent theoretical re-assessment. Clearly, task allocation is a highly socio-political issue and although there may appear to be well-defined gender divisions of labour, these are constantly being challenged and re-negotiated between and within communities in response to the bargaining power of different individuals. It is also important to distinguish agro-ecological knowledge from agricultural work as women's participation in the latter is often a very poor indicator of their role or interest in agricultural management: a situation confirmed by the fact that twelve of my eighteen in-depth interviewees had absolutely no role in agricultural decision-making.

Although WED's rather simplistic and undifferentiated 'add women to environment and stir' (Braidotti et al, 1994) approach has obvious appeal for donor agencies and policy-makers, therefore, it is important to consider the implications of using it as a basis for development planning. As the Ambatoli data illustrates, significant inter- and intra-gender differences in agro-ecological knowledge possession exist even in areas of relatively high female equality. Patrilocality and patrilinearity are particularly important in influencing women's control over land as well as their abilities to develop and put into practice their agro-ecological knowledge. Prevailing cultural restrictions on women's mobility also constrain their access to public places such as markets in which agro-ecological knowledge is articulated. Wider class and community-related factors, meanwhile, can be significant in discouraging women in better-off households from undertaking field-based agricultural work and thereby obtaining agricultural expertise. In addition, more specific factors including marital status, age, education, 'economic value' and individual personalities have an important influence on intra-household task allocation and the opportunities that different women get for putting their ecological knowledge into practice.

So, while eco-feminist/WED discourse has been extremely valuable in challenging traditional male-dominated views, it seems to work best as a springboard for local investigation into gender-environment relations rather than as a representation of reality. Unfortunately, the idea that women's empowerment can be achieved simply through the provision of greater opportunities for their participation is unrealistic in that it fails to acknowledge the socio-cultural barriers that most women face in trying to articulate their views and develop their knowledge bases. Real gender equality is likely to be a long-term process involving major structural changes to land ownership and inheritance patterns (Agarwal, 1994a; 1994b). In the meantime, locally-grounded research into the micro-political ecology of gender-environment relations is required as a means of revealing more effective ways to promote the participation of both men and women in environmentally-oriented local development initiatives.

Notes:

[1] The idea of investigating such transfers came up in conversation with Anil Gupta in 1992. I am grateful to him for encouraging me to pursue this part of my research. See Appendix II for a list of respondents with whom discussions took place on this topic.

[2] Much of the data used in this section was first published in Jewitt, S. (2000b), 'Unequal Knowledges in Jharkhand, India: De-Romanticizing women's Agroecological Expertise', *Development and Change* vol. 31 (5), pp. 961-985.

[3] Out of the other fifteen women who had been married for over five years, four still felt unable to make any suggestions about agriculture. Their reticence was due to the fact that they had either married young or grown up in towns. As a result, they had learned most of what they knew about crop growing in Ambatoli.

[4] In many joint households, by contrast, homestead gardens are controlled by the oldest married women (and their daughters) who delegate the more irksome tasks to their daughters-in-law.

[5] It was not possible to find out the botanical names for these plants.

[6] There is also some doubt about Silamuni Orain's and Nauri Orain's claim to introduce *gongra* and *jhingi* to the village as both vegetables are mentioned in S.C. Roy's book 'The Oraons of Chota Nagpur' which was published in 1915.

10 Forest Use and Management in the Research Area

Introduction

The dependence of many rural Indians (particularly *adivasis*) on forests for both subsistence-related and socio-cultural purposes has been acknowledged for some time (Fernandes and Kulkarni, 1983; Guha, 1983; 1989; Shiva, 1988). In addition, the idea that forest-dependent populations possess detailed silvicultural knowledges and have potential for (re)-establishing environmentally sustainable systems of community-based resource management has received much attention from India's populist writers and forest intellectuals (see, for example, Fernandes, 1983; Guha, 1983; 1989; Joshi, 1983; Kulkarni, 1983; Prabhu, 1983; Shiva et al, 1983; Shiva, 1988; Gadgil and Iyer, 1989; Alvarez, 1994; Pathak, 1994; Gadgil and Guha, 1995). With the exception of the studies done by Kelkar and Nathan (1991) on gender and forests, by Corbridge (1991a; 1995c) on forest struggles, and by Mehrotra and Kishore (1990) on forest protection committees, however, little work has been done on the relationship between rural people and forests in the once *adivasi*-dominated 'land of forests': Jharkhand. Yet in many ways, Jharkhand is a particularly interesting area for research on local people's interactions with forests.

As in many other areas of India (Guha, 1983; 1989; 1993; Shiva, 1988; Pathak, 1994; Sivaramakrishnan, 1999; Karlsson, 2000; Rangan, 2000a), forest-dependent villagers in Jharkhand found means of opposing the introduction of scientific forestry. Although the state's response was often repressive, villagers still devised a variety of 'peasant resistance strategies' (Scott, 1985) to counter official attempts to de-legitimise their traditional forest use and management practices. Sometimes they did so by destroying commercially valuable forests in the way that peasants in Uttarakhand did in the 1920s (Guha, 1989), but in other cases they had the opposite response and tried to protect local forests from illegal exploitation by forest contractors. In later years, their efforts were influenced and strengthened by the Jharkhand movement's emphasis on local people's resistance to state-imposed development; notably the imposition of commercial forest management and its criminalisation of local forest users.

The main aim of this Chapter is to examine the history of forest management in the research area and outline the different ways in which local people depend on forests for subsistence and socio-cultural purposes. More specifically, the first section traces changes in forest management practices from the pre-colonial to the post-Independence period and the impacts that these have had on local forest users. The second section examines the main causes of forest decline in the research area drawing particular attention to how villagers' loss of interest in state-owned forests (many of which had traditionally been managed *de facto* as common property resources) contributed to deforestation in the research area (Guha, 1983; 1989).

Continuing Chapter Eight's emphasis on the interlinkages between socio-cultural and technical knowledge systems (Bebbington, 1990a; 1992), the third section of this Chapter focuses on the socio-cultural and magico-religious importance of forests in the research area and how this has helped to prevent their destruction. Particular attention is drawn to the central role of forests in village culture and the influence of the Jharkhand movement on local attitudes towards them. This is followed in the fourth section by an examination of the more practical aspects of local forest use and of the pivotal role that forest produce plays in many villagers' household survival strategies. In the concluding section, an attempt is made to assess the extent to which traditional forest uses and silvicultural management systems have been displaced by scientific forestry and later forest decline.

Forest Management History in the Research Area

The Pre-colonial Period   Although few concrete data actually exist on forest management in the fieldwork area prior to the British colonial period, Kelkar and Nathan suggest that 'the village community ... seems to have been the acknowledged owner of the forest around it in the sense that its use of the forest, whether to gather/hunt forest products or to bring forest land under cultivation, was an unhindered right, and did not involve any payment to or permission of a superior' (Kelkar and Nathan, 1991, p. 121).

Although village communities did pay a tribute to the Chota Nagpur Raja, this was a fixed amount and did not relate to the extent either of forest or agricultural land (Kelkar and Nathan, 1991). This situation changed slightly when the Chota Nagpur Raj was made a tributary of the Mughal Emperor and the right to convert forests into agricultural land was restricted. In all other senses, local people's use of forests remained unrestrained until the onset of British commercial forest management.

This description of unrestricted yet apparently ecologically sustainable forest use in the pre-colonial period has been echoed by other writers on South Asian forestry (Guha 1983; 1989; 1993; Prabhu, 1983; Tiwari, 1983; Kulkarni, 1983; Arnold and Campbell, 1986; Gupta 1986; Shiva, 1988; Gadgil and Iyer, 1989; Mehrotra and Kishore, 1990; Kant et al, 1991; Ghate, 1992; Pathak, 1994). Indeed, their accounts have been instrumental in creating a populist vision of traditional forest management sustained by 'customary limitations on users' (Guha, 1989,

p. 31) that highlights what has been lost as a result of the development process (Jeffery, 1998b; Sivaramakrishnan, 1999; Rangan, 2000a). In Jharkhand, however, the ecological reality of this 'golden age' of forest management may not have been quite so idyllic as some populist accounts suggest. In particular, the widespread use of fire both to drive game and attract it to patches of fresh grass within the forest caused substantial damage that may well have resulted in localised ecological crises (Nathan and Kelkar, 1991). To what extent these problems would have affected the longer term ecological sustainability of community-based forest management systems is impossible to comment on, however, as many of them declined with the onset of British scientific forestry.

The British Colonial Period During the British colonial period, many Indian forests were brought under state ownership and local people's traditional forest rights were suddenly reduced to mere privileges granted by the colonial state. In the research area, control over most of the non-Reserved forests was placed in the hands of local *zamindars*. According to the Annual Progress Report of Forest Administration in the Province of Bihar 1936-7 (the first to be published after the separation of Bihar from Orissa), only around 4000 square kilometers out of a total of 29,000 square kilometers of forest land in Bihar had been classified either as Reserved or Protected forest and placed under scientific forestry by the end of 1937.

The *zamindars* themselves usually undertook little or no forest management and the freedom of local people to collect forest products such as fuel wood, timber and non-timber forest products (NTFPs) depended to a large extent on whether they were held under the *rakhat* or the *katat* systems mentioned in Chapter Six. According to Kelkar and Nathan, the nineteenth century uprisings helped to preserve many of the forests in Ranchi District under *rakhat* systems. This meant that forests were held as common property resources by the village 'with regulated user rights for its residents' (Kelkar and Nathan, 1991, p. 126).

The details of these rights were recorded in the Kattiyan Part II and were issued to local people by the *zamindars* in the form of 'plain papers'.[1] If villagers had additional requirements for forest produce, they could make a request to the *zamindar* or a village-level representative - usually the *Pahan*. The *Pahan* was also often responsible for deciding which trees should be cut to meet villagers' customary concessions and additional requirements for forest produce. Any disputes over unfair distribution of fuel wood, construction timber and so on tended to be dealt with through the *Parha* system.

A number of the older villagers in Ambatoli and Jamtoli remember the *zamindari* period well and talk fondly of how forests throughout Jharkhand were well stocked with mature *sal*, *asan* (*Terminalia tomentosa*), *bar* (*Ficus bengalensis*), *bhelwa*, *jamun*, *karonda* (*Carissa carandas*), *karam*, *karanj*, *kend*, *kusum*, *mahua*, mango (*Mangifera indica*), *parsa* and other trees. They also maintain that their forefathers had traditionally had the free use of local forests and that they used to protect them to ensure that nobody went short of fuel wood or house construction timber.

Simon Oraon, for example, mentioned that villagers used to abide by specific rules about the collection of fuel wood, the cutting of trees for timber and the ways in which coppicing or pollarding should be done. Most villagers used to collect dead branches or waste wood (from trees cut for timber) as their major source of fuel wood and only if there were none left did they take lateral branches from living trees. Many people also used to collect branches and twigs from 'weed' bush species like *putus* (*Lantana camera*) and *tetair* (*Ipomoea cornea*) for use as fuel. Unlike in later years, young trees were rarely felled and anybody cutting a healthy fruit tree would have been scolded by other villagers.

Sal has long been the preferred timber species as its wood is strong and it regenerates quickly through coppicing. Many villagers tried to promote this by cutting the tree at an angle to prevent rotting by allowing water to drain off. They also used to leave a fair amount of stump above the ground to ensure that the new growth was too tall to be trampled by cattle or stunted by weeds. Villagers sometimes used to pollard trees to prevent browsing by cattle, promote vertical growth and provide a source of fuel wood (discussion with Tila Oraon, 1993). Interestingly, and unlike the situation in Mathwad Forest Range, Madhya Pradesh (Pathak, 1994), emphasis in the research area was (and largely still is) placed on the need to protect forests rather than convert them to agricultural land.

Nevertheless, during the 1930s, the colonial Forest Department became very dissatisfied with the condition of many *zamindari* forests and their criticism (and de-legitimisation) of local forest management practices helped to justify attempts to bring them under scientific forestry. The main means of achieving this was under the Bihar Private Forests Act, which gave the colonial government power to take over the management of private forests. As a result of *zamindari* opposition, it was not possible to implement the Act until 1946. In the mean time, however, a number of private landlords were persuaded to place their forest land under Forest Department management in exchange for a fixed rent and a share of the profits (Government of Bihar, 1941). Between 1937 and 1947, the area of privately-owned Reserved Forest under Forest Department management increased from 40,453 acres (16,370 hectares) to 200,314 acres (81,064 hectares) and the Protected Forest area increased from 13,501 to 31,623 acres (5,464 to 12,797 hectares) (see figures 10.1 and 10.2).

Unlike the situation in Uttarakhand described by Guha, very few of Bihar's forests were completely 'divested of existing rights of user' (Guha, 1989, p. 38) during the British colonial period. This was mainly because most (83 per cent) of Bihar's forests were notified as Protected rather than Reserved Forests. The major factor determining forest classification was the ability to satisfy both commercial and local

Figure 10.1   Reserved Forest acreage 1935-1956

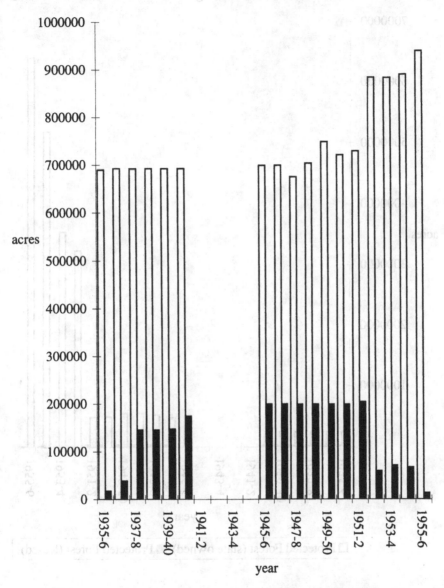

Figure 10.2   Protected Forest acreage 1935-1956

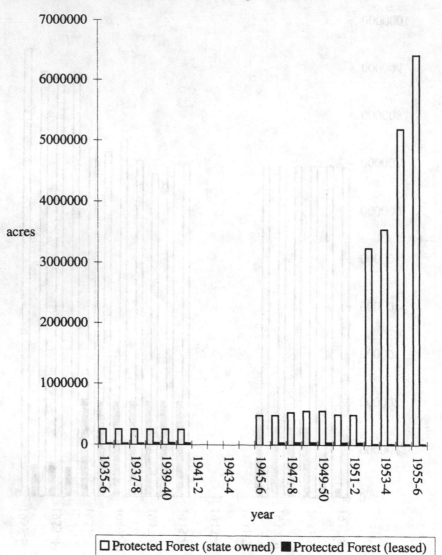

requirements. In Protected Forests where everything is permitted unless specifically prohibited, right-holders' requirements had to be satisfied before timber could be cut for commercial purposes. Reserved Forests, by contrast, are usually much less 'right-burdened' (everything is prohibited unless specifically permitted) and have greater scope for commercial timber production.

Although some forests were reserved 'free from rights' in the early 1920s (Society for Promotion of Wastelands Development and Forest and Environment Department, Bihar, 1994), around 90 per cent of the Bihar *zamindari* forests that were handed over to the Forest Department for management between 1937 and 1947 were notified as Protected Forests: mainly because they were too heavily right-burdened for commercial management to be feasible (interview with Sanjay Kumar, 1995).[2] During the 1930s and 1940s, certain state-owned forests that were large or productive enough to be exploited commercially were notified as Reserved Forests. Those like Jamtoli that had previously been used by local people, however, were 'reserved with rights' and had to satisfy right-holders' requirements before they could be felled for commercial use. Bihar's most intensive commercial timber production, meanwhile, took place in the large yet sparsely populated Reserved Forests of Singhbhum and Palamu where management disagreements with local people were fairly uncommon.

Part of the reason for this relative lack of conflict could have been that the British colonial forest management system in Bihar did not 'close' forests to same the extent as it did in Uttarakhand. This was particularly true of the ex-*zamindari* Protected Forests and the smaller right-burdened Reserved Forests like Jamtoli where right-holders were expected to 'coppice the annual coupe area, protect the forests from fire, report the commission of forest offences and help forest officers in preventing the commission of offences and in apprehending the offenders' (Society for Promotion of Wastelands Development and Forest and Environment Department, Bihar, 1994, p. 13).

With regard to villagers' rights in local forests, there was relatively little difference between the Protected Forests and the right-burdened Reserved Forests. The first 'Annual Progress Report' to be published after the reservation of Jamtoli forest in 1944,[3] for example, stated that 'there are no rights and privileges in the Government Reserved forest except the following:- *Bona fide* tenants living in the neighbourhood of either a reserved, protected or private estates' forest have been allowed the privileges of taking timber and fire wood, grazing cattle and collecting mohua flowers and fruits for their genuine domestic use but not for sale or barter' (Government of Bihar, 1947, p. 10).

As Guha (1989) rightly points out, local people's use of both Reserved and Protected Forests was certainly curtailed compared to the pre-colonial situation. Essentially, the Forest Department was (and is) a profit-making institution and villagers without rights in local forests had to pay concessionary rates for forest produce that they had formerly obtained free (see figures 10.3, 10.4, 10.5 and 10.6). Even right-holders sometimes had to pay for forest produce when they exceeded their annual allowances. As in other areas of South Asia, therefore, official misunderstandings about indigenous systems of property ownership and management (Singh, 1986; Guha, 1989; Shepherd, 1992b), plus the sudden imposition of forest regulations must have had severe psychological and socio-cultural impacts on local people (Elwin, 1964). Many of the Forest Department's new rules also had significant implications for the satisfaction of villagers' subsistence requirements (Singh, 1986).

On the whole, however, the colonial Forest Department's forest management system seems to have been less burdensome for local people in Jharkhand than in Uttarakhand. This was largely due to the gradual transition from the free use of forests during the pre-colonial period to the *zamindari* system and then state-based forest management. In addition, villagers in some areas of Jharkhand were able to participate in forest management through the *'taungya'* system which enabled them to grow food crops in the forest in exchange for planting (or coppicing) and protecting commercially valuable trees (Government of Bihar, 1938; 1939; 1947; Sivaramakrishnan, 1999; Karlsson, 2000). As a result, levels of local detachment from forests and the associated temptation to over-exploit or even destroy them were less pronounced and cases of major local 'unrest' over forest issues were much less common than in Uttarakhand (Guha, 1989).

Certainly in Jamtoli, a number of the older villagers now look back nostalgically upon the British period when the village population was smaller and there was enough wood for all in spite of the semi-commercial scientific management of Jamtoli forest (interviews with Simon Oraon, Vishvanath Bhagat and Phulchan Tirkey, 1993). Many villagers also commented that timber, fuel wood, NTFP and grazing allowances were perfectly adequate for their needs, so they rarely had to pay concessionary rates to obtain additional wood.

Echoing Karlsson's (1999) findings in northern West Bengal, many Jharkhandi villagers recall relationships with colonial Forest Department staff as being cordial and remember the forest guards as 'honest men' who did not take bribes and were sympathetic to villagers' requests for additional forest produce in times of need (interviews with Jittu Bokta and Buddhu Bhagat, 1993). From a more general administrative point of view, too, the lack of British interference with the *Parha* system (Roy, 1936; Kelkar and Nathan, 1991) and their 'respect for local customs' (interviews with Tila Oraon, Somra Bhagat, Budhu Bhagat and Simon Oraon, 1993) is fondly recalled in both Ambatoli and Jamtoli.

In 1946, the Bihar Private Forests Act was finally implemented and the Forest Department started to take control of the private forests that made up nearly 80 per cent of Bihar's total forest land so that they could be managed 'along proper and scientific lines' (Government of Bihar, 1947, p. 18). The eventual target was to place around 25,900

Figure 10.3   Free and concessional timber grants

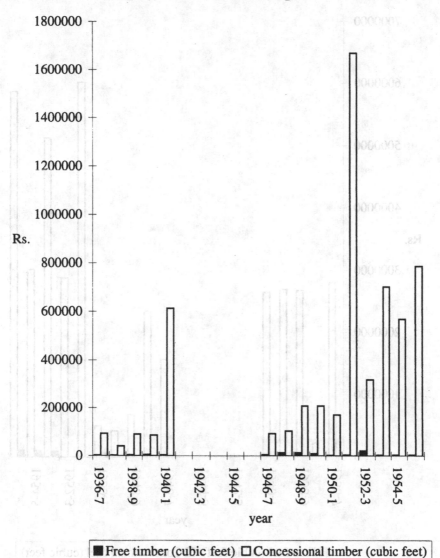

Figure 10.4   Free and concessional fuel wood grants

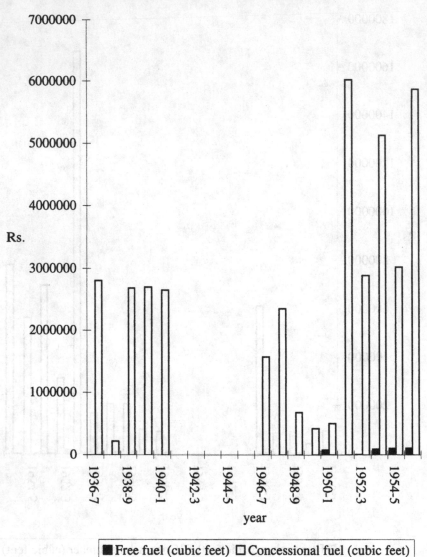

Figure 10.5   Free and concessional bamboo grants

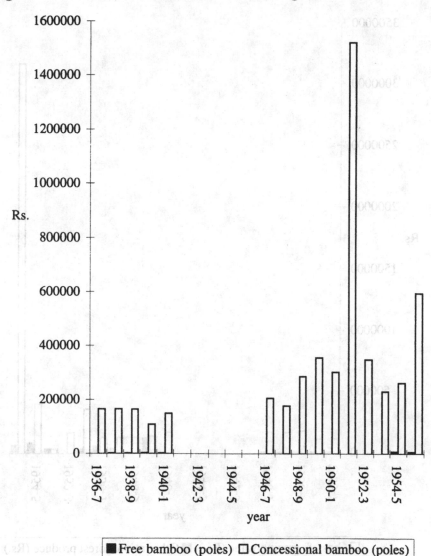

Figure 10.6   Free grazing and minor forest produce

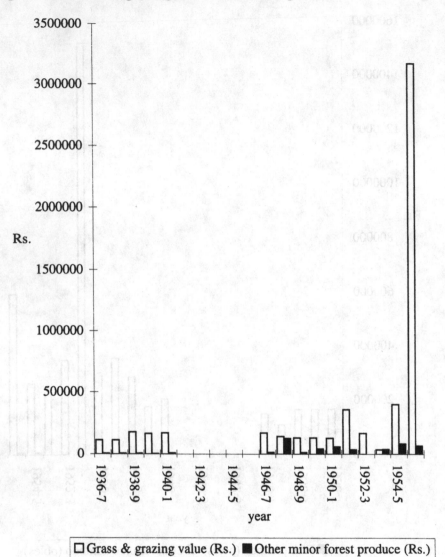

square kilometers of private forest under Forest Department management to prevent it from being 'destroyed at the hands of the private owners who paid little attention to this asset' (Government of Bihar, 1947, p. 18).

Working plans were soon devised by the Forest Department to improve the condition of the most 'badly destroyed and spoilt' (Government of Bihar, 1947, p. 18) *zamindari* forests. In most cases, these consisted of improvement fellings to stimulate the growth of the larger trees and coppice coupes to promote regeneration in badly degraded areas (Government of Bihar, 1948). The replanting of 'blank' areas was also carried out in many forests using local and exotic species including *gamhar* (*Gmelina arboria*), *semar* (*Bombax malabricum*), teak (*Tectona grandis*) eucalyptus (*Eucalyptus citriodora*) and pine (*Pinus longifolia*) (Government of Bihar, 1948).

The actual process of demarcating the boundaries of the ex-private forests and determining whether they should be re-notified as Reserved, Protected or Unclassed Forests was a very slow one that stretched staff resources to the limit (Government of Bihar, 1948). It was also a very costly process and expenditure on the demarcation, improvement and general management of these forests resulted in a significant dip in Forest Department profits during the late 1940s (see figure 10.7).

The financial implications of the Bihar Private Forests Act were also a source of concern for many local people who feared that the former *zamindari* forests might be re-scheduled as 'right free' Reserved forests. This situation was exacerbated by some of the large landlords who had lost control over their forests under the Bihar Private Forests Act and saw local opposition to the Act as an ideal opportunity to stir up trouble. To allay these fears, the Forest Department organised 'forest publicity weeks' to clear up misunderstandings over the concept of reservation and to explain that local people would not lose their rights in the ex-private forests.

The Post-Independence Period As mentioned in Chapter Four, there was a strong degree of continuity between British and post-colonial systems of forest management (Guha, 1983; 1989; Jewitt, 1995a; Pathak, 1995). Indeed, one of the main differences between the two systems was the 'greater zeal' (Singh, 1986) with which the post-colonial Forest Department de-legitimised local forest management practices to justify a more modernising and commercially-oriented system of scientific forestry.

Soon after Independence, India implemented a process of land reforms. In Bihar, the *zamindari* system was abolished in 1950 with the introduction of the Bihar Land Reforms Act which vested in the state the ownership of all agricultural land that was not privately owned (i.e. tilled) by *zamindars* or their tenants. It also brought just under 28,000 square kilometers of *zamindari* forest land under Forest Department ownership to enable the 'malformed and pallarded (*sic*) crop of the

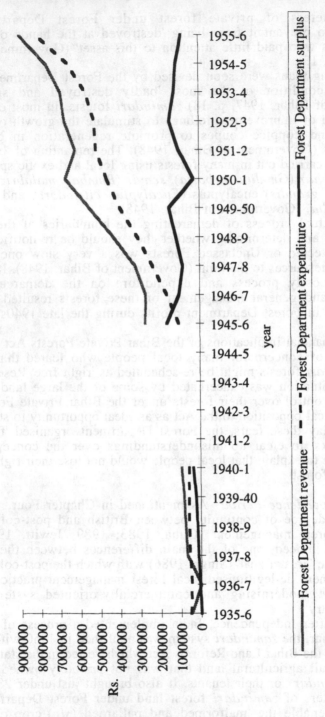

Figure 10.7   Forest Department revenue and expenditure 1935-1956

zamindari regime' to be 'quickly improved' (Government of Bihar, 1952).[4]

The majority of the ex-*zamindari* forests that had been designated as Private Protected forests prior to 1950 were re-notified as Protected Forests. One reason for not reserving them was that reservation was a very lengthy process (involving 15 sections of the Indian Forest Act, 1927) and mass reservations were simply not practical at that time. In addition, many of the ex-*zamindari* forests were small and heavily right-burdened and could not therefore support commercial as well as local use (interview with Sanjay Kumar, 1995).

There were some exceptions. Ambatoli forest is large and was well stocked with large trees at Independence according to both official and local sources (interview with Sanjay Kumar, 1995). Backwards extrapolation from the first village-based census data for Ambatoli and Jamtoli in 1971 also shows that Ambatoli had a significant amount of forest land (0.7 acres or 0.29 hectares) per right-holder in 1951. This meant that it was more than capable of supplying timber for commercial production in addition to fulfilling the requirements of its right-holders in both Ambatoli and the neighbouring village of Patratoli, which has little forest land of its own.

Prior to the implementation of the Bihar Land Reforms Act, many local forests also suffered at the hands of local *zamindars* who were determined to plunder as much timber as they could. As a result, many of the new Protected Forests were in a degraded state when the Forest Department took them over. Ambatoli's *zamindar*, the Ratu Maharaja, for example, cut 14 acres of mango orchard before Ambatoli forest was re-designated as a Protected Forest in 1953.

To make good this type of damage and improve the condition of the new Protected Forests, the Forest Department undertook large scale *sal* coppicing and re-planting operations using mainly local species such as *asan, arjun (Terminalia arjuna), khair (Acacia catechu), gamhar, semar* and *sisam (Dalbergia sissoo)* (Government of Bihar, 1950). Forest protection by local villagers was also expected in exchange for their rights to dry branches for fuel wood and NTFPs which were listed, as during the colonial period, in the Kattiyan Part II.

In Ambatoli and Jamtoli, for example, all villagers are right-holders and their freedom to collect NTFPs (with the exception of *kend* leaves for which a licence is required)[5] and dead branches for fuel wood did not change either after Independence or the land reforms. Their rights to cut forest trees as timber for house construction and agricultural implement making also remained the same with the exception that they had to first obtain permission from Forest Department officials rather than the *zamindar* or his village level representative.

Non right-holders, on the other hand, were supposed to buy forest produce at concessionary rates from the Forest Department. As in many other areas of India, however (Guha, 1989; Gadgil and Guha, 1992; 1994; Colchester, 1994; Pathak, 1994), the pilferage of such products from nearby woodlands were common 'everyday forms of peasant resistance' (Scott, 1985) to commercial forest management.

The management of each forest in Bihar was carried out according to detailed working plans and differed only slightly from the scientific forestry principles of the British. In the ex-Private Protected Forests, some unregulated fellings were initially carried out to meet right-holders' requirements (Government of Bihar, 1950; 1952), but by 1964, most forests were being managed according to twenty year working plans which aimed to improve the growing stock using short rotation coppice coupes (Government of Bihar, 1952). Because most of the ex-*zamindari* forests were heavily right-burdened, their management plans were either 'non-commercial' or 'quasi-commercial' in nature.

Non-commercial management was usually restricted to the smaller, heavily right-burdened Protected Forests which needed substantial improvement in the form of replanting and regeneration through coppice. Initially, at least, 'right-holder only' coupes were the only form of exploitation possible in many of these forests (Government of Bihar, 1953). Quasi-commercial management was aimed at forests that were 'in surplus to the requirements of the right-holders' (Government of Bihar, 1951; 1953) and was usually undertaken in the larger Protected Forests like Ambatoli and some of the smaller, right-burdened Reserved Forests like Jamtoli. In theory, local needs were supposed to be met first from annual coupes and any surplus was to be sold to contractors at auction or to non right-holders at concessionary rates. In practice, the contractors often cut far more of the forest than they had paid for in order to sell off the surplus wood to charcoal-makers for their own profit. In Jamtoli, villagers used to mark out the coupes with arrows, but the contractors often ignored them and felled much larger areas (discussion with Simon Oraon, 1993).

In Ambatoli, villagers were sometimes employed by forest contractors to help them to illegally exploit the forest, but because the contractors had been 'sent by the government', nobody felt that they could object either to the low wages (villagers were paid less than the minimum wage) or to the over-exploitation of their forest. The contractors frequently also took the lateral branches and twigs left over from the commercial fellings that they were supposed to leave for the right-holders to use as fuel wood.

During the 1950s and 1960s, the activities of forest contractors in both Ambatoli and Jamtoli left many right-holders unable to meet their subsistence needs for forest products. As a result, they decided to cut the additional trees that they needed without asking the Forest Department's permission. Unfortunately, but in keeping with the situation in other areas of India (Guha, 1983; 1989; Blaikie, 1985; Mehrotra and Kishore, 1990; Shepherd, 1992b; Pathak, 1994; Karlsson, 2000), many forest guards saw this type of pilferage as an opportunity to make money for themselves and threatened forest 'thieves' that they would take out court cases against them unless they paid the requisite bribe. Before long, financially-based 'understandings' started to develop between forest guards and some of the wealthier local inhabitants: a situation that caused severe inter- and intra-village tension as right-holders saw their traditionally equal rights of access to forests disappear (discussion with Fulla Lohra, 1993). It also provoked a widespread feeling of resentment towards local forest staff who could 'afford not to care if the forest disappeared as they would get their salary whatever happened' (discussion with Tila Oraon, 1993).

Throughout Bihar, these types of situation sparked off increasing hostility towards the Forest Department, often accompanied by a vicious cycle of illegal fellings followed by conflict over who received punishments and who did not. The number of 'forest offences' increased rapidly after 1946-48, coming to a peak in 1952-53 when local people were forced to come to terms with the rules, regulations and corruption (from contractors as well as local forest staff) associated with the new system of forest management (see figures 10.8 and 10.9). Interestingly, data on forest guard dismissals, warnings, financial penalties and suspensions show a good deal of similarity with these patterns, perhaps indicating that higher-level forest officials were just as unsatisfied with the situation as were the villagers themselves (see figure 10.10).

Forest Decline in the Fieldwork Area

Although increases in village (human and animal) populations certainly put pressure on local forest resources, this 'official view' on the cause of forest decline (Government of India, 1976) is not the only explanation for deforestation in the research area. As has been the case elsewhere in South Asia (Blaikie, 1985; Arnold and Campbell, 1986; Guha, 1989; Pathak, 1994), a decline in local people's access (and rights) to forest produce in Jharkhand was accompanied both by a decline in their willingness (and sense of duty) to protect forests and an increase in everyday forms of resistance to commercial forest management.

Like the villagers of Uttarakhand described by Guha (1989), many villagers in the Bero area felt that when 'their' forests were nationalised and exploited by ex-*zamindars*, forest contractors and forest guards, their 'relationship with the jungle was broken' (discussion with Simon Oraon, 1993). Also, like the villagers of Mathwad Forest Range, Madhya Pradesh studied by Pathak (1994), many people saw the Forest Department's dealings with them as a form of 'harassment' which divided local communities internally. They were particularly resentful that people they did not respect (notably corrupt forest officers) could 'keep them in line' by threatening them with prosecution for cutting trees that their ancestors had once owned (discussion with Tila Oraon, 1993). Typically, however, it was local people and not forest officials or contractors who initially got most of the blame for forest decline (Ghate, 1992).

Figure 10.8   Illegal forest felling 1936-1956

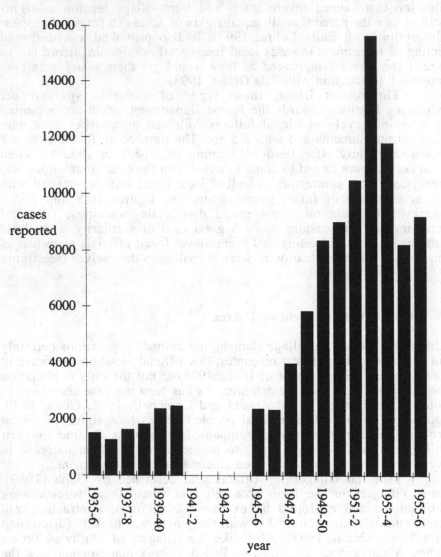

Figure 10.9   Forest offences

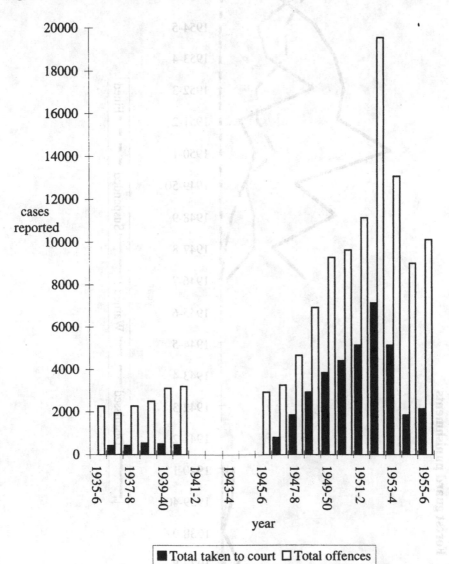

Figure 10.10   Forest guard punishments

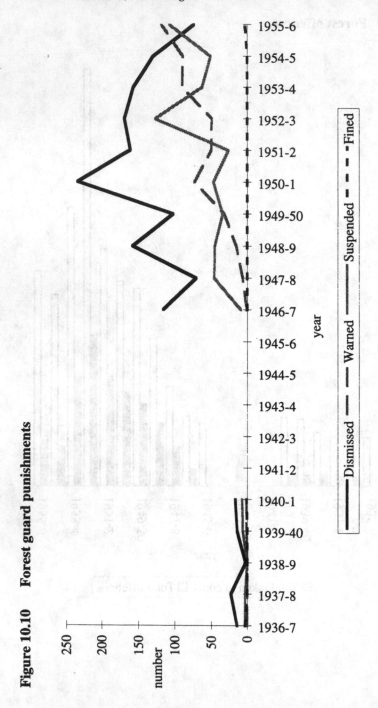

Although few villagers in the research area became so 'alienated' from local forests that they wanted to destroy them completely (Guha, 1989) or convert them to agricultural land (Pathak, 1994), many did take the attitude that as 'the people working for the Forest Department were exploiting forests for their own profit, why shouldn't we' (discussion with Phulchan Tirkey, 1993). Even Forest Department staff now acknowledge that the corruption of the contractors and forest officials 'left an impact on villagers' minds' (interview with Anand Jha, Forester for Bero Block, 1993). As a result, many people lost respect for the Forest Department and started taking the view that they would be fools to protect rather than exploit the state's resources. As Chapter Eleven will show, however, not everybody undertook this particular form of resistance. Villagers from Jamtoli, for example, started, in the 1960s, to oppose scientific forestry in a more positive way: by protecting 206 acres (83.36 hectares) of Jamtoli Reserved Forest from further commercial exploitation.

In many other villages, however, the deliberate over-exploitation of forest produce by local people had severe implications for long-term forest regeneration because villagers who had formerly taken care to coppice forest trees abandoned this practice to hasten their 'getaways' with stolen timber. In addition, many villagers were aware that despite the re-designation of most ex-*zamindari* forests as Protected forests, there were plans to re-classify some of the most productive of these as Reserved Forests at a later date. Consequently, right-holders of large forests like Ambatoli were particularly wary about encouraging forest regeneration through careful coppicing in case this encouraged the state to reserve the forest.

Interestingly, this shift in attitudes towards forests coincided with the breakdown of most traditional *Parha* administration systems following the introduction, in 1949, of the Gram Panchayat system of local government. According to many villagers in Ambatoli and Jamtoli, the formal recognition of elected Panchayat leaders by the state, plus the higher levels of political power and funding available to them (compared to what could be mobilised at the local level by the *Parha* system), were major factors in the decline of many *Parhas* to little more than social institutions (interviews with Simon Oraon and Indru Swansi, 1994).

As the *Parha* system had formerly dealt with conflicts over access to forest resources, its decline meant the loss of an important safety valve in a situation of growing tension. Illiterate villagers who had traditionally relied on *Parha* leaders to act as intermediaries between them and the Forest Department were increasingly forced to deal directly with forest officials. Misunderstandings and antagonism over forest issues spiralled as villagers became suspicious that Forest Department staff were trying to exploit their lack of familiarity with legal documents in order to claim unreasonable fines for forest crimes. Interestingly, many villagers still feel that if the *Parha* system had been maintained, 'local people would not have felt so resentful about the

Forest Department's power over them and the forest would not have suffered so much' (discussion with Simon Oraon, 1993).

By the late 1970s, antagonism between local people and the Bihar Forest Department came to a head as attempts to assert traditional forest rights were greeted with increasingly repressive responses from the state. In 1978, police opened fire on *adivasis* from Simdega, Ranchi District, who had started a movement to oppose the nationalisation of NTFPs (Fernandes, 1983). Similarly, on the eighth of September 1980 in the town of Gua, Singhbhum District, a peaceful *adivasi* protest against 'police terror, government forestry policies and state employment policies' (Corbridge, 1995c, p. 5) ended in bloodshed when a scuffle broke out and the police opened fire, killing eight *adivasis*. A further nine wounded *adivasis* were shot dead by injured policemen at the hospital when they arrived for treatment (Corbridge, 1995c; Corbridge and Jewitt, 1997).

By the early 1980s, most forests in Ranchi District were in very poor condition and some had been lost altogether. In Jamtoli, for example, the forest area declined from 609 acres (246.45 hectares) to 206 acres (83.36 hectares) between 1961 and 1991. The main reasons for this were that during the early 1960s, commercial timber exploitation coupled with right-holders' requirements for wood and illegal felling by forest contractors resulted in the complete failure of large parts of Jamtoli forest to regenerate: a situation that helped to stimulate the establishment of a village-based forest protection system. Ambatoli forest, meanwhile, became so degraded that no new management plan was drawn up to replace the 1964 plan which expired in 1984 and coupe cutting was totally abandoned.

In many respects, Singh's scenario in which local people who have their 'natural resources usurped' (Singh, 1986, p. 5) are forced into stealing, poverty or migration is a very accurate representation of the situation in Jharkhand during this period. From the mid-1970s, struggles between local people and forest contractors became increasingly common (Fernandes, 1983; Kelkar and Nathan, 1991; Corbridge, 1995c). In the late 1970s, poor harvests set off a trend of seasonal migration to brickfields in Uttar Pradesh and West Bengal that increased as forests declined still further. Of those that stayed behind, the worst off were villagers whose local forests had become badly degraded as they had to travel long distances and plead (or fight) for access to less degraded forests.

In many forests, a kind of open access situation prevailed as both local and official rules of forest access and exploitation broke down. Increasingly, local people stopped thinking about whether the young trees that they cut would re-grow or whether cattle would browse or trample regenerating coppice seedlings (discussion with Simon Oraon, 1993). Many villagers also stopped caring about whether their forests were being exploited by non right-holding outsiders, as everyone knew that they had nowhere else to go for their forest produce and most could not afford to pay for it.

By the late 1970s, the widespread degradation of state-owned forests brought an end to most coupe systems and forced people to cut fuel wood illegally to satisfy their subsistence requirements. As the situation worsened, it became clear that efforts would have to be made to address the problem of forest decline and alleviate tensions between local people and the Forest Department. Social forestry schemes were promoted as an ideal solution to these problems and throughout Bihar, farm forestry programmes were accompanied by attempts to replant 'blank areas' in Protected Forests situated close to areas of habitation (interview with Anand Jha, 1993). In Ambatoli, the Forest Department paid villagers 15 rupees per day to plant trees on five acres of 'blank' forest land near Pokaltikra hamlet.[6]

Although social forestry schemes did go some way towards improving villager-Forest Department relations, many of the gains that they made were eroded by the rapid rate of population increase in the region (see figures 10.11 and 10.12).[7] After the early 1980s, villagers in Jharkhand, as elsewhere in the developing world (Eckholm, 1984; Blaikie, 1985; Shepherd, 1992a; Holmgren et al, 1994; Tiffen et al, 1994), started to become very aware of the pressure that increasing human and animals populations were putting on declining forest resources. In Ambatoli and Jamtoli, for example, the subject of population increase was brought up by villagers in twelve out of fourteen in-depth interviews about forest use and management.

According to village level census data, Ambatoli's and Jamtoli's populations increased from 1480 to 1751 and 880 to 1339 respectively between 1971 and 1991 (see figure 10.13) and the forest area to right-holder ratios declined in response. Ironically, it was Ambatoli with its Protected Forest (and double burden of right-holders) rather than Jamtoli with its Reserved Forest that maintained the greater area of forest per right-holder ratio after the mid-1960s. Between 1971 and 1991, Ambatoli's forest area to right-holder ratio declined from 0.54 to 0.46 acres (0.22 to 0.19 hectares) per person while Jamtoli's plummeted from 0.46 to 0.15 acres (0.18 to 0.06 hectares) per person (see figure 10.14). The main reason for this rapid fall in Jamtoli's ratio was the decline in forest area caused by commercial forest exploitation coupled with illegal felling in the 400 acres (161.8 hectares) of forest that were not protected by the forest protection committee.

By the mid-1980s, the high levels of demand for forest products caused by widespread deforestation encouraged some households to start selling fuel wood. Initially, moral restrictions limited this type of forest exploitation to 'welfare' poor villagers (Reardon and Vosti, 1995) who could not afford not to exploit forests if they were to succeed in meeting their household subsistence requirements. Before long, however, the fuel wood market became so buoyant that other villagers started to see it as an easy way of making money; many of them ignoring erstwhile unspoken restrictions about the cutting of fruit and non-coppicing trees in favour of making a quick profit (discussion with Sila Lohrain, 1993). The Forest Department responded to this trend by

Figure 10.11   Population in Ranchi District 1911-1981

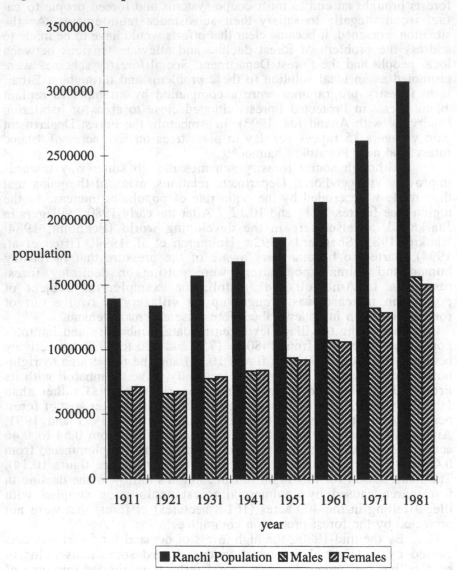

■ Ranchi Population   ☒ Males   ☑ Females

Figure 10.12 Population in Bero Block 1921-1991

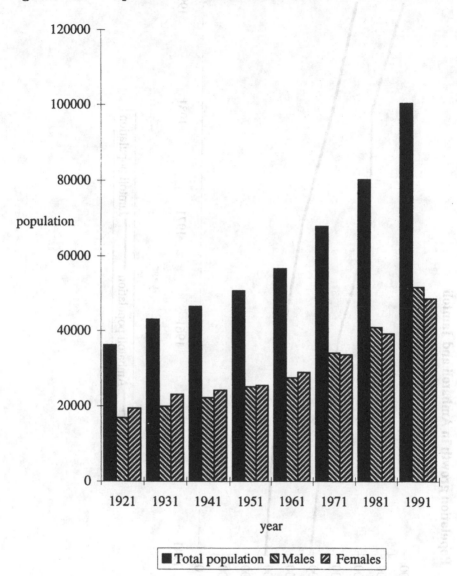

Figure 10.13 **Population growth in Ambatoli and Jamtoli**

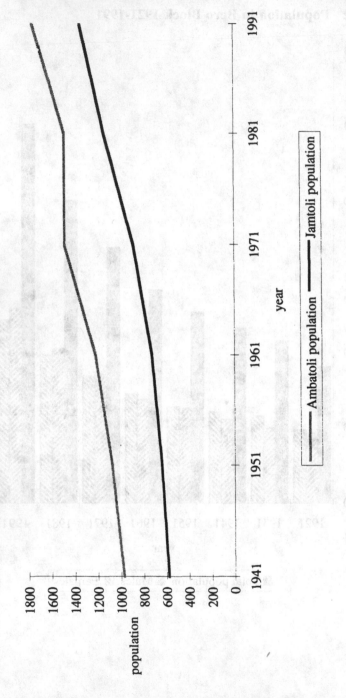

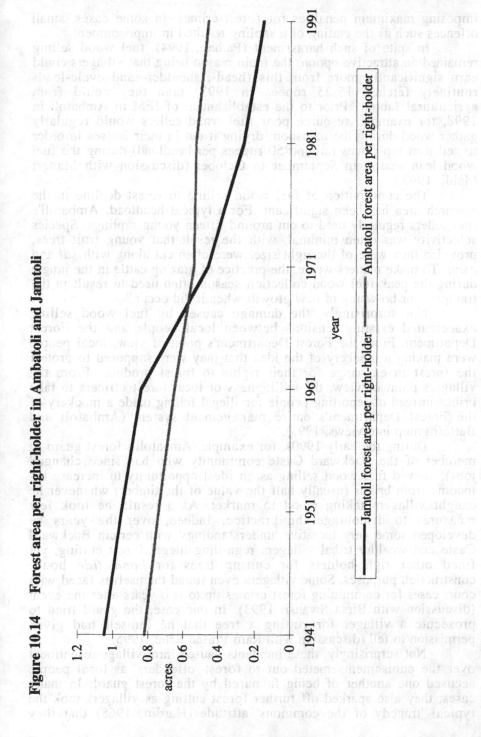

Figure 10.14 Forest area per right-holder in Ambatoli and Jamtoli

imposing maximum penalties for forest crimes. In some cases, small offences such as the cutting of a sapling resulted in imprisonment.

In spite of such harassment (Pathak, 1994), fuel wood selling remained an attractive option; the main reason being that villagers could earn significantly more from this (head-, shoulder- and cycle-loads routinely fetched 15-25 rupees in 1993) than they could from agricultural labour.[8] Prior to the establishment of JFM in Ambatoli in 1998, for example, resource poor fuel wood sellers would regularly gather wood during the monsoon; drying it out in their houses in order to sell it at top prices (around 50 rupees per headload) during the fuel wood lean season in September to October (discussion with Mangri Mahli, 1993).

The contribution of fuel wood selling to forest decline in the research area has been significant. For a typical headload, Ambatoli's fuel sellers regularly used to cut around fifteen young saplings. Species selectivity was often minimal, with the result that young fruit trees, provided they were of the right size, were often cut along with *sal* and *asan*. To make matters worse, the practice of grazing cattle in the jungle during the peak fuel wood collection season often used to result in the trampling or browsing of new growth when it did occur.[9]

Not surprisingly, the damage caused by fuel wood selling exacerbated existing tensions between local people and the Forest Department. From the Forest Department's point of view, local people were making a mockery of the idea that they were supposed to protect the forest in exchange for their rights to forest produce. From the villagers point of view, the willingness of local forest officers to take bribes instead of reporting people for illegal felling made a mockery of the Forest Department's entire management system (Ambatoli and Bartoli group interviews, 1993).

During the early 1990s, for example, Ambatoli's forest guard (a member of the Backward Caste community who has since changed jobs), viewed fuel wood selling as an ideal opportunity to increase his income from bribes (usually half the value of the timber) whenever he caught villagers taking wood to market. As a result, he took few measures to discourage the practice. Indeed, over the years he developed some very lucrative 'understandings' with certain Backward Caste and wealthy tribal villagers regarding illegal forest cutting, yet fined other right-holders for cutting trees for *bona fide* house construction purposes. Some villagers even found themselves faced with court cases for committing forest crimes up to two years after the event (discussion with Birsa Swansi, 1993). In one case, the guard tried to prosecute a villager for cutting a tree that he himself had given permission to fell (discussion with Ram Prasad Sahu, 1995).

Not surprisingly, these incidents caused intra-village resentment over the punishments meted out to forest 'offenders' as local people accused one another of being favoured by the forest guard. In many cases, they also sparked off further forest cutting as villagers took the typical 'tragedy of the commons' attitude (Hardin, 1968) that they would be silly not to cut extra wood if other villagers were doing it. In spite of these problems, however, the centrality of Ambatoli's forest to villagers' culture and livelihoods ensured that their interest in protecting it was never completely overshadowed.

The Socio-cultural Importance of Forests

The socio-cultural importance of forests to Indian people is a topic that should be much more central to both the academic literature on Indian forests and more policy-oriented guides to the development of participatory forms of resource management like JFM. This is especially true given the ways in which local silvicultural knowledges and forest management practices have been de-legitimised by the Indian state (to justify the imposition of scientific forestry - Shepherd, 1992b) and re-defined, often negatively, by local people who are forced to come to terms with commercial forestry (Guha, 1989).

    Confirming Bebbington's (1990a; 1992) point that the technical aspects of local knowledge systems are far more widely studied than their socio-cultural aspects, most references to forest-based sites of socio-cultural importance are subsumed within wider critiques of state-imposed development and forest management (see, for example: Fernandes, 1983; Corbridge, 1986; 1991a; 1995c; Shiva, 1988; Guha, 1989; 1993; Imam, 1990; Kumar, 1991; Ghate, 1992; Ramakrishnan and Patnaik, 1992; Pathak, 1994). The main exceptions to this are the studies carried out by anthropologists such as Elwin (1939; 1964), von Furer-Haimendorf (1989) and Roy (1912; 1915; 1928) who spent many years investigating the wider socio-cultural environments of India's tribal populations.

    As Roy's work shows, Jharkhand provides much scope for research into the socio-cultural aspects of local silvicultural knowledges as forests play a central role in Jharkhandi village culture and religion. Yet to my knowledge, no recent village-based studies have been done on this topic. The aim of the following sub-sections, therefore, is to fill in some of this gap by emphasising the value of knowledge about the magico-religious importance of forests for understanding their use, management and (increasingly nowadays), protection. The Jharkhand movement's emphasis on the centrality of forests to the region's cultural integrity, future environmental stability and economic development is also outlined.

Sarna *Religion and Forests*

As indicated in Chapter Eight, forests play a central role in some of Jharkhand's major *Sarna* festivals. The most important religious event of the year is *Sarhul* which must be celebrated before villagers can collect fruits or flowers from the forest. The reason for this is thought to be that *Sarhul* originated when the Oraons were hunters and gatherers and acted as the main opportunity for villagers to pray for future prosperity (Roy, 1928). Central to this festival is the *sal* tree which is usually also found in *Sarna* sacred groves and which is essential for villagers' subsistence as it provides edible flowers, strong timber, leaves that can be made into leaf plates and bowls and twigs that are used as tooth brushes.

The *Karam* festival is also closely linked to the forest both as a place of worship and for the provision of the three *karam* branches used in both the *Burhia Karam* and main *Karam* festivals. Most village forests also have a special *karam* tree where *Sarna Burhia* is worshipped during the *Karam* festival. In addition, forests (and the *karam* tree, in particular), are associated with good rainfall and many local people are convinced that forest loss has caused a decline in monsoon rainfall because forests 'provide food for the clouds and soak up the rain when it falls' (discussion with Simon Oraon, 1993). Indeed, some villagers commented that when the forest was well stocked with large trees, it rained so much that they always used to carry umbrellas during the monsoon, whereas nowadays there is not enough rain to warrant carrying an umbrella (interviews with Tila Oraon and Simon Oraon, 1993).

Other trees that are particularly important in *Sarna* religious festivals are the *bel* (*Aegle marmelos*), *bhelwa*, *kend* and *sinduar*, the leaves or twigs of which are used in *Danda Katta* ceremonies. *Sal* leaf plates and bowls are used at most major festivals and other important occasions such as births, deaths and marriages. Many villagers also use *sal* branches and leaves to construct a canopy under which people can dance during weddings. Another part of the *sal* tree that is used in magico-religious ceremonies is the sap that exudes from the root of the tree. When dried and burned, this smells rather like incense (*dhuwan*) and is commonly used by *bhagats* (ghost-finders) and occasionally village *Pahans* to propitiate malevolent *bhuts*.

The Importance of Forests for non-*Sarna* Worshippers

In keeping with the importance of tree worship in Hindu theology (Chandrakanth et al, 1990), Hindu villagers in Ambatoli and Jamtoli use individual trees, notably the sacred *pipal* (*Ficus religiosa*) as important sites for *puja* (worship). In contrast to some other areas of India where Hindu sacred groves and temple forests are revered and protected by local people (Guha, 1989; Khiewtam and Ramakrishnan, 1989; Chandrakanth et al, 1990; Gadgil and Subash Chandran, 1992; Pereira, 1992; Ramakrishnan and Patnaik, 1992; Vatsyayan, 1992; Pathak, 1994; Jeffery, 1998a), however, most Hindu households in Jharkhand do not feel that forests have as much religious importance for them as for *Sarna* worshippers. Nevertheless, this does not stop Hindu villagers (along with Christians and Tana Bhagats) from participating, to some extent, in the main forest-based *Sarna* festivals.

In addition, Ambatoli, which is one of the few Oraon villages where a village deity called Mahadeo is worshipped, has a shared Hindu and *Sarna* religious festival. Every year in July, villagers go to a place deep within the forest called Ghargari where they offer sacrifices and worship the Mahadeo deity (who is represented by a round stone situated between two tridents) in a Hindu-style temple. According to Roy, the Mahadeo stone's phallic shape represents the 'procreative force of nature' (Roy, 1928, p. 54) and is often worshipped by childless Hindu and *Sarna* villagers who pray to the Mahadeo deity for sons.

Hunting

Another important tradition that is carried out in the jungle by both *adivasi* and non-*adivasi* men is hunting. According to Roy, most Oraon-dominated villages have three major hunts per year: the *Phagu-Sendra* which is held in March, the *Bisu Sikar* which is held in April or May and the *Jeth Sikar* which is held in June (Roy, 1915). Nowadays, by contrast, most villagers in the research area only go on two annual hunts: the *Phagu-Sendra* which involves an entire village and the *Bisu Sikar* which involves a whole *Parha*. As forest decline has also meant a decline in habitats for game animals such as deer, wild pigs, wild dogs, bears and even tigers, however, these annual hunts have become little more than social occasions in most areas.

Malevolent Forest Bhuts *and Spirits*

In addition to the religious importance of forests, many villagers also believe them to be the abode of evil spirits (*bhuts*) which must be regularly propitiated to prevent them from bringing harm to the village. One of the most feared parts of a village is the place where people who die from unnatural deaths are buried. Where possible, these burial grounds are situated within the forest at the boundary with a neighbouring village (Roy, 1916; 1922; 1915; 1928).

Women who die in childbirth or during pregnancy are particularly feared as they are thought to become malevolent spirits called *churils* which chase any man who passes near to the burial place because their 'carnal appetite remained unsatisfied in life' and they still long 'for a mate of the opposite sex' (Roy, 1928, p. 187). In keeping with Roy's observation that villagers often mutilate the bodies of women who die during pregnancy or childbirth to prevent them from becoming *churils*, Oraons in Ambatoli and Jamtoli bury such women deep in the forest with their eyes, nose and mouth stitched up with thorns and sometimes with their arms and legs broken (interviews with Jitan Manjhi and Ram Prasad Lohra, 1993).

There are also a number of places within village forests that are associated with particular *bhuts* that must be propitiated at regular intervals. In Jamtoli, *adivasi* villagers sacrifice a cow every twelve years to a *bhut* known as Goindi Burhia which lives in the forest on Jamtoli hill. Another *bhut* that lives in the jungle close to Kaxitoli hamlet is propitiated with a buffalo once every twelve years (discussion with Budhwa Oraon, 1993). In Ambatoli, meanwhile, there is a *bhut* which lives in a place known as *Bar Pahari* (*bar* tree hill) and must be propitiated with a goat once every three years to prevent it from bringing illness to the village (discussion with Gandra Barla, 1993). In addition, (and providing an interesting contrast to Baines's (1989) work on the disturbance of sacred sites by logging companies in Melanesia), there is a place in Ambatoli forest known as Porhosokra from which even forest contractors keep away unless they have first offered sacrifices to the resident *bhut* known as Darha (discussion with Sukra Bhagat, 1993). If these sacrifices are not made, local people believe that the *bhuts* could cause the monsoon rains to fail or bring widespread crop failures, famine and death to the village.

The Importance of Forests in Jharkhandi Politics and Culture

The widespread belief amongst Jharkhandis that much of the region's wealth (generated by natural resources) has been taken away by Bihari *dikus* was extremely important in maintaining support for the Jharkhand movement. Many villagers have direct experience of forest exploitation by contractors and were broadly sympathetic to the movement's attempts to preserve and strengthen traditional cultural events and systems of community organisation.

Nagpuri Music and the Jharkhand Movement

An important (although not widely studied) source from which both *adivasis* and non-*adivasis* in the research area draw their sense of Jharkhandi cultural identity is from Nagpuri songs. Some of these are traditional or folk songs that are non-political but are known to villagers throughout the research area.[10] Others, written mainly since the 1970s by professional musicians, contain a strong political message, often highlighting forest issues

Some of the most influential of these have been 'Jharkhand awareness songs' written by Ranchi artists who worked closely with Jharkhand leaders such as Dr. Ram Dayal Munda and Dr. B.P. Kesari at the Tribal and Regional Languages Department of Ranchi College (see Appendix III).[11] One of the best known examples of this is a song called '*jagu jawan*' which, although written by a non-*adivasi*,[12] raises exactly the same issues as those mentioned by villagers in Ambatoli and Jamtoli: namely the exploitation of Jharkhand's resources by outsiders and the lack of education and employment opportunities for Jharkhandi people.

In many senses, this type of 'awareness song' has been a much more efficient vehicle for informing rural people (of all communities) about the Jharkhand movement than political rallies: a fact that is not lost on cultural leaders like Ram Dayal Munda who retain strong links with their own villages and recognise how central song and dance is to both *adivasi* and non-*adivasi* village culture (discussion with Mukund Nayak, 1993). Throughout my fieldwork in the Bero area, for example, I interviewed far more people who knew *'jagu jawan'* than had participated in a political rally.

Songs like *'jagu jawan'*, *'murga chap'* and *'chala re Jharkhand raikhe mangay'* were also influential in both pinpointing the cause of forest decline in Chota Nagpur and helping to link this issue to the Jharkhand movement without re-enforcing its traditional *adivasi* bias. Their popularity during the mid-1970s also helped to bring forests onto the agenda of a movement that had been dominated previously by the overriding interest in land issues of its elite *adivasi* leaders (discussion with Mukund Nayak, 1993). Indeed, it was not until the late 1970s when the JMM supported villagers in Singhbhum District in their struggles against timber contractors, that local people (of all communities) really started to see the Jharkhand movement as an ally against this particular form of outside exploitation (Corbridge, 1991a; Devalle, 1992). It did not take long, however, before forests were added to most people's lists of 'things that would be managed more fairly and effectively in a separate Jharkhand State' (discussion with Nicholas Lakra, 1993).

In response, the Jharkhand movement tried to consolidate local people's anger about the extent to which outsiders had gained from exploiting their 'land of forests' (discussion with Ram Prasad Lohra, 1993). Forests were also emphasised by the socio-cultural wing of the Jharkhand movement as part of a wider revival of community music and dance (Kelkar and Nathan, 1991). In particular, Dr. Ram Dayal Munda and Dr. B.P. Kesari in conjunction with several Nagpuri music and dance artists worked hard to stress the importance of traditional *adivasi* festivals like *Sarhul* to Jharkhand's cultural integrity.

These themes have, in turn, been picked up and re-emphasised at the village-level by local political and *Parha* leaders such as Simon Oraon and the local Member of the Legislative Assembly, Vishvanath Bhagat. During the annual *Parha Jatra* at Bero which follows the traditional *Parha*-based *Bisu Sikar* on the 3rd of June, Vishvanath and Simon have often made speeches emphasising the importance of local *Parhas* in ensuring the survival of traditional cultural events such as the *Bisu Sikar* and *Sarhul*.[13] Both men are very interested in wider development issues and have emphasised the need to maintain 'the social development of the local area through the *Parha* system' (interviews with Vishvanath Bhagat and Simon Oraon, 1993). And as neither man likes to miss an opportunity to talk about the importance of forests, a second element of their speeches has usually focused on the centrality of forests to Jharkhandi culture and the need to protect them to ensure future supplies of fuel wood and house construction timber (interviews with Vishvanath Bhagat, Simon Oraon and Pothwa Bhagat, 1993). Indeed, in 1993, so many villagers were seeking Vishvanath's and Simon's help with forest-related problems that they were finding it difficult to visit everybody who requested their assistance (interviews with Vishvanath Bhagat and Simon Oraon, 1993).

The Practical Importance of Forests

As in so many other areas of India (Elwin, 1936; 1939; 1964; 1986; von Furer Haimendorf, 1989; Guha, 1989; Ghate, 1992; Pereira, 1992; Kakat, 1993; Pathak, 1994; Sivaramakrishnan, 1999; Karlsson, 2000; Rangan, 2000a) forests in Jharkhand are absolutely essential for providing vital subsistence products such as fuel wood, timber, food and medicinal herbs. Fernandes (1983), for example, suggests that the Oraon, Munda and Char tribes depend on forests for more than 40 per cent of their 'survival needs'. Even degraded forests and wastelands can provide important subsistence resources (such as twigs and leaves for fuel wood, cattle fodder and potential agricultural land) as well as wider 'ecological services' (Kumar, 2001).

In Jharkhand where local cultural practices as well as subsistence-strategies are strongly related to forests, most local people possess fairly well developed silvicultural knowledge systems. In addition, there are a few villagers who have special skills in forest management, knowledges about forest plants, or even the ability to fuse the magico-religious and practical values of forests through indigenous medicine (IDS Workshop, 1989a; Pereira, 1992; Fairhead, n.d.). The extent of these knowledges regarding forest use and management in the research area and their importance for helping to prevent further forest decline will be examined below.

The Use of Forests for Subsistence-related Purposes

As can be seen from table 10.1, many forest tree species have a variety of practical and socio-cultural uses. For the majority of people in the research area, however, the most important forest products are fuel wood, construction timber and wood for plough-making. Looking at fuel wood collection first, all households in Ambatoli and Jamtoli use the village forest as their primary source of supply. The main difference is that in Jamtoli, villagers only collect wood during an annual coupe held in January or February by the village's forest protection committee: the details of which will be provided in Chapter Eleven. The situation in Ambatoli between 1993 and 1997 is described below as it is characteristic of most non-JFM villages in the Bero area during this period. The changes that occurred after 1998, when Ambatoli's JFM committee banned fuel wood selling, the cutting of green trees and unrestricted cattle grazing are described in Chapter Eleven.

Fuel Wood Collection Like many other rural people in India and throughout the developing world (Joshi, 1983; Eckholm, 1984; Agarwal, 1986b; Dankelman and Davidson, 1988; Henshall Momsen,

Table 10.1   The different uses of forest trees

Tree*	Fuel	Plough/ house timber	Furni-ture	Oil	Medicine	Religion	Fodder
Asan	*	*					*
Acacia	*	*	*				
Amla	*	*			*		*
Bahera	*	*		*	*		
Bar	*		*			*	
Ber	*				*		*
Bel	*				*	*	
Bhelwa	*		*	*			
Dahu	*	*	*		*		
Eucalyptus	*	*					
Footkal	*						
Gamhar	*	*	*				
Jackfruit	*		*		*		
Jamun	*		*		*		
Jirhul	*				*		
Karam	*					*	
Karanj	*		*	*			
Karonda	*				*		*
Kend	*	*	*				
Koinar	*						*
Kusum	*		*	*			
Mahua	*		*	*			
Mango	*		*			*	
Neem	*			*	*		
Parsa	*						
Pipal	*					*	*
Piyar	*						*
Putus	*						
Sal	*	*	*				
Simbal	*		*				
Imli	*				*		
Teak	*						
Tetair	*						

* For the botanical names of these trees, see Appendix I.

Source: Jewitt, 1996.

1991; Pathak, 1994; World Resources Institute, 1994; Singh, 1999), villagers in Jharkhand spend a considerable amount of their time gathering fuel wood. In 1993, most Ambatoli households collected fuel wood at least twice a week during the agricultural 'slack season' between November and May when their work loads were lower and the wood in the forest was fairly dry and easy to carry. At the start of the collecting season, most villagers collected dry twigs and branches that had fallen on the ground since the previous collection season. Head-loaders usually stripped off any lateral branches, twigs and leaves to make bundling up and carrying the wood easier. To prevent wastage, these discarded branches and dry leaves were often used as filling for the headload bundle, or were picked up later by other villagers carrying baskets instead of headloads.

Later on in the season when most of the dry branches had already been collected, villagers used to cut green wood in the form of small trees or branches from larger trees. Because large branches are hard both to reach and to cut, most people preferred to fell small trees of less than 10 cm diameter at breast height. Preference rankings showed *sal* and *asan* to be the most favoured species as their wood makes a hot fire without too much smoke (see table 10.2). These trees frequently also produced good coppice growth, despite the fact that many villagers used to cut them in ergonomically convenient ways rather than in a manner designed to stimulate coppice shoots.

Having said this, some people, especially the older villagers, were careful to cut only crooked or poor quality *sal* and *asan* trees to promote a future supply of straight, good quality timber for house construction and plough-making purposes (discussion with Dhosia Orain, 1993). Others avoided cutting green wood altogether if they could meet their fuel wood requirements using dry leaves and woody weed species such as *putus* and *tetair* (discussion with Gandri Bhaktain, 1993). A majority of villagers were also quite careful to avoid cutting fruit trees because there were 'plenty of other trees' that could be used for fuel wood and they wanted to 'preserve as many fruit trees as possible' (interviews with Gandri Bhaktain and Silamuni Orain, 1993).

Villagers usually used an axe (*tangiya*) or a sickle shaped tool known as a *dowli* to cut the trees and strip them of lateral branches. To tie up bundles and baskets of fuel wood, many people used the root of a jungle creeper known as *dhudiya*, although some also used rope (see plate 10.1). Most of the bundles taken from Ambatoli forest in 1993 weighed over 20 kilograms and even children of ten to twelve years often carried bundles weighing over 15 kilograms.

Large Timber Throughout the research area, most people consider *sal*, which is strong and rot-resistant, to be the best wood for both house construction and agricultural implement-making, although *asan* is also well liked. As can be seen from table 10.2, however, preference rankings also identify certain fruit trees as providing good quality large timber. *Jamun*, mango, *ber*, *dumar* (*Ficus glomerata*), *kusum*, *kend*, and *piyar*

Table 10.2   Tree preference rankings[+]

Tree[*]	Fuel	Timber	Fruit nuts leaves	Oil	Medicine	Fodder
Asan	2	2			11	3
Acacia	9	13				
Amla				2	1	
Bahera		19	14	5	2	5
Bar		17	15			16
Ber		10	12			18
Bel			10		6	10
Bhelwa		20	4		9	
Bokla						
Dahu		18	13			13
Dhaunta		21				11
Dumar		11	11			
Eucalyptus	10	12				
Footkal			9			2
Gamhar		5				
Harra					4	20
Jackfruit		4	3			
Jamun		7	6			9
Karam		9				14
Karanj	7			1	5	
Kend		8	5		10	15
Koinar						17
Kusum		14				20
Mahua		16	8	4	7	19
Mango	8	6	1			8
Neem		22			3	6
Parsa	6					
Piyar		15	2			1
Putri						4
Putus	3					
Sal	1	1	16	3	12	12
Tamarind	5		7		8	7
Teak		3				
Tetair	4					

[+] The highest ranking is 1.
[*] For the botanical names of these trees, see Appendix I.

Source: Jewitt, 1996.

Plate 10.1   **The use of *dhudiya* creeper to tie fuel wood bundles**

(*Buchanania latifolia*) are particularly well liked as house construction timber and jackfruit, because of its wide girth, is favoured for door-making. It is usually only wind thrown or over-mature fruit trees that are cut for this purpose, however.

Other species that are commonly used as construction timber include *bahera* (*Terminalia bellirica*), *bar*, *bhelwa*, *dahu*, (*Artocarpus lakoocha*), *dhaunta* (*Anogeissus latifolia*), *gamhar*, *karam*, *mahua*, *neem* and *simbal* (*Bombax ceiba*). The best species for furniture-making according to local villagers are *sal*, *bahera*, *bar*, *ber*, *dahu*, *dhaunta*, *gamhar*, *jamun*, *kend*, *kusum*, *mahua*, mango, *neem*, *piyar* and *simbal*. In addition, villagers' preference rankings rate eucalyptus and acacia wood as highly as many local species for furniture-making and house construction (particularly for use as roof rafters): a situation which supports Eckholm's (1984) view that existing critiques of the use of these species in social forestry projects may be somewhat of an over-reaction.

In Jamtoli, members of the forest protection committee obtain the large timber that they need for house construction, agricultural implements and furniture from the annual coupe. In Ambatoli, by contrast, it was becoming increasingly difficult by 1993 for right-holders to satisfy their large timber requirements from Ambatoli forest. As a result, many villagers wanting to undertake house construction work often had to buy at least some of the wood that they needed. But with a roof costing up to 2000 rupees in timber alone, the temptation to fell trees illegally was very great.

Between January and June 1993, for example, at least three groups of villagers who needed to extend their houses got together, felled enough trees for their requirements and brought the timber home in a bullock cart. One group from Bartoli and another from Nehalu worked solidly for two days cutting the timber up into usable pieces to reduce the risk that the forest guard would catch them. The members of the third group (all of whom were Backward Caste villagers like the forest guard) were less concerned as they were on good terms with the guard and had bribed him to 'look the other way' while they felled and cut up their tree.

Apparently, this was also a common practice amongst the wealthier *adivasi* villagers, especially those with jobs in government service, as they preferred (and could afford) to bribe the guard rather than risk having a court case taken out against them by the Forest Department. The least wealthy villagers, particularly the Lohras, Mahlis, Manjhis, Swansis and some of the poorest Oraons usually tried to minimise their timber requirements by making alterations to their houses only when absolutely necessary. When they did need timber they usually felled and quickly sawed up a tree when they thought the forest guard was elsewhere. The consequences for being caught could be very severe, however. Banwa Swansi from Panrtoli was put in gaol for seven months for felling a mature *sal* tree and Somra Oraon from Bartoli was fined 300 rupees for cutting a tree that was little bigger than a sapling.

Forest Food Although I did not attempt to assess the overall calorific contribution of forest food to household diets in the research area, these items are, as Kelkar and Nathan (1991) suggest, of great importance as they are mostly available between March and August when *kharif* season food stocks are starting to dwindle (see table 10.3). Forest food is also very popular as it provides a welcome change from the monotony of a rice-based diet. More importantly, forest fruit provides an important source of vitamins for the many villagers who suffer from rickets. It is also an important additional source of calories for the poorest and near-landless villagers (particularly the *sadans*) whose staple food stores rarely last for more than nine months.

In Ambatoli, 75 per cent of the sample households regularly collected forest fruits in 1993 and their access to these items (along with other NTFPs) has not been restricted since the JFM committee was formed. The most popular fruits are *kend*, *piyar*, mango, *jamun*, *ber*, *dumar* and *bel*, all of which ripen between late February and August. Many villagers also eat the dried, newly formed leaves of the *footkal* (*Ficus retusa*) tree which are collected in late-February. In Jamtoli, by contrast, only about half of the villagers actually obtain fruit or seeds from the village forest. This is mainly because Jamtoli has a high average number of homestead trees per household (87 per sample household compared to 12.5 in Ambatoli) and villagers prefer to collect fruit and seeds from their own (nearby) trees rather than from the forest.

In both villages, a number of the most food deficit households bake a hard unleavened bread out of dried and crushed *sal* flowers (which come out in March) and many people eat roasted *sal* seeds. Mushrooms are another forest food that are popular throughout the research area and villagers from virtually every household go to the forest during the mushroom season (June to August) to collect enough for a mushroom curry. In addition, there are a number of forest plants that villagers sometimes collect between January and July for use as *sag* (green vegetables) in curry dishes. Many of these have fallen from use as they are not as palatable as some of the newer *rabi* season vegetables, but plants that are still commonly collected include '*kattay sag*', '*chimti sag*' and '*matak sag*'.[14]

It is not only humans that depend heavily on forests as a source of food, however. In Ambatoli, prior to the formation of the JFM committee, most villagers used to pay a child or youth to take their cattle and goats to graze and browse in the forest between November and May, when they were not required for agricultural duties. Unlike Guha's (1989) account of cattle grazing in Uttarakhand, however, setting fire to the forest to get a fresh crop of grass was uncommon. In Jamtoli, cattle grazing has traditionally been restricted to wastelands and fallow fields. When grazing is poor, however, households in both villages feed their cattle on rice straw and, like the villagers in northern India mentioned by Shiva (1988), supplement their diet with leaves and branches from the forest.

Table 10.3   Forest food

Tree*	Plant parts	Time available
Amla	Fruit, Leaves	November-December
Bar	Fruit	April-May
Ber	Fruit	July-August
Bel	Fruit	June-July
Dahu	Fruit, Flowers	March-April
Dumar	Fruit	March-April
Footkal	Leaves	February-March
Jackfruit	Fruit	February-March
Jamun	Fruit	July-August
Jirhul	Flowers, Leaves	June-July
Kachnar	Flowers, Leaves	May-June
Karonda	Fruit	September
Kend	Fruit	May-June
Koinar	Leaves	May-November
Kusum	Fruit	April-May
Mahua	Flowers	March
Mango	Fruit	April-June
Neem	Leaves	June-July
Pithod	Fruit	September
Piyar	Seeds	May-June
Sal	Flowers	March-April
Tamarind	Fruit	May-June

Herbs and mushrooms+	Plant parts	Time available
Chimti Sag	Leaves	January-February
Kattay Sag	Leaves	May-June
Matak Sag	Leaves	May-June
Sataur	Fruit	November-December
Suga Sag	Leaves	July-August
Mushrooms	All	February-May

* For the botanical names of these trees, see Appendix I.
+ It was not possible to find out the botanical names for these plants.

Source: Jewitt, 1996.

Liquor As in some other *adivasi*-dominated areas of India (Hardiman, 1987), country liquor, many of the ingredients for which come from the forest, is another important source of 'lean season' calories for rural people in Jharkhand (Corbridge, 1986; Kelkar and Nathan, 1991). In 1993, 48 per cent of Ambatoli's sample villagers compared to 33 per cent in Jamtoli admitted to the collection of liquor-making ingredients including *mahua* flowers (for distillation into *mahua* liquor) and *ranu*, a collection of herbs and forest products which is used in the fermentation of *hanria* (rice beer). As some villagers, particularly the women, were embarrassed to admit that they collected *ranu*,[15] these percentages may well be underestimates.

The main collection period for *mahua* flowers is March and April although many are dried for later use. *Ranu*, likewise, is collected mainly in the spring and early summer and processed into a tablet form that can be used throughout the year. According to Kelkar and Nathan (1991), the most important time of year for liquor as a source of calories is in July and August when grain stocks are running low. In the research area, however, the actual consumption of country liquor is probably just as high during the festival and wedding season between November and May.

The major collectors of liquor-making ingredients in Ambatoli are, unsurprisingly, the *Sarna*-worshipping Oraons and Mundas and the semi-Hinduised Lohra and Mahli families. Indeed, a number of Oraons and Mundas frequently set fire to the grass around their own *mahua* trees (as well as some of those in the jungle) to make the flowers easier to collect (discussion with Ram Prasad Sahu, 1995). Amongst Ambatoli's Hindu population, the only villagers who do not drink alcohol on a regular basis are the better-off Backward Castes. Nevertheless, this does not stop a number of Sahu and Pradhan households from bringing *ranu* and *mahua* back from their trips to the jungle. Indeed, for the Sahu liquor sellers, the collection of *ranu* and *mahua* is a major cost-cutting activity.

Amongst many of Ambatoli's Scheduled Caste households, the consumption of country liquor is fairly widespread and some of the Manjhi households in Pokaltikra and the Swansi households in Panrtoli rival the *adivasis* in terms of the volume of alcohol that they brew and drink. In Jamtoli, by contrast, only the Hindu Ghops and a few of the Oraon *Sarna* worshippers actually admitted to collecting liquor-making ingredients for their own use, although several other households collected small quantities for sale.

Other Non-timber Forest Products (NTFPs) As emphasised by writers such as Fernandes and Kulkarni (1983), Nathan (1988), Shiva (1988), Pathak (1994), Shah (1995) and Poffenberger (1996), in addition to providing many wood and food items that are vital for rural subsistence, India's forests also provide a number of other NTFPs that are used by local people on a day-to-day basis. This is certainly the case in Ambatoli and Jamtoli where villagers put NTFPs to many different uses.

Taking the *sal* tree as an example, the leaves are often stitched together with bamboo splinters to make *sal* leaf plates or bowls and the young straight twigs are used as tooth brushes. *Sal* seeds can be pressed for oil and the sap, as mentioned earlier, is dried to form *dhuwan* for religious and *bhut*-propitiation ceremonies.

Many people also collect the seeds of *amla* (*Emblica officinalis*), *karanj*, *bahera* and *mahua* trees to use for oil. *Karanj*, *bahera* and *mahua* oils are used in the manufacture of soap and *sal* oil is used for cooking. *Amla* is sometimes used as a hair dye and is often found along with *karanj* oil as one of the main ingredients in hair styling and conditioning oil. Another useful seed is that of the *simbal* tree which contains a fluffy cotton-like substance that is used to fill pillows and mattresses.

Other household items that are made essentially from forest produce include a sweeping brush known as a *jarru* and mats used for sitting or sleeping on. *Jarrus* are made from *sabai* grass (*Eulaliopsis binata*) that grows in open jungle and mats are usually made either from *dhub* grass (*Cynodon dactylon*) which also grows in open jungle or the plaited leaves of the *kujur* (*Phoenix acaulis*) tree. Bamboo, in addition to being split to make roof battens and woven into baskets and winnowing fans, is also used to make fish traps that are placed in the gaps where water enters and exits paddy fields during the monsoon.

Indigenous Medicine Another group of very important NTFPs are the medicinal plants used throughout the world by indigenous healers to cure and protect people from many different types of ailment (Dankelman and Davidson, 1988; Nathan, 1988; Gupta, 1989; Pereira, 1992). In Kerala, for example, tribals in the Attapadi valley use the roots of thirty nine species, the fruits of fifteen species and the leaves of thirty species for medicinal purposes (Fernandes, 1983). Yet the medicinal knowledges of these local experts are not particularly well researched and understood, despite the fact that many drugs in allopathic and ayurvedic medicine are made from plant extracts. Also poorly understood and documented are the 'ritual symbolic' aspects of indigenous medicine (Fairhead, n.d.); particularly the linkages (or lack of distinction) that many indigenous healers make between what outsiders would call infections or diseases and the mischievous activities of *bhuts* and *dains*.

In Ambatoli and Jamtoli, education and development more generally has brought about a major shift in many local people's attitudes towards traditional and allopathic medicine and their understandings of the various causes of illness. Most villagers have a fairly rudimentary knowledge about the use of forest products to cure minor ailments and about the allopathic treatments available from local 'compounders' or doctors. At the same time, many people still rely heavily upon local herbalists who are considered to have expertise in the curing of 'natural' illnesses (that can also be treated with allopathic or ayurvedic medicines) as well as those caused by supernatural powers.

In Ambatoli, a participatory rural appraisal 'timeline' indicated that twenty five years ago, only a quarter of the village population used allopathic medicine for humans and only around two per cent used it for animals. After the early 1980s when a compounder came to live in Patratoli and veterinary treatments became more widely available in Bero, the number of allopathic medicine users increased rapidly and now stands at around 75 per cent for humans and 25 per cent for animals.

This trend does not indicate a displacement of herbal treatments by allopathic medicine, however, as these systems currently co-exist in a very complementary manner. In 1993, 33 per cent of Ambatoli's sample population claimed to possess some knowledge about herbal remedies for minor 'natural' illnesses (such as coughs, mouth ulcers, fever, dysentery and pain in humans or coughs, sore tongues and sore feet in cattle). Moreover, in-depth interviews with twenty five villagers in Ambatoli showed that only five used allopathic medicine exclusively in 1993. Two of the five said that they 'did not believe in herbal medicine' (interviews with Gandra Bhagat and Binita Manjhi, 1993) and the other three said that they did not use it as they had grown up in towns where few people were experts on herbal medicine (interviews with Satan Lohrain, Rina Sahu and Pairo Mahli, 1993).

Three of the other twenty interviewees had never used allopathic or ayurvedic medicine for illnesses in their families (interviews with Etwa Oraon, Hiramuni Orain and Fagni Pradhan, 1993). All of these villagers had relatives or friends who were knowledgeable about herbal medicine and did not see the point in paying for allopathic medicine when herbal remedies were free. The remaining seventeen interviewees all said that they usually tried herbal medicine first (if the problem was perceived to be a minor one) and then went to see the compounder if it did not work.

Although most of these seventeen villagers believed that allopathic medicine was effective, its relatively high cost (75 rupees for a tetanus injection and 10 rupees for ten aspirin tablets) was one of the major factors that encouraged them to try herbal medicine first. A second factor was one of convenience. But as only three interviewees were familiar with more than four different herbal remedies, most of the complex cases were referred to village herbalists.

In Ambatoli, the best known and most respected of these is Jittu Bokta, who regularly uses twenty six different forest plants to treat 'natural illnesses' in both humans and cattle. More importantly, Jittu is regarded as an expert at curing and preventing illness caused by malevolent forces through the propitiation of *bhuts* and the distraction of the 'evil eye': skills that allopathic doctors simply cannot compete with. Jittu also works on an 'in kind' rather than a cash-based system of payment for his treatment (usually a few bottles of liquor) and this suits the pockets of the poorer villagers, in particular.

Detailed information on the use of allopathic as opposed to herbal medicine was not sought in Jamtoli, although the sample villagers were all asked about their knowledge of herbal remedies. Most of their answers showed that the co-existence of allopathic and herbal medicine is probably even greater in Jamtoli than in Ambatoli as a slightly higher percentage of villagers from the sample survey (38 per cent) had some knowledge about forest-based medicinal herbs. In addition, allopathic drugs were both readily available and significantly cheaper from chemist shops in Bero than from the compounder in Ambatoli.

One of the main experts on herbal medicine in Jamtoli is Simon Oraon who regularly uses nineteen different forest plants to cure fever, mouth ulcers, toothache and dysentery and to speed up the healing of broken bones: treatments that he makes no charge for. Being a Christian rather than a *Sarna*-worshipper, however, Simon has no knowledge of how to cure illnesses caused by malevolent *bhuts*. As a result, many of Jamtoli's *Sarna* worshippers go to their local equivalent of Jittu Bokta for the treatment of 'supernatural' illnesses. Others prefer to visit professional herbalists (*Badhyam*) who sell many different (and quite cheap) herbal remedies for both 'natural' and *bhut*-induced illnesses at Bero market.

The Financial Importance of Forest Produce

In many areas of India, the sale of forest produce is a very important source of cash income for villagers who do not undertake seasonal migration (Pathak, 1994; Saigal, 1995; Shah, 1995). In Ambatoli and Jamtoli, by contrast, the existence of forest protection committees largely prevents the sale of wood and few villagers make much of an income from selling NTFPs. In 1993 when Ambatoli villagers regularly used to sell fuel wood, however, a number of households made quite significant profits, although strong disapproval of the practice helped to restrict it to particularly resource poor villagers. Incomes from fruit and oil, meanwhile, were (and remain) rather insignificant in both villages (see tables 10.4 and 10.5).

Fuel Wood Selling In Ambatoli, the twenty sample households that regularly sold fuel wood in 1993 earned an average of 1375 rupees. Nine of these owned less than 2.5 acres of land and had great difficulty producing enough crops to last them through the year. A further eight owned between 2.5 and 5 acres of land but were unable to meet their subsistence requirements either because of poor land quality, high household size to land ratios or, in the case of a female-headed household, a dependence on others to work her land.

The remaining three fuel wood selling households all owned over 12 acres of land, but one of them, (Silamuni Orain's husband, Bandhna), had had to mortgage out most of his fields to pay compensation for his mother's 'witchcraft' (discussion with Bandhna Oraon, 1993). Another had 14 acres of unirrigated land and sold fuel wood irregularly to enable him to buy *rabi* season vegetables. His annual income from fuel wood selling was only around 300 rupees. The

Table 10.4   Sales of forest produce by Ambatoli's sample population

Community sub-group	Fuel wood income	Forest food income	Oil income	Total income
Oraon	6680	1380	0	7060
Munda	2020	20	40	2140
Mahli	8010	0	180	8130
Lohra	0	10	0	10
Bokta	3450	0	0	3450
Pradhan	850	0	0	850
Manjhi	0	0	0	0
Swansi	5500	0	0	5500
Chik Baraik	0	0	0	0
Sahu	0	0	0	0
Thakur	0	0	0	0
Teli-Sahu	1000	0	0	1000
Total (Rs.)	27510	1410	220	28140

Source: Jewitt, 1996.

Table 10.5   Sales of forest produce by Jamtoli's sample population

Community sub-group	Fuel wood income	Forest food income	Oil income	Total income
Oraon	150	150	0	300
Munda	0	0	0	0
Lohra	0	0	0	0
Ghop	0	60	0	60
Pradhan	0	20	0	20
Total (Rs.)	150	230	0	380

Source: Jewitt, 1996.

third 'wealthy fuel seller' had no good excuse for his activities, but could afford both to bribe the forest guard to 'turn a blind eye' and to ignore other villagers' criticisms of his actions.

In Jamtoli, by contrast, only one person from the twenty one sample households admitted to selling fuel wood, but claimed to do it only very irregularly to meet short term cash deficits. He also claimed to have earned only 100 rupees from this source in 1993. Interestingly, the wood that he sold was taken from a neighbouring village forest rather than from Jamtoli forest, as he is a member of Jamtoli's forest protection committee (discussion with Budhram Lohra, 1993). Group discussions in all three hamlets indicated that the incidence of fuel wood selling was very low overall; mainly as a result of the forest protection initiative, but also because regular annual deficits in staple crops were less of a problem in Jamtoli than elsewhere (interviews with Ram Prasad Lohra, Sukra Bhagat and Budhwa Oraon, 1993).

Forest Food and other NTFPs Although the sale of NTFPs in Ambatoli and Jamtoli does not provide anything like half of the village's annual income, as is the case in some other areas of Jharkhand (Nathan, 1988), the sale of forest fruits, leaves, seeds and mushrooms at local markets does provide a small but important income for some of the poorest villagers. In 1993, eight of Ambatoli's sample households regularly sold forest food and two households also sold *sal*, *mahua* and *karanj* seeds for oil. The average income for forest food sales was just over 51 rupees and the two oil selling households earned 40 and 180 rupees each.

In Jamtoli, most villagers sell fruit and seeds from their own household trees, so it is usually only cash deficit households that sell NTFPs from Jamtoli forest. In 1993, the four sample households that sold fruit and seeds from forest trees included three of the four food deficit households plus a particularly business-oriented Oraon household. Their average income from these sales was just over 57 rupees.

Perhaps surprisingly, I came across nobody in either Ambatoli or Jamtoli who regularly sells items such as *sal* leaves for leaf plates and bowls, *sal* twigs for tooth brushes, *kend* leaves for making cigarettes or other NTFPs that are commonly sold elsewhere. From time to time, villagers make leaf plates and bowls for other households if they are having a celebration of some kind, in the expectation that the favour will be returned. Nevertheless, compared to other areas of India (Fernandes, 1983; Gadgil, 1983; Joshi, 1983; Agarwala, 1990; Gadgil and Guha, 1992; Pathak, 1994; Saigal, 1995; Saxena, 1995; Shah, 1995; Poffenberger, 1996), villagers' financial returns from NTFPs seem relatively small.

Large Timber Sales Because of the scarcity of large trees in Ambatoli after the late-1980s and the forest protection system in Jamtoli, very little large timber was cut illegally for sale between 1993 and 1997.

Conclusion

In many ways, forest decline in the research area reflects India-wide patterns of deforestation resulting from fundamental conflicts between the state's need to conserve forests, exploit them commercially and satisfy villagers' subsistence requirements. Since the 1950s, 4000 square kilometers of Bihar's Protected Forest has been converted into agricultural land and out of the 24,000 square kilometers of Protected Forest remaining, 14,000 square kilometers is officially classed as 'degraded' (interview with Sanjay Kumar, 1996). While increasing human and animal populations have undoubtedly compounded existing competition for forest resources in Jharkhand, there is also empirical support for the idea that local people became 'alienated' from commercially managed forests (Guha, 1989) or developed a 'free for all' mentality towards them (Bahaguna, 1991).

   In many cases, the widespread nature of illegal felling by forest contractors and the frequent corruptibility of forest guards undermined villagers' respect for the 'official' system of forest management. Throughout Jharkhand, relationships between local people and the Forest Department became particularly antagonistic during the 1970s when everyone started to feel the pinch of forest decline. Indeed, the previous accounts of deforestation, landuse intensification, population increase and loss of traditional institutional structures would appear, according to the policy-oriented community-based forest management models discussed in Chapter Two (Shepherd, 1993a; 1993b), to represent an irreversible situation.

   Whilst not exactly providing support for contrasting populist emphases on the persistence of peasant moral economies (Guha, 1989) and environmental management strategies, however, the situation in Ambatoli (and other villages like it), never became quite as bad as some local accounts suggest. During my main period of fieldwork in 1993, Ambatoli forest was in very poor condition and had few large trees left, but was still far too extensive to confirm some villagers' estimates of a 50 percent decline since the mid-1980s. In addition, many species were managing to regenerate in spite of villagers' failure to coppice them (or to restrict cattle grazing) and there were few areas that did not have at least some remaining rootstock with potential for rapid regeneration when protected. And although mature *jamun*, mango, *amla*, *piyar*, *dumar* and *kend* trees had become fewer in number and the availability of forest fruit (and the country liquor made from it) was declining, certain other useful forest plants were becoming more widespread. Indeed, some species of forest *jaree bootee* (herbs), grasses and weed bushes (notably *putus* and *tetair*) were thriving in response to increased light availability within the forest.

   Moreover, evidence to suggest that traditional forest usage patterns had been eroded as a direct result of forest decline was fairly limited. As with agriculture, villagers' enthusiasm for tapping into both traditional and modern knowledge systems has provided a continual source of demand for a variety of different forest-based items. With respect to everyday items made from forest products, for example, in-depth interviews with village elders such as Jittu Bokta, Tila Oraon and Tuttu Oraon suggested that the only things no longer obtained from the forest are raincoats made from *gungu* leaves. This is due not to a decline in *gungu* bushes, however, but to the replacement of *gungu* raincoats (which are time-consuming to make, heavy and hard to store) with umbrellas and plastic mackintoshes. In certain cases, forest products have actually replaced other traditionally used items; a widespread example being the use of *karanj* oil soap (instead of ash and 'nagera' mud from the river) for washing clothes and hair. Although scented toilet soap and shampoo have started to displace *karanj* oil soap for bathing purposes, most villagers still use it for washing clothes and the sale of *karanj* seeds provides a regular source of income for several households.

Nor is there any real evidence of a replacement of traditional silvicultural knowledges by modern technology. Allopathic medicine, for example, seems to co-exist well with herbal medicine and has shown few signs yet either of de-legitimising or displacing it. Indeed, most of the forest-based ingredients for herbal remedies are still widely available in Ambatoli forest and as Jittu Bokta is training two young men how to practice herbal medicine, it seems likely that these systems will continue to co-exist for some time to come. Proof that villagers have not lost their coppicing and pollarding skills, meanwhile, can be found in the careful manner with which many men tend their own trees.

Whether most people in villages like Ambatoli are too keen on individual property ownership and conventional development goals to embrace populist alternatives centred on community-based co-operation and traditional environmental knowledges is rather more difficult to address, however. Amongst Oraons, agricultural land has long been owned individually and villagers' feelings about 'their' forests must certainly have changed with the introduction of British forestry policy and/or *zamindari* systems of forest management. The post-Independence exploitation of forests by contractors also distanced many people from traditional systems of autonomous forest management.

On the other hand, commercial forest exploitation did make local people aware of broader environmental problems that they have come to associate with forest decline: notably reduced water availability and more widespread air pollution near urban areas. It also made many people angry about the de-legitimation of local forest use and management practices by a scientific forestry system which made no allowances for the centrality of forests to village culture. This, in turn, strengthened their belief that forests should be brought back under local control, as rural people could not only manage them better than the Forest Department, but would have the incentive to do so, if they were making the decisions about how and by whom they were to be used (interviews with Tila Oraon, Ram Prasad Sahu and Jittu Bokta, 1993).

Indeed, few villagers willingly gave up their control over local forests in the first place and many employed strategies ranging from the deliberate exploitation of state forests to the forcible eviction of timber contractors to resist the imposition of scientific forestry. From the late 1970s, the Jharkhand movement helped to strengthen opposition to Forest Department control by supporting local protests against contractor-cutting and the planting of teak monocultures (Kelkar and Nathan, 1991; Corbridge, 1991a). In view of India's recent emphasis on participatory forest management, it could be argued that this type of resistance has finally paid off for local people.

To make the most of these new developments, however, it is necessary to know where, when and how initiatives like JFM are most likely to succeed. And despite the value of populist discourse in highlighting the plight and potential of India's 'ecosystem people' (Gadgil and Guha, 1995), it offers little practical help in identifying the most important pre-conditions for successful autonomous or joint resource management. Instead, I would argue that broad 'enabling criteria' for successful forest management (Ostrom, 1990; 1999) coupled with village-level research into local institutional ecologies, different stakeholder priorities and the linkages between technical and socio-cultural knowledges is a more effective means for development planners to identify the most promising areas for (as well as the best means of promoting) sustainable and participatory resource management.

My own research, for example, indicates that so long as local people maintain their belief in the power of malevolent *bhuts*, they will continue to protect village forests in places like Pothalgudha where *churils* and other spirits are believed to live, or Porhosokra where sacrifices are offered to the forest dwelling *bhut*, Darha. Forest cover in religious places like Ghargari or Karamatungri where the *Pahan* gives *puja* under a *karam* tree will also be retained, so long as villagers continue to worship in these places. The fact that traditional healers give forest-based medicinal herbs for the treatment of both 'natural' and *bhut*-inflicted illnesses also means that the detailed knowledges about jungle *jaree bootee* possessed by people like Jittu Bokta will co-exist with allopathic medical knowledges as long as the threat of witchcraft remains.

On the basis of this, it is suggested that in areas where forests have particular cultural or magico-religious importance, local people are often more willing than villagers elsewhere to fulfil the demands necessary to develop autonomous or participatory forest protection. Indeed, Jamtoli, which was mentioned very little in the previous account of forest decline in the research area, is an excellent example of successful autonomous forest management established by villagers with a strong socio-cultural as well as practical dependence on forests.

Ambatoli, meanwhile, has shown a consistent desire (if not always the ability), to address the problem of forest decline through village-based forest protection. Unfortunately, it has been hampered by a lack of favourable pre-conditions for community-based resource management as well as the Forest Department's lack of interest, until recently, in participatory schemes. But after the introduction of the Bihar JFM programme, Ambatoli did show itself capable, eventually, of establishing village-based forest management. Stimulating and maintaining something as fragile and potentially volatile as participatory forestry, however, is a task that requires substantial commitment, patience and flexibility from both parties. The extent to which this has been achieved in the research area will be taken up in Chapters Eleven and Twelve.

Notes:

[1] The Kattiyan is a village-level record of rights that contains details of local land holdings (Kattiyan Part I) and allowances for forest produce (Kattiyan Part II).

[2] At the time when I did my main period of fieldwork, Sanjay Kumar was the Divisional Forest Officer for Ranchi East Division. Since then, he has been promoted to Conservator of Forests and embarked on a programme of doctoral research at the University of Cambridge. I have worked closely with Sanjay since 1993 and am deeply indebted to him for all the help that he has given me.

[3] The publication of the Annual Progress Reports of Forest Administration in the Province of Bihar was suspended between 1940 and 1946 as a result of staff shortages during the Second World War.

[4] Some of this forest land has since been lost as a result of conversion to agriculture.

[5] *Kend* (*Diospyros melanoxylon*) leaves are used in *bidi* (local cigarette) making and villagers are supposed to register with the Divisional Forest Officer in Ranchi if they want to collect *kend* leaves for sale.

[6] In support of the criticism that social forestry programmes have received for their emphasis on exotic trees such as eucalyptus (Shiva et al, 1983; Eckholm, 1984; Foley and Bernard, 1985; Shiva, 1988; Arnold, 1991; Ghate, 1992), Ambatoli's villagers expressed a preference for fruit or fast-growing fuel trees. The Forest Department, by contrast, wanted to plant eucalyptus and acacia as these are unpalatable to cattle and their leaves can be burned for fuel. As it turned out, the Forest Department's instinct was correct as most of the young fruit trees that were planted were browsed within a month whereas the eucalyptus and acacia are still growing.

[7] Per capita fuel wood requirements are around 0.65 kilograms per day and Ranchi District alone currently has a human population of well over three million. Cattle populations are also very large (over two million in Ranchi District) and depend heavily on forests for browse material as well as grazing (interview with Sanjay Kumar, 1995).

[8] In 1993, the daily rate for agricultural labour in Ambatoli was 15 rupees for one man and two bullocks.

[9] Even villages with successful forest protection systems have often had problems in preventing tree felling by non-local fuel wood sellers from heavily deforested areas. To reduce their chances of being caught, such people usually head for the most remote parts of the forest. When challenged, they often point out that the

forest is owned by the government, not by the right-holders, so nobody except the government can stop them from taking wood.

10 In Ambatoli, for example, the main singers and musicians are all from the Mahli community and one Mahli man in particular, Asant, writes many of his own lyrics for both traditional and modern tunes. He has drawn upon his family's experience of forest decline in Ambatoli; describing the beauty of the forest and the importance of preserving it. But this desire to preserve forests often conflicts, in practice, with the need to exploit them for day-to-day subsistence purposes: a contradiction that Asant Mahli is particularly well aware of because in 1993, his own family could not meet their household subsistence requirements without the cash that they obtain from fuel wood sales.

11 Translations of three of these songs can be found in Appendix III.

12 '*Jagu jawan*' was written by Mukund Nayak who belongs to the Scheduled Caste Ghasi community. Like a large number of other so-called Scheduled Castes in Jharkhand, Mukund's family have strong socio-cultural linkages with *adivasis* and shared their reasons for wanting a separate Jharkhand State.

13 *Sarhul* is also organised within a *Parha* as the festival travels from village to village within a set time period.

14 It was not possible to find out the botanical names for these plants.

15 Those villagers who did admit to collecting *ranu* often referred to it laughingly as '*dawai*' or medicine.

11 Autonomous and Joint Forest Management in Jharkhand

Introduction

After the mid 1980s, the Indian government's failure to halt deforestation, ease fuel wood shortages or significantly further its goal of bringing 33 per cent of India's land under forest cover promoted a re-thinking of forest policy. In the aftermath of the furore caused by the 1981 Draft Forest Bill, India's forest intellectuals managed to bargain successfully, on behalf of forest dependent populations, for more sensitive and people-friendly forest policy-making (Fernandes and Kulkarni, 1983; Pathak, 1994). They also helped to bring to light the impact of tension between forest officials and local people on India's forests (Palit, 1993).

At the same time, wider shifts in development thinking coupled with successful attempts by forest officers in Arabari, West Bengal to initiate joint forest management helped to bring participatory forest management to the attention of the Indian central government, external donor agencies and NGOs (Palit, 1993; Saigal, 1995). Also important was a growing awareness of communities throughout South Asia which, instead of destroying local forests, managed them as common property resources through the establishment of autonomous forest protection committees (Arnold and Campbell, 1985; Gadgil and Iyer, 1989; Mehrotra and Kishore, 1990; Kant et al 1991; Messerschmidt, 1995).

The main aim of this Chapter is to examine India's recent interest in community-based and participatory forest management in light of detailed village-level data on forest management from Jharkhand. In the sections that follow, an attempt will be made to assess the practicalities of establishing autonomous and joint forest protection systems in forest-dependent villages which, following the logic of India's forest intellectuals, should be enthusiastic and ready to re-develop community-based resource management. In particular, 'kindly' populist discourse (Corbridge and Jewitt, 1997) about rural people's forest management skills is compared with empirical findings from locally-based research in Bihar and Orissa (Mehrotra and Kishore, 1990; Kant et al, 1991; Kelkar and Nathan, 1991). It is also contrasted with Shepherd's (1993b) policy-oriented community-based forest management model which suggests that even with assistance from the state (through JFM schemes) 'area type three' villages like Ambatoli and Jamtoli would be unlikely to succeed in (re)-developing village-based forest management.[1]

An underlying emphasis is the value of supplementing wider resource management theories with village-level research on the micro-political ecology of people-forest-agriculture interactions as a means of developing flexible and locally appropriate institutional systems for the management of degraded forest areas. An investigation is also made of the conditions likely to promote successful forest protection through participatory forestry as opposed to a more demand side populist approach in the form of autonomous forest protection. Particular attention is placed on the Bihar Forest Department's changing response, especially after the introduction of JFM, to the idea of involving local people in forest management.

More specifically, the first section tries to highlight factors that assist in the development of community-based resource management. The second section draws a broad comparison between Ambatoli and Jamtoli and attempts to classify them according to Shepherd's model. The third and fourth sections examine Jamtoli's success at establishing and maintaining autonomous forest protection and Ambatoli's efforts to follow suit, paying particular attention to the circumstances under which other 'area type three' villages might succeed in (re)-establishing community-based forest management. The remaining sections trace the establishment of the Bihar JFM programme in the research area and attempt to evaluate its success in providing villages like Ambatoli with the incentives that they need to establish community-based forest management and giving villages like Jamtoli the support they need to maintain forest protection.

Conditions Favouring Community-based Management

To make sense of why local people have had such different reactions to the privatisation of India's forest resources, it is helpful to examine the situation in a narrower context. Looking at Jharkhand, Kelkar and Nathan (1991) identify two types of response to a loss of local control over forest resources. The first is 'to cut down or otherwise destroy forests' (Kelkar and Nathan, 1991, p. 140); a reaction that centres around the assertion of local people's traditional subsistence rights to convert forests into agricultural land (Guha, 1989; Pathak, 1994). Examples of this include the felling of commercially valuable timber plots by Santhals in Midnapur during the 1920s and the widespread felling of *sal* trees in the 1950s, prior to the re-notification of *zamindari* forests by the Forest Department (Kelkar and Nathan, 1991). As mentioned in the previous Chapter, conflict between local people and timber contractors was also common throughout the 1950s and often sparked off rounds of deliberate forest exploitation. Opposition to the plantation of commercially (but not locally) valuable teak monocultures was also prevalent in Singhbhum District during the 1970s (Kelkar and Nathan, 1991).

The second response is to 'preserve and defend forests against attempts to cut them down' (Kelkar and Nathan, 1991, p. 140) and centres around an emphasis on local silvicultural management strategies as a means of maintaining access to forest produce. Instances of this vary from isolated attempts to protect local forests from exploitation by outsiders to more sustained systems of autonomous forest protection. Examples of isolated incidents include the obstruction of Forest Department tree felling by local people in Gumla District in 1978 and attempts to stop the sale of a local coupe (by taking bows and arrows to the timber auction) by people from Jhagram village in 1985 (Kelkar and Nathan, 1991).

More prolonged efforts to oppose state forest management and re-establish local silvicultural knowledges and management systems include a movement led by Fetal Singh in Palamu District which demanded both a halt to forest cutting by contractors and the right to obtain timber and other forest produce free of charge (Singh, 1982c). Although this movement was crushed with the use of police force (Kelkar and Nathan, 1991), many villages in Jharkhand have been successful in preserving both local forests and local people's rights to their produce (Mehrotra and Kishore, 1990; Kant et al, 1991). Jamtoli and the village-based forest protection committees in Bihar and Orissa studied by Mehrotra and Kishore (1990) and Kant et al (1991) are good examples of this.

The specific reasons why certain communities react to a loss of control over common property resources by damaging them while others make efforts to protect them can only really be examined with the help of local-level research (Wade, 1988; Ostrom, 1990; 1999; Agrawal and Sivaramakrishnan, 2000b). Data from Mehrotra and Kishore's study suggest that a common concern about local forest loss, a perception of negligence by the Forest Department, the existence of legal rights to forest produce, the presence of a strong leader or traditional resource management institution and the influence of 'an uprising against social and economic deprivation of the masses' (Mehrotra and Kishore, 1990, p. 2) can all help to encourage the establishment of autonomous forest protection. Similar 'conditions for success' plus an emphasis on the need for a unified and trusting user group, a well defined resource and a clear set of rules (with sanctions for breaking them) can be found in much of the mainstream literature on common property resources discussed in Chapter Two (Blaikie et al, 1986; Wade, 1988; Gibbs and Bromley, 1989; Gadgil and Iyer, 1989; Ostrom, 1990; 1999; Stevenson, 1991; Singh, 1991a; Baland and Platteau, 1996; Kaufman, 1997a; Fernandez, 1999). Factors that deter village-based resource management, on the other hand, include external elements such as powerful agents that are either indifferent to (or stand to gain from preventing) it and internal matters such as high population densities, intensive agriculture and a loss of communal land (Ostrom, 1990; Shepherd, 1993a; 1993b).

Comparison of Ambatoli and Jamtoli

An underlying assumption of this section is that temporally and spatially specific institutional ecologies; notably an ability to overcome intra-community tensions and seek greater access to forest produce for all, are responsible for Jamtoli's successful forest protection system and Ambatoli's many false starts. Before launching into a comparison of the fieldwork villages, however, it is important to first establish that there are no major ecological or management-related differences that could account for the disparities between them. It is also necessary to confirm the 'area type' of the two research villages with reference to information on population to resource ratios, cultivation intensity and the extent of common land.

Looking first at the ecology of the two forests, there are very few differences to be found. Both forests belong to a continuous belt of moist deciduous *sal* forest and their species composition would, if their management systems and levels of exploitation had been roughly the same, be expected to be very similar (discussion with Sanjay Kumar, 1993). The soils in both villages are lateritic interspersed with sandy loam, but the soil in Ambatoli is of an overall better quality as the relief in the village is more undulating, giving a greater average depth of soil than in Jamtoli, which is rather hilly and rocky to the east with quite shallow soils. The forest on this rocky area is rather degraded now, but used to have a similar species composition to the lower-lying areas (interview with Sanjay Kumar, 1993).

According to existing forest records and local oral histories, the condition of Ambatoli and Jamtoli forests were very similar during the late 1950s, prior to the establishment of a formal forest protection system in Jamtoli. Although fourteen acres of Ambatoli's orchard land were felled by the local *zamindar* in the wake of the Bihar Land Reforms Act 1950, there are no records of any damage to Ambatoli forest proper and attempts were made by the Forest Department to improve the quality of the forest after it came under state management in 1953. Jamtoli, being a non-*zamindari* village, was not affected by the *zamindari* exploitation of forests and was managed on a quasi-commercial basis from 1944. As it had spent longer under Forest Department 'improvement fellings' than Jamtoli, it probably had a slightly higher proportion of large trees by the late 1950s, but the actual crop density would have been very similar (interview with Sanjay Kumar, 1995).

Similarities in the commercial exploitation of the two forests and their forest area to right-holder ratios helped to ensure that their condition remained fairly similar throughout the 1950s and 1960s. As mentioned in Chapter Ten, however, after the mid-1960s, Jamtoli's forest area to right-holder ratio declined below that of Ambatoli when forest degradation (resulting from commercial timber exploitation and illegal felling by forest contractors and local people) became so severe that parts of Jamtoli forest never regenerated (interview with Sanjay Kumar,

1995). By 1971, Jamtoli's forest area had declined to approximately 400 acres (161 hectares) and by 1991, little forest area remained outside the 206 acres (83.36 hectares) protected by the village-based forest protection committee.

In terms of Shepherd's model, it is a long time since population density has been so low and agriculture so extensive that villagers were able to practise swidden cultivation. Indeed, the fact that all of the cultivated land is held in permanent registered plots places both villages firmly in 'area type three'. Agricultural intensity is also sufficiently high to justify an 'area type three' classification as almost all villagers who have access to irrigated land grow two crops per year and some villagers in Jamtoli manage to grow three. Fertiliser is also widely used in addition to dung, although sample survey data showed it to be more common in Jamtoli than in Ambatoli.

According to Shepherd's third criterion, common land, both villages fit the 'area type three' classification. With the exception of the five acre state-owned '*maidan*' that is used as a recreation ground, the forests are the only 'commons' in Ambatoli that can be used for purposes such as cattle grazing. These are not real commons, however, as they are owned by the state which has its own (albeit largely disregarded) rules about when and what types of animals can be grazed there. Jamtoli has no true commons and although cattle grazing is not permitted in the 206 acres of forest protected by the forest protection committee, villagers routinely allow unregulated grazing in non-protected areas of degraded forest or wasteland.

In summary, then, it seems that Ambatoli and Jamtoli forests were in a fairly similar condition at the time when forest protection was established in Jamtoli. There appear not to be any ecological factors that could account for current differences between the two forests in the absence of forest protection. With regard to Shepherd's second model, both villages fit the criteria set out in the 'area type three' classification, although the lower forest area to right-holder ratio and higher degree of agricultural intensity in Jamtoli suggest that this village is slightly closer to 'area type four' than Ambatoli (see figure 11.1).

Autonomous Forest Protection in Jamtoli

When autonomous forest protection was established in the early 1960s, Jamtoli would still have been classed as an 'area type three' village with no true commons, no swidden and all agricultural land held in permanent registered plots. Prior to its reservation in 1944, however, Jamtoli forest (being a *zamindari* forest) was managed by the village community as a kind of commons and this may well have been influential in the later development of forest protection. The village also had many other features that were favourable for the establishment of community-based forest management.

Figure 11.1   The position of Ambatoli and Jamtoli within Shepherd's 'second model'

Area-type 1	Area-type 2	Area-type 3	Area-type 4
Geographical site characteristics			
remote low pop. density	----->-------> ----->------->	----->-------> ----->------->	near to town high pop. density
extensive --->------> Type of land use --->------> intensive			
high -----> Extent of Common Property Resources -----> low			
high -----> Likely villager interest ----> low			
Estimated position of Jamtoli and Ambatoli in Shepherd's matrix			
		AMBATOLI JAMTOLI	

Source: Adapted from Shepherd, 1993b.

Firstly, the fact that the village is home to Simon Oraon, one of the area's most influential *Parha Rajas*, certainly supports the emphasis in the common property management literature on the importance of a strong leader and prior organisational experience as catalysts for establishing forest protection committees (Blaikie et al, 1986; Wade, 1988; Berkes, 1989; Mehrotra and Kishore, 1990; Kant et al, 1991; Ostrom, 1999; Stevenson, 1991; Mukherji, 1998; Singh, 1999). Simon's efforts to maintain the role of his *Parha* as a local administrative and developmental as well as a social institution also fits with Mehrotra and Kishore's opinion that traditional decision-making institutions are important pre-conditions for community-based forest protection and management.[2] Indeed, it was primarily around the issue of forest protection that Simon started to reorganise his *Parha* following the introduction of the Gram Panchayat system. Such recent control over the use and management of Jamtoli forest made its reservation very hard to tolerate and was undoubtedly an important factor behind villagers' willingness to challenge the imposition of scientific forestry and its criminalisation of local forest users.

A major catalyst for the change to a more formal system of forest protection was the pressure put on local forests by population growth coupled with commercial timber exploitation. To make matters worse, many of the forest contractors who undertook felling operations refused to give local right-holders their share of the coupe and forest guards sometimes prosecuted villagers who cut timber for *bona fide* house construction purposes. Like the forest dwellers of Uttarakhand (Guha, 1989) and Mathwad Forest Range (Pathak, 1994), villagers in Jamtoli gradually lost interest in protecting the newly reserved Jamtoli forest and started to exploit it for their own profit.

According to Simon Oraon, many people formed the opinion that 'this forest is not ours, it belongs to the government, so why shouldn't we cut it? The contractors are cutting timber without any restrictions, so we will do the same' (discussion with Simon Oraon, 1993). Simon himself, along with several other *Parha* officials took a different view on the matter saying that although the coupe system was 'destroying the forest' (discussion with Sukra Bhagat, 1993), it was local people who were suffering the most both from forest decline and from the court cases taken out against them by the Forest Department. Consequently, they argued that efforts should be made to protect, not destroy Jamtoli forest.

The Establishment of Autonomous Forest Protection in Jamtoli

Autonomous forest protection actually started in Jamtoli in the late 1960s and was stimulated by the arrival of forest contractors with instructions to undertake routine felling operations in Jamtoli forest. The villagers were watching these activities closely and, on seeing the contractors fell more than the allocated area, Simon called together a number of local people to challenge them. Armed with bows and arrows, they succeeded in frightening away the contractors and confiscating the timber that they had cut.

Simon then decided to use his influence as *Parha Raja* to convince local people that as they depended so heavily on forests, they would not be able to survive unless they undertook long-term forest protection. Not surprisingly, many villagers objected, at first, to the idea of protecting a state forest from which the Forest Department (or forest contractors) would reap all of the benefits. Nevertheless, they recognised that what Simon said made sense, and made moves to divide Jamtoli forest between the three *tolas* (Jamtoli, Bertoli and Kaxitoli) and decide upon how forest protection would be administered. Kaxitoli, the hamlet in which Simon Oraon lives, was allocated the small (38 acres) but good quality Letteguttu forest which is situated close to the hamlet. A further 162 acres of poorer quality forest were divided up amongst Jamtoli and Bertoli.

After a few weeks, each of the three *tolas* had established a committee which met every fortnight and consisted of one adult male from each household. At first, the committees decided that forest protection would be best achieved through the employment of a villager as a 'guard' to patrol the forest. Each household had to give one kilogram of paddy per year in payment to the guard. When he caught anyone cutting trees without permission, that person was brought before the committee which would decide on whether a fine (usually of about five rupees) would be levied.

Local forest officers, meanwhile, were less than impressed by the Jamtoli villagers' decision to resist the imposition of scientific forestry and assert their customary subsistence rights by protecting a Reserved Forest from commercial exploitation. Predictably, the Forest Department threatened to prosecute the Jamtoli villagers but after a few years, Simon's forest protection scheme was so obviously successful that a number of forest officers started to show an interest in his initiative.[3]

By the early 1970s, Kaxitoli's forest had regenerated rather more successfully than Jamtoli's and Bertoli's. The main reasons for this were the poorer quality of Jamtoli's and Bertoli's forests (both of which lie on rather infertile, rocky land) coupled with their accessibility to outsiders which resulted in a fair amount of illegal felling. Unfortunately, this put a lot of stress on the resources actually available to villagers in Jamtoli and Bertoli, causing a 'loss of unity' which resulted in widespread tree felling. Kaxitoli, however, maintained its guard over the forest and Simon refused to let the Jamtoli and Bertoli villagers cut any timber from Letteguttu forest until their own committees had been re-established. In the meantime, a number of neighbouring villages were so impressed by the regeneration of Letteguttu forest that they decided to establish forest protection systems of their own.

This 'demonstration effect' had serious implications for the villagers of Jamtoli and Bertoli, however, as it curtailed their access to neighbouring forests from which they had previously been able to steal wood. As a result, they decided to re-establish their forest protection committees. At first, three men were employed by the villagers of both *tolas* to guard the forest and every household gave five kilograms of paddy per year for their services. A few years later, this form of protection was stopped as villagers felt that it was too much of a drain on their resources. Instead, they moved to the current system where four people go daily on a rota basis to check on the forest. Kaxitoli, by contrast, has continued to maintain the guard system (discussion with Ram Prasad Lohra, 1993).

Since the autonomous forest protection system was established in Jamtoli, villagers have organised an annual coupe within the protected area during which fuel wood and timber is collected by individual households. To ensure that enough good timber trees are cut for house building purposes, the committee usually marks them prior to the coupe. Poor quality timber trees or 'single crop' species such as *asan* that can only be used for wood are usually selected for cutting as fire wood. Fruit trees are almost never felled unless they are unhealthy and most of the other species that are cut are chosen for their ability to regenerate rapidly through coppicing.

When the coupe takes place, each household usually collects a specified number (usually three or four) cart loads of fuel wood which must last them all year. If they do run short, they have to cut their own trees or search in the forest for fallen branches or weed species such as *putus* or *tetair*. Villagers who have unforeseen requirements for house-building or plough-making timber must make a request to the committee which decides whether the need is justified. If villagers are caught felling trees in the forest without permission, they are usually fined by the committee.

Often, villagers undertake re-planting operations immediately after the coupe: an activity that has increased in scale recently as the Forest Department has made funds available for such activities from its afforestation budget. Most of the species selected by the committee members have been fruit trees as the most important timber species (*sal* and *asan*) regenerate very quickly from coppice growth (see plate 11.1). Mango, guava (*Psidium guajava*), *mahua*, *jamun*, *bel* and jackfruit have been particularly popular although *ber* is also liked for its fruit and as a host tree for lac. To ensure that the young trees are not damaged by grazing cattle, villagers usually plant a living fence around them (made of fast-growing *tetair* or thorny *putus*) which can later be cut for fuel.

Interestingly and in corroboration with the many critiques of the use of exotic trees in social forestry programmes (Shiva et al, 1983; Foley and Barnard, 1985; Shiva, 1988; Arnold, 1991; Ghate, 1992), the committee has not opted for exotic eucalyptus and acacia trees because, as Simon puts it, 'they take space and give nothing. Their leaves cannot be eaten by people or cattle and their wood will last for only one day'. Moreover, they 'drink a lot of water' and can only be used for their wood; unlike fruit trees which can provide a 'double or triple crop' in the form of food, wood, fodder and even lac (discussion with Simon Oraon, 1993).

A rather more unusual element that has been integrated recently into Kaxitoli's forest protection strategy consists of a form of 'welfare' to help some of its poorest villagers. Surrounding the *tola*, there are eight acres of individually owned orchard land. When mangos and other fruits or seeds become ripe on trees within this orchard, the poorest members of the village have the first pick of the fruit. This can be used for their own consumption or for sale at market.

The Resolution of Internal and External Conflicts over Forest Protection in Jamtoli

Jamtoli's intra-village relationships have not always been so trouble-free, however. In 1989, five families from Kaxitoli actually became estranged from the rest of the community as a result of conflict over forest issues. A man whom I shall call 'Rakia Oraon' was expelled from the forest protection committee for stealing money from its central funds. He tried to retaliate by attempting to organise a gang to cut Kaxitoli's forest on *Fagun* (a festival) day, when the other villagers would be celebrating. Simon heard about this plan and managed to stop it so Rakia Oraon threatened to take out a court case against Simon. Since then, Simon has

Plate 11.1   *Sal* **coppice regeneration**

not allowed him, or the four households that supported him, to have any wood from the annual coupes.

With regard to Simon's relationship with the Forest Department, there was a fairly long period of tension following the occasion when Simon chased the contractors away from Jamtoli and confiscated their timber. As time went on, however, many forest officials came to respect the work that Simon was doing in Letteguttu forest. Indeed, in 1978 when villagers objected to the auction of a coupe for Jamtoli forest worth 84,000 rupees, the Chief Conservator of Forests wrote to the government in support of the villagers' plea that another forest should be found from which the coupe could be taken (interviews with Simon Oraon and Sanjay Kumar, 1995).

The successful outcome of this incident encouraged the villagers of Kaxitoli to extend their area of protected forest by promoting coppice regeneration in 6 acres of very degraded forest known as Chanjkapari. Despite the fact that there was only rootstock left in Chanjkapari, regeneration has been very successful: a situation that helped to increase the Forest Department's support for the village's forest protection initiative. According to Simon, the Forest Department now comes to him for advice on how to settle forest-related disputes. Indeed, local rumour has it that when Rakia Oraon tried to prosecute Simon for refusing to let him have any wood from Jamtoli forest, the Deputy Commissioner in Ranchi said to him: 'Don't complain to us. As far as I'm concerned Simon Oraon is the Divisional Forest Officer of Jamtoli forest and you would do well to go back there and learn from the work that he is doing' (discussion with Sukra Bhagat, 1993).

The Key to Jamtoli's Success

While Jamtoli would not have been picked out by Shepherd's (1993b) model as a likely candidate for common resource management, its success could be interpreted as providing support for populist emphases on community-based forest protection (D'Abreo, 1982; Fernandes and Kulkarni, 1983; Kulkarni, 1983; Guha, 1983; 1989; Shiva, 1988; Gadgil and Guha, 1995). It also illustrates how, given the right conditions, local forest protection systems can survive not only the privatisation of commons and but also a transition, in terms of increasing population densities and agricultural intensity, to a definite 'area type three' (approaching 'area type four') village.

Although the existence of community-based forest management in Jamtoli prior to 1944 must have been one of the most important factors in the establishment of autonomous forest protection, many of the other pre-conditions were also present. Examples include a strong cultural attachment to forests coupled with a desire to prevent forest loss, a perception of Forest Department over-exploitation of forests, the existence of legal rights to forest produce (the Kattiyan Part II), the presence of a strong leader (Simon Oraon) and a strong traditional decision-making institution (the *Parha*) to mobilise support for forest protection and wider village development (Mehrotra and Kishore, 1990; Kumar, 1991). There has even been an 'uprising against social and economic deprivation of the masses' (Mehrotra and Kishore, 1990 p. 2) in the form of the Jharkhand movement which helped, particularly after the early 1980s, to increase local people's awareness of their customary rights to use forests. Jamtoli also meets many of the more general conditions for the establishment of forest management institutions outlined in by Ostrom (1999): notably a well-defined resource, high levels of common understanding of and dependence on it, a clear set of rules, autonomy from the state, prior organisational experience and an unified user group.

According to Simon Oraon and a number of other village elders, Jamtoli's success in village-based co-operation has more to do with 'the high proportion of non-drinking Tana Bhagat and Christian households' than the high percentage of Scheduled Tribe villagers (interviews with Simon Oraon, Etwa Bhagat and Sukra Bhagat, 1993). Indeed, Simon Oraon puts high levels of alcohol consumption at the top of his list of hindrances to village unity because it 'leads to conflict and jealousy which, if not controlled, make any type of village development impossible' (discussion with Simon Oraon, 1993). As a result, Simon has tried to discourage drinking within his *Parha*, which he claims is, as a result, 'very unified'.

Simon's *Parha* has been very influential in encouraging forest protection within its boundaries. Partly as a result of Jamtoli's demonstration effect, the neighbouring villages of Basananda, Hutar and Saheda started to protect their forests in the late 1960s and the nearby villages of Khurhatoli and Baridih followed suit soon afterwards. Simon Oraon has also set up a tree nursery to supply seedlings to villages where replanting work is necessary and claims to have established forest protection systems in another twenty nine villages within Bero Block.

An important forum for addressing the issue of forest protection is the *Parha Jatra* at which an interesting mix of local politics and wider development issues are discussed. Although Simon became disillusioned with the Jharkhand movement many years ago because of its corruption and lack of unity, he remained broadly sympathetic to its aims and, like Vishvanath Bhagat, has supported its emphasis on the need for local people to claim forests back from outsiders and protect them for their own use (discussion with Simon Oraon, 1993). Unlike Vishvanath, Simon is not a member of any political party as he feels that this could compromise his position as *Parha Raja*, but in his speeches at the *Jatra*, he has often referred to the *Parha*-based development and forest protection work that he and Vishvanath have done together. Both men emphasised that 'like fish out of a pond, local people cannot live without forests and if they cut the jungle, they are cutting away one of their most important resources' (discussion with Vishvanath Bhagat, 1993). On the other hand they felt that the unity to protect the region's forests on a sustained basis would not be achieved until 'people stop caring so much about drinking' (discussion with Simon Oraon, 1993).

In his approach to village development, Simon has always emphasised the need to get to the root of a problem and has argued that successful forest protection, in particular, must be accompanied by what he calls 'village social uplift' to remove the fundamental causes of forest cutting. As much recent development literature points out, environmental degradation is often associated with poverty (Blaikie, 1985; Blaikie and Brookfield, 1987b; World Commission on Environment and Development, 1987; Shepherd, 1992b; World Bank, 1992a; Reardon and Vosti, 1995; DfID, 1997; 2001; Agrawal and Sivaramakrishnan, 2000b). In recognition of these links, Simon has tried to alleviate Jamtoli's underlying poverty by promoting irrigation and agricultural intensification. He has also tried to assist the poorest villagers by helping them to obtain government-owned (*garmajurwar*) land for cultivation and with 'welfare' schemes like the forest fruit collection system mentioned earlier.

Indeed, Simon's success in promoting increased levels of livelihood security, resource distribution and forest conservation is a good illustration of how community-based resource management can help to achieve some of the wider aims of rural development (Berkes and Taghi Farvar, 1989). It is also an important reminder that what many local people seem to want is more, not less development; but development over which they have command, rather than supply side populist or 'anti-development' alternatives which could take away this local control.

Within Jamtoli, the lack of intra-village tension and the widespread recognition of Simon Oraon as an 'expert' in forest management, agriculture and *Parha* leadership are seen as the prime factors behind the success of forest protection and wider village development. Most people are also quick to point out Simon's personal virtues; notably his role as an 'intelligent, honest and trustworthy leader who always tries to help people, no matter what the cost is to himself' (discussion with Buddhu Bhagat, 1993). Local people also respect him for the fact that he really cares about village welfare and development. One man said of him: 'Even now that Simon is an old man, whenever he hears that people are cutting the forest unnecessarily, he rushes to stop them, telling them that if they protect the forest one tree will be sufficient for four people, but if they continue to cut it, every person will have to cut twenty young trees unnecessarily' (discussion with Somra Bhagat, 1993).

The fact that Simon is over sixty years old begs the question of whether forest management in Jamtoli will shift to an 'area type three' open access pattern upon his death. When I raised this issue with the members of the forest protection committee with whom I had been working most closely, they said that even without Simon Oraon's leadership, there was sufficient unity within Jamtoli to maintain village-based forest management (interviews with Ram Prasad Lohra and Budhwa Oraon, 1993). According to Ram Prasad Lohra, everybody in Jamtoli is very aware of how crucial forests are to both everyday subsistence and village life more generally, and would maintain their guard over the forest, 'no matter what happened' (discussion with Ram Prasad Lohra, 1993).

Autonomous Forest Protection in Ambatoli

In contrast to Jamtoli, Ambatoli is typical of most villages in the research area in that its tradition of community-based forest management is much more remote and its internal unity is less pronounced. In addition, the village's 'forest citizens' are significantly less active than those described by India's forest intellectuals and there is no strong leader to help address the poverty that lies at the root of their forest exploitation. To compound matters, forest protection has been hindered by the existence of dominant elites who can afford not to care about the forest plus a group of resource poor villagers who cannot afford not to exploit it.

In spite of these problems and Ambatoli's status as an 'area type three' village with little potential for re-establishing common property resource management, a number of attempts have been made to initiate autonomous forest protection within the village. Although many of these have been unsuccessful, their existence does illustrate how in villages where forests are important for practical and socio-cultural reasons and local people are concerned about forest decline, there may be scope for autonomous forest management if a sensitive and flexible strategy can be devised.

Ambatoli's Attempts to Establish Autonomous Forest Protection

As mentioned in Chapter Ten, after Independence, villagers in Ambatoli, as in Jamtoli, started to encounter difficulties in obtaining forest produce as a result of illegal felling by forest contractors. A major shift in local attitudes occurred when forest contractors started to pay local villagers to help them to exploit Ambatoli forest illegally. Many people showed the classic reaction to resource privatisation by the state (Berkes and Taghi Farvar, 1989; Stevenson, 1991), and started to look upon Ambatoli forest as a government forest that they would be foolish not to try to profit from. One villager's recollection was 'we started thinking that if outsiders were making a profit from our forest, why shouldn't we?' (Panrtoli group interview, 1993).

Before long, however, many people became concerned about how they would meet their subsistence needs if forest decline continued and started to think that they could and should protect Ambatoli forest for the sake of their grandchildren. In the late 1960s, a particularly severe coupe cut by forest contractors stimulated the establishment of a village-wide system of forest protection.

From what villagers could remember, forest protection was organised by a committee made up of representatives from each of the main *tolas*. Ambatoli forest was divided up between the seven *tolas* according to boundaries established during the British period and a patrol rota of representatives from each household was drawn up for each *tola*. The main aim was to keep all outsiders, especially contractors, out of the forest. The committee also decided upon which species and trees should be cut by villagers for timber and fuel wood and whether requests for additional timber should be granted.

Unfortunately, the committee collapsed not long after its establishment as a result of personal conflicts between villagers which were fuelled, according to one villager, by 'too much drinking and quarrelling' (Ambatoli group interview, 1993). As elsewhere in India (Shiva et al, 1983; FAO, 1985; Shepherd, 1985; 1992b; Arnold and Campbell, 1985; Falconer, 1987; Noronha and Spears, 1988), the issue of land tenure also caused concern as many villagers were wary of protecting a state-owned Protected Forest which the Forest Department had the right to cut at any time.

An additional area of conflict stemmed from right-holders in the neighbouring village of Patratoli who had never been allocated their own area within Ambatoli forest proper and had traditionally obtained their timber and fuel wood from 'Patratoli forest' which is situated between Ambatoli and Patratoli. As Patratoli forest is very small, however, protecting it from the Forest Department and taking timber or fuel wood for local use without being caught by the forest guard was very difficult.

In addition, Patratoli forest has traditionally been used to supply Ambatoli with large timber for use at *jatras* and during religious festivals and has been protected for socio-cultural reasons to promote the growth of large trees. This made it even more difficult for the villagers of Patratoli to obtain their timber and fuel wood requirements without coming into conflict with the need to maintain a supply of 'festival timber'. As no concessions were made to set aside an area of Ambatoli forest for the Patratoli villagers to use, they refused to protect it and continued to make themselves unpopular with other right-holders over the exploitation of 'festival timber'.

By the early 1980s, the demands of increasing village populations upon declining forest resources produced a kind of open access situation where the cutting of fuel wood in Ambatoli forest was carried out on a daily basis by right-holders and outsiders alike. In an attempt to control this situation and to force non right-holders to regenerate their own forests, the villagers of Ambatoli started to think about re-establishing village-wide forest protection. To some extent, local people re-examined their attitudes towards forests as a result of the Jharkhand movement's emphasis on villagers' rights to forest produce and its involvement in struggles against forest contractors in Singhbhum District. This encouraged villagers to believe that they 'could manage the forest better than the government' (Panrtoli group interview, 1993), and helped to stimulate an assertion of local forest rights and management systems in opposition to state corruption and commercial forest exploitation.

During the mid-1980s, however, a number of attempts to re-establish autonomous forest protection collapsed as a result of intra-village tensions and conflict between Patratoli and Ambatoli: a situation not helped by the increasing incidence of seasonal migration after the late 1970s. Fuel wood selling was also recognised as a source of tension as well as a major factor in forest decline, but the fact that resource poor households depended on this as a source of income made other villagers uneasy about clamping down on it. Fuel selling by wealthier villagers, on the other hand, has been hard to control because of the forest guard's reluctance to discourage the practice.

As a result, uncontrolled forest exploitation continued and in spite of the tree planting work undertaken in Pokaltikra by social forestry programmes, the effects of deforestation were clearly visible. In addition to the problems of obtaining trees large enough for house construction or plough-making, many villagers also believed that deforestation was making it increasingly difficult to find underground water for well construction.

In the late 1980s, several villagers from Nehalu and Panrtoli hamlets were again stimulated into protecting their forest areas from exploitation by villagers from other *tolas*. One of the main leaders of this initiative was Indru Swansi, a Panchayat member and respected village elder. Indru's interest in forest protection stemmed initially from his attendance, during the 1960s, at a training course in Panchayat administration which had a strong emphasis on forest management (discussion with Indru Swansi, 1995).

In 1990, Indru and several Backward Caste men from Nehalu hamlet got together to discuss what Simon Oraon and Vishvanath Bhagat had said in their speeches at the *Parha Jatra*. Soon afterwards, they started to attend meetings about forest protection held by Simon in Bero (Ram Prasad Sahu, 1993). This gave them the encouragement that they needed to establish a joint Nehalu and Panrtoli forest protection committee. The committee was set up on the basis of one adult male per household and made decisions about the worthiness of villager's requests for large timber and the types of punishment to be meted out (usually a fine) for unauthorised tree felling. In order to guard the forest, all members of both hamlets were expected to 'keep an eye out' for intruders and call for assistance if any were seen.

The demonstration effect of Nehalu's and Panrtoli's initiative, plus the fact that their area became out of bounds to other *tolas*, helped to stimulate the establishment of forest protection in the other five *tolas*. The Ambatoli, Bartoli, Deepatoli and Kaparia committees lasted only a short time due to conflicts over fuel wood selling, 'guard duty' and the rights of Patratoli villagers to use Ambatoli forest. Nehalu's committee also collapsed in 1992 as a result of personal conflicts between members. Panrtoli, however, managed to maintain its unity and the protection of its forest area until the JFM committee was formalised in 1998.

Villagers of the seventh *tola*, Pokaltikra, did not establish a forest protection committee until 1992, but were successful in maintaining their unity until the advent of JFM. The main person behind the *tola* committee's establishment was Jageshwar Ram Manjhi, a fairly well-off Scheduled Caste man who became involved in forest protection after participating in an education programme sponsored by the Xavier Institute of Social Services and held at Simon Oraon's house. Jageshwar's concern about forest decline stemmed mainly from the fact that his wife's job as a teacher (and the high social status that her education and income brought) made it undesirable for her to collect fuel wood with the result that Jageshwar had to go to the forest himself or rely on either his mother or paid labourers to collect it for him.

As a result of Simon's influence, Jageshwar has been strict in his organisation of Pokaltikra's forest protection system. The committee was made up of one adult male per household, although several women from female-headed households were allowed to become committee members. All members were expected to attend the weekly committee meetings or face a fine of five rupees. Every day, six villagers were expected to go out on 'guard duty' to patrol Pokaltikra's 50 acres of forest land and any trespassers that they caught had their axes confiscated and were made to pay a fine of 50 rupees per person.

Despite Pokaltikra's and Panrtoli's success in protecting their areas from outsiders, their efforts were unable to halt the decline of Ambatoli forest as committee members imposed few restrictions upon their own use of the forest. Equally damaging was the lack of restrictions on grazing animals. But the absence of any rules about this coupled with wider intra- and inter-village tensions over the issue of forest protection made the resolution of such problems very difficult. Pokaltikra's forest protection system, for example, made it very unpopular amongst villagers from other *tolas*, who said that the Pokaltikra villagers were only protecting the forest so that they could make more of a profit from fuel wood selling (Bartoli, Ambatoli and Nehalu group interviews, 1993). Villagers from Patratoli, meanwhile, viewed Pokaltikra as a 'thief *tola*' and accused its inhabitants of greed (Patratoli group interview, 1993).

The Reasons for Ambatoli's Failure to Achieve Village-wide Autonomous Forest Management

In contrast to populist discourse about knowledgeable and active forest communities poised for the re-establishment of village-based resource management systems, therefore, field data from Ambatoli hints at a more complex and less promising reality. Indeed, the village's firm 'area type three' status suggests that a move back to village-based forest management ought to be impossible, even with state assistance. For a start, Ambatoli's fast fading tradition of community-based forest management, declining forest area to right-holder ratios and shift towards more intensive agriculture have all helped in undermining its efforts to re-establish village-wide autonomous forest protection. In addition, the absence of a strong leader and a long history of village-, *tola-* and community-based tension have made the development of village-level unity and co-operation especially problematic. To exacerbate matters, Ambatoli's *Mukhia* (who lives in Patratoli) has shown little interest in forest protection; partly because he is afraid that Patratoli's villagers will lose their access to Ambatoli forest.

These factors, plus the absence of wider village development programmes of the type found in Jamtoli, have made it difficult for Ambatoli to overcome the poverty that lies at the root of many villagers' over-exploitation of the forest. Another issue that is perceived to contribute to Ambatoli's lack of unity is the high level of alcohol consumption amongst many of the village's *Sarna* worshippers. According to one villager, *adivasis* 'care only for drinking. They cut the forest and sell the wood to buy liquor and then they get drunk and fight over who is causing forest decline' (discussion with Fakir Oraon, 1993).

But although Shepherd's model is broadly correct in its assessment of Ambatoli's likely success in establishing autonomous forest management, it underestimates the extent of local people's dependence on (and attachment to) forests and the potential of JFM in harnessing this. For in spite of its problems, Ambatoli has a number of characteristics favouring community-based forest management; notably a strong sense of the practical and socio-cultural importance of forests, concern about forest loss and a perception that the Forest Department has seriously neglected both local forests and the rights of local people to use them (Mehrotra and Kishore, 1990).

As a result, there are grounds for hope that certain 'area type three' villages (Shepherd, 1993a; 1993b) in areas where forests have an important role in village culture (and where local people want greater control over them) can be nudged towards successful village-based resource management with appropriate and flexible state intervention. Indeed, Shepherd herself has pointed to the persistence of local management systems in small areas of forest that have particular socio-cultural significance to local communities (Shepherd, 1993a). These issues will be discussed more fully below with reference to with Ambatoli's efforts to establish JFM and Jamtoli's attempts to build it into the existing forest management system.

The Shift to Joint Forest Management in India

During the late 1980s, there was a definite India-wide shift in attitudes towards forest management when local people started to be seen as part of the solution to rather than the main cause of forest decline (Ghate, 1992; Palit, 1993). At the same time, there was growing recognition that the imposition of scientific forestry had often provoked local people into asserting their silvicultural knowledges and forest rights as means of either destroying commercially-valuable forests or protecting local forests from state management (Guha, 1989; Kelkar and Nathan, 1991).

The need for a more sensitive approach to forest management was taken on board in the 1988 National Forest Policy and consolidated by specific resolutions on JFM (Government of India, 1990; 2000) which acknowledged the importance of forests for local subsistence and made recommendations for involving local people in forest management. While accepting local people's rights to forest produce, the state made clear its expectation that in return, right-holders would help to protect the forest areas from which they derived benefits (Indurkar, 1992). Indeed, a more cynical interpretation of the Indian state's interest in JFM could point to its potential for overcoming existing conflicts over forest management by promoting forest conservation and satisfying local requirements for forest produce whilst at the same time freeing up the state's best forests for commercial use. By relying heavily on local people's time, effort and institutional capital, it also promises to be cheap to run.

The Bihar Joint Forest Management Programme

As mentioned in Chapter Four, the Bihar JFM programme has claimed to be revolutionary in its provisions for returning 'almost 100 per cent of the benefits of forest protection to local people' (interview with Sanjay Kumar, 1993). Although later interpretations of Bihar's Resolution no. 5244 on joint forest management envisaged a 'sharing of benefit among the people, the government and the village on a 1:1:1 basis' (Society for Promotion of Wastelands Development and Forest and Environment Department, Bihar, 1994, p. 16), the financial rewards to participating households could still, in theory, amount to a doubling of their disposable incomes (Corbridge and Jewitt, 1997).

In addition, the Bihar Forest Department seemed (initially, at least) to realise that unless the poverty that forces local people to over-exploit forests in the first place could be alleviated, it would be difficult to sustain JFM in the long-term. In an attempt to create 'employment opportunities for rural poor' and promote the 'overall development of villages' (ibid., p. 9), therefore, the Forest Department worked closely with the World Food Programme. The main outcome of this was an emphasis on JFM schemes that also aimed to improve local infrastructure (roads, drainage, check dams and irrigation canals), village facilities (potable water, education and medical facilities) and local income generating potential (tree, medicinal plant and oilseed plantations).

Although local participation in forest management may not, in principle, have been 'a new concept for Bihar' (ibid., p. 13), the reality has been rather different. Indeed, it has taken major financial incentives (potentially for both parties) plus a significantly more bottom-up and sensitive approach to even start to change the attitudes of local people and Forest Department staff towards one another. But there is some evidence that this has borne fruit. A number of articles published in the 'Workshop on Joint Forest Management in Bihar' (Society for Promotion of Wastelands Development and Forest and Environment Department, Bihar, 1994), for example, emphasised the importance of allowing local people to exploit forest products for subsistence purposes and as a source of cash: a significant departure from the restrictions on local forest use envisioned in the Draft Forest Bill of 1981. This document also indicated a shift away from previous forestry practices with its emphasis on the need to utilise indigenous silvicultural knowledges for the development of forest working plans.

In practical terms, the Bihar JFM programme has made it necessary for both the Forest Department and local people to carry out specific duties and to shoulder certain responsibilities. As stated in the 1988 National Forest Policy, villagers are expected to help the Forest Department to protect local forests from grazing, fire and illicit felling. The role of local people in forest management is much greater under JFM, however, as in addition to undertaking manual forest work, village forest protection and management committees (VFPMCs) are also supposed to help prepare forest management plans and determine how access to forest produce will be regulated. In return for these efforts, members of the VFPMCs are allowed to take forest produce according to their needs and should expect to receive forest produce 'free of cost at the time of thinning and final harvest' (ibid., p. 13).

The responsibilities of the Forest Department, meanwhile, are to 'provide administrative and technical guidance ... assist and train committee members' (ibid., p. 26), participate in the formation of forest management plans, record the minutes for committee meetings, maintain a record of expenses, take action to punish forest offenders and dissolve VFPMCs if their actions are 'found to be detrimental to protection and management of the forest' (ibid., p. 26). In return for these administrative and management duties, the Forest Department takes a set amount of timber as a royalty. The exact amount of this is fixed by the Commissioner and Chief Conservator of Forests, but may vary between the different forest divisions.

In Ranchi East Division, 20 per cent of the final timber and bamboo harvest will be taken by the Forest Department and the villagers will be entitled to take the other 80 per cent for their own use or for sale. The Forest Department's royalty will not always be taken in the form of cut timber, however. In coppice-with-standards management plans, the Forest Department is likely to request VFPMCs to leave 20 per cent of the large trees as 'standards' that will eventually provide good quality timber but will in the meantime be useful for regeneration from seed and soil conservation purposes (interview with Sanjay Kumar, 1995).

To date, only Protected Forests are eligible for JFM and the Bihar JFM Resolution is aimed primarily at improving the quality of the more degraded right-burdened forests in this category. The Resolution does not apply to right-burdened Reserved Forests, although where these forests are becoming degraded, it is suggested that 'action by local officials can be initiated for forming VFPMCs and the government's approval may be obtained subsequently' (Society for Promotion of Wastelands Development and Forest and Environment Department, Bihar, 1994, p. 16). This has not happened to date, but the situation may change in view of the attention given to the possibility of JFM in 'good forests' by the most recent central government resolution on JFM (Government of India, 2000).

When Bihar's JFM programme was first initiated, it was projected to embrace around 5000 villages and bring around 800,000 hectares of degraded forest 'under protection and rehabilitation' (Society for Promotion of Wastelands Development and Forest and Environment Department, Bihar, 1994, p. 27). The intention was to regenerate 85,000 hectares of forest land with viable root stock, plant 85,000 hectares and improve the remaining 630,000 hectares through 'protection and management' (ibid., p. 27).

For a region dominated by 'area type three' villages (Shepherd, 1993a; 1993b), this was obviously a very ambitious target, so to increase the chances of success in the early stages of JFM implementation, the Forest Department established certain village selection criteria. The most desirable villages were those that had relatively small, ethnically homogeneous populations that were eager to participate in JFM and which also had relatively good quality forests that did not require a long lag time before they could start to generate an income (interview with Sanjay Kumar, 1993). Interestingly, little attention was paid to the socio-cultural importance of forests and how this might influence villagers' willingness to implement and sustain JFM in the long-term.

Before attempting to formalise VFPMCs, the Forest Department tried to make people aware of what JFM involved by bringing existing VFPMC members to the target village or taking villagers to see a working JFM area (Society for Promotion of Wastelands Development and Forest and Environment Department, Bihar, 1994). This was sometimes followed by a visit from a 'JFM promotion team' and/or a local NGO to help the villagers to form a VFPMC and an executive committee.

On certain occasions, the Forest Department has allowed an NGO member to join the executive committee if it was thought that this would assist the initiation of JFM in a target village. The standard executive committee model (current and former *Mukhia*, *Sarpanch*, *Pahan*, school teacher, Forester, four Scheduled Castes or Tribes and three to five women) outlined in Chapter Four has also occasionally been adjusted to include more women in villages with high levels of female interest or involvement in forest protection (ibid., 1994).

One of the main practical difficulties encountered during attempts to establish JFM committees has been the failure of the required 50 per cent of village right-holders to attend the general body meeting that must be held before a VFPMC can be established. Other problems have included the development of vested interest groups (especially in economically or ethnically heterogeneous villages), a lack of interest in JFM by the forest guard (or resistance to sharing authority with local people) and the long lead time between the establishment of JFM and the receipt of financial rewards (ibid., 1994; Mukherji, 1998).

Indeed, the issue of delayed gratification presents a major hurdle for people who have problems in meeting their subsistence needs (Corbridge and Jewitt, 1997). In villages with pronounced wealth inequalities and inter-community tensions, the temptation to suddenly fell trees in the protected area to meet short term cash needs is likely to be very high. This is especially true in view of the substantial long-term commitments of time and energy that villagers are expected to make coupled with the constant threat of free-riding by others.

To reduce the likelihood of brute economic realities undermining local people's interest in JFM, the Bihar program placed a lot of initial emphasis on employment provision, the sale of NTFPs and wider development initiatives. Indeed, the programme's early efforts to address the poverty at the root of forest decline through wider income generating projects and its attempts to use indigenous knowledges in the development of forest management plans looked like a perfect example of appropriate development. It also seemed to have potential for overcoming traditional tensions between local people and forest officials as well as for helping the state out of its dilemma over forest exploitation versus forest conservation. A few years later on, however, the situation seems less positive as only 123 VFPMCs had been formally recognised in the Ranchi East Division by mid 1999, the Bihar Forest Department's enthusiasm for JFM has shown signs of dwindling: a situation that bodes ill for the programme's long-term success.

The Establishment of JFM in the Research Area

Although none of the villages that I visited during the course of my main period of fieldwork in 1993 had actually obtained formal recognition of their VFPMCs, the programme had generated great initial excitement. Indeed, Simon Oraon was so optimistic about the possibility that JFM could, if implemented properly, play a major role in rebuilding Jharkhand's forests that he tried to encourage all non-protecting villages to overcome 'the intra-village conflicts preventing them from making the most of the programme' (discussion with Simon Oraon, 1993). As a result of his enthusiasm, a number of villages that were actively protecting their forests but had no official status as JFM villages formed VFPMCs in preparation for formalisation by the Forest Department. One such case was Jamtoli which, as a Reserved Forest, did not come under the initial remit of Bihar's JFM Resolution.

JFM in Jamtoli

Apart from the status of its forest, Jamtoli met all of the JFM 'target village' criteria with its high degree of unity and socio-economic homogeneity, long-standing interest in forest decline and well-stocked forest which could provide immediate financial benefits. Local forest officers therefore capitalised on these advantages and encouraged the Jamtoli villagers to form a VFPMC in the hopes that government approval would be given for its formal recognition under the Bihar JFM Resolution.

To all intents and purposes, Jamtoli was treated just like any other 'target village' where the establishment of JFM was being attempted. In 1991, local forest staff from Bero went to talk to the villagers about the Bihar JFM scheme and persuaded them to re-arrange their existing autonomous forest protection committee structure to conform with the regulations set out in the Bihar JFM Resolution. The formation of an eighteen-member VFPMC caused surprisingly few difficulties and was particularly welcomed by villagers in Jamtoli and Bertoli hamlets as their existing committees were starting to show signs of stress as a result of persistent thieving by villagers from outside.

Jamtoli's VFPMC is made up of fifteen Scheduled Tribe members and three Backward Castes and includes the *Mukhia*, *Sarpanch*, *Pahan* and school teacher, the local Forester and five Scheduled Tribe women. Committee meetings are held once per month and are attended by the executive committee plus the forest guard who is a 'specially invited member'. Unlike the ex-forest guard in Ambatoli, Jamtoli's guard is regarded as an 'honest man' (Jamtoli group interview, 1995). He helps the villagers to protect their forest from illegal felling and does not appear threatened by JFM, as has often been the case elsewhere (Shepherd, 1992b).

When I asked Simon Oraon whether the superimposition of JFM onto his existing system of forest management had caused any problems, he said that he was pleased with how smoothly the new system had been implemented and accepted (discussion with Simon Oraon, 1995). The amalgamation of the three *tola*-based forest protection committees into one VFPMC went unopposed as the main leaders of these committees have all been elected onto the executive committee and the other ex-committee members are entitled to a say in the new system of forest management courtesy of their 'general body' membership.

One thing that the Forest Department (quite sensibly) has not tried to change, is the annual coupe system which has proved successful over a number of years in providing villagers with their yearly allocation of fuel wood and timber. Interestingly, and in spite of the fact that the Bihar JFM programme does not recognise the *tola*-based division of Jamtoli forest (and would therefore support any claims made by the villagers of Jamtoli and Bertoli hamlets to take wood from the better-protected Letteguttu forest), local forest officers have not tried to amalgamate the areas protected by each *tola*.

To help to compensate villagers in Jamtoli and Bertoli for their lack of mature forest, the Forest Department gave them a lot of personal and financial attention at the start of the project. In the summer of 1993, Forest Department 'afforestation funds' were used to provide seedlings and to pay villagers 25 rupees per day to undertake weed clearing, tree planting and forest maintenance work: employment that was welcomed by the village's poorer inhabitants. In total, around 170,000 trees were planted in the more degraded parts of Bertoli's and Jamtoli's forest areas. To prevent cattle from getting into the forest and damaging the young trees, the villagers made a trench around this area.

The species chosen for planting by the VFPMC consisted mainly of 'double crop' species such as mango, *jamun*, *karanj*, *mahua* and *ber*. To provide for longer term timber requirements, a few teak and *sisam* trees were also planted. In addition, the Forest Department encouraged the VFPMC to plant fast growing eucalyptus, acacia and *gamhar* trees to minimise the time before villagers could benefit financially from JFM. In addition to increasing support for the scheme, this helped to reinforce earlier efforts to provide short term (but well paid) employment opportunities for the most needy villagers.

To maximise its forest protection and regeneration efforts, Jamtoli's VFPMC selected a guard from each *tola* to patrol the forest, control cattle grazing and supervise the village's tree nursery. These efforts were supported by the official forest guard who continues to work closely with the *tola*-based guards and inform them if he finds evidence of illegal felling or loose cattle. In addition, the villagers themselves help to guard the forest on a rota basis with four of five people (men and women) making daily patrols. Any offenders that are caught have their axes confiscated and are made to pay a fine, the amount of which is decided by the executive committee.

Given the past history of distrust between local people and the Forest Department, I was surprised at villagers' lack of suspicion that JFM might be a trick to get them to protect forests that would be brought back under state control as soon as they became productive. The issue of insecure tenure is one that dominated much of the early social forestry literature (Shiva et al, 1983; FAO, 1985; Shepherd, 1985; Arnold and Campbell, 1986) but is noticeable by its absence in much of the official literature on JFM. Yet in effect, the payment of a royalty to the Forest Department underlines the fact that the forests managed under JFM programmes remain under state ownership.

Most villagers have been reassured by the fact that a document specifying the procedures for the distribution of forest produce between the villagers and the Forest Department must be signed by both parties at the time of VFPMC formalisation. But despite initial hopes to extend the JFM programme to include small Reserved Forests, Jamtoli's VFPMC still has not been formalised. Nevertheless, Simon Oraon remains unconcerned that the Forest Department might try to reclaim the trees protected by local people and feels sure that the JFM programme is 'genuine' as the Forest Department has made great efforts to provide trees and employment for Jamtoli's villagers. If any conflict does arise, however, he is certain that the villagers of Jamtoli will 'stand up and fight against the Forest Department, just as they have done in the past' (discussion with Simon Oraon, 1995).

Nevertheless, many villagers have been keen to hedge their bets by planting trees on land that the Forest Department (and, better still, the state) has no jurisdiction over. Tree planting on individually-owned land has become especially popular over the last twenty years as villagers have come to realise the profits that can be made from planting the free tree seedlings provided by the Forest Department under 'farm forestry' and (sometimes) JFM programmes. When I visited Jamtoli in April 1995, for example, I found Simon Oraon holding a meeting about the financial and ecological advantages of growing fruit and fast-growing timber trees on individually-owned poor quality *tanr* land that is hard to cultivate. To see whether local people were more willing to grow eucalyptus on private land than in the forest, I asked the group if they would consider growing the species if there were enough local demand (such as a paper mill) for it. Everybody was adamant that they would do so as 'eucalyptus is fast growing and coppices almost as well as *sal*' (Kaxitoli group discussion, 1995). It is also profitable, with ten year old poles fetching 200 rupees each at the timber yard.[4]

Indeed, over the last five years, this emphasis on tree planting on privately owned land seems to have displaced the villagers' interest in the young trees planted at the start of the JFM programme; 50 per cent of which had become choked by weeds in 1996. Since then, villagers have actually cleared some of the weaker saplings on the lower slopes to make more room for cultivation: a situation that reflects their sense of disillusionment with the Forest Department's failure to officially formalise Jamtoli's VFPMC.

JFM in Ambatoli

Ambatoli's experience of JFM has been rather different as although VFPMC formalisation took a long time, most villagers remain very optimistic about the programme's achievements to date and its potential for long-term success. The process started as early as May 1993 when several villagers attended meetings on JFM held by the Forest Department in Bero. In the enthusiasm that followed the discussion of the programme at the *Parha Jatra* in June of that year, there was hope that JFM would be implemented quite quickly. Vishvanath Bhagat was approached to assist right-holders in Ambatoli and Patratoli to overcome their lack of unity and establish a VFPMC. The main reason that he was chosen was that he had previously helped to prosecute a group of outsiders who were caught felling *sal* trees in Patratoli forest.

By the end of June 1993, most villagers had a clear idea of what JFM involved and how the membership of the executive committee would be organised. As the revenue village (Ambatoli plus Patratoli) was the unit around which the VFPMC would be formed, it was decided that to minimise intra- and inter-village conflict, one person from each of Ambatoli's seven *tolas* should be members of the executive committee and the *Mukhia* would act as Patratoli's representative. The villagers had also decided which women they would elect as executive committee members.

By the end of December, however, little further progress had been made. In Patratoli, a number of people were suspicious that Ambatoli villagers would take all of the benefits of forest protection, yet the *Mukhia*, who had attended JFM meetings and knew that Patratoli's right-holders would get the same shares as Ambatoli's right-holders, seemed unable (or unwilling) to reassure them. Help from outside was also not very forthcoming and at least two meetings between village leaders, Vishvanath Bhagat and the Forest Department were cancelled between June and December 1993.

Furthermore, divisions between villagers who were very keen on forest protection and those 'willing to think only of today and too busy drinking to be bothered about making plans for the future' (discussion with Bineswar Thakur, 1995) acted as a major hindrance to the establishment of a village-wide consensus regarding JFM. An additional problem was the absence of many of Ambatoli's poorest and most forest-dependent villagers (who were undertaking seasonal migration) when forest-protection meetings were held: a situation that tended to reinforce the say of better-off villagers.

By the end of 1994, little had changed in this regard as Vishvanath Bhagat was concentrating hard on his campaign for the Bihar State elections and did not get around to visiting Ambatoli. A number of villagers also felt let down by Forest Department staff who (they say) never bothered to come to Ambatoli to talk to them about JFM (interviews with Bineswar Thakur, Indru Swansi, Ram Prasad and Kristo Sahu, 1994). While villagers accepted that Vishvanath Bhagat was very busy with his political work, they were angry about the failure of Forest Department staff to visit Ambatoli because 'they get their petrol for free, so why can't they make the effort to come here, talk to us and help us to form a forest protection committee?' (discussion with Ram Prasad Sahu, 1995).

From the Forest Department's point of view, Ambatoli was not viewed as a high priority 'target village' for early JFM implementation because of its large size, non-unified (and heterogeneous) population and the limited scope of its forest to provide immediate financial benefits to the VFPMC (interview with Sanjay Kumar, 1995). Nevertheless, Anand Jha claims to have visited the village several times during 1993 because he knew that a number of villagers were very keen to establish a VFPMC. He also pointed out that when Forest Department staff tried to hold a 'general body' meeting in Ambatoli in early 1994, considerably less than 50 per cent of the village right-holders attended, making it impossible to move on to the next stage of VFPMC formation (interview with Anand Jha, 1995).

Obviously, there were two sides to this story, but there were certainly a lot of missed opportunities due to a lack of communication between local people and the Forest Department. When I talked to Anand Jha in detail about the practicalities of JFM establishment in Ambatoli, it became clear that the scheme could be tailored to benefit villagers from Patratoli in a way that might put an end to their conflict with Ambatoli villagers over the issue of forest protection. For a start, he pointed out that the areas of Ambatoli forest traditionally protected by individual *tolas* have no legal standing as far as the Forest Department is concerned. Moreover, Patratoli's right-holders have the same rights in both Ambatoli and Patratoli forests as do Ambatoli's right-holders.

In addition, Anand Jha suggested that if Patratoli villagers wanted a nearby source of fuel wood but did not want any conflict over the socio-cultural significance of Patratoli forest, JFM would enable them to build up their own woodland on the village's common land. The Forest Department could provide free tree seedlings (chosen by the villagers) and would pay villagers 25 rupees per day to plant them. For villagers who were still wary about the land tenure implications of planting trees on government land, meanwhile, free seedlings and some financial assistance were available for 'farm forestry' tree planting on individually-owned land (interview with Anand Jha, 1995).

When I returned to the village in April 1995, most villagers in Ambatoli had overcome their concerns about JFM as they had heard about the formal agreement between the Forest Department and a formalised VFPMC in the neighbouring village of Hatu. Nevertheless, there were still a number of communication problems between villagers and Forest Department staff that prevented the *tola*-based committees, in particular, from making the most of their forest protection initiatives.

In 1993, for example, Anand Jha had told me that the *tola*-based forest protection committees in Pokaltikra and Panrtoli were entitled both to financial priority over non-protecting *tolas* and to assistance from the Forest Department in drawing up forest management plans. But as a result of the forest guard's failure to pass on this information, both *tolas* were trying to struggle on in the face of opposition from non-protecting hamlets. The forest guard's concern about losing bribes from fuel wood sellers was a major factor in his unwillingness to encourage village-wide forest protection.

In spite of such problems, most villagers retained a great deal of enthusiasm for JFM and remained confident that a VFPMC would soon be established. One of the most influential people in nurturing this interest was the resident KRIBHCO 'village motivator', Ms. Mary Tete, an Oraon woman who has been involved in the Rainfed Farming Project since its initiation and has earned much respect amongst the Ambatoli villagers since moving there in 1994. Using her popularity (and bargaining power) within the village to good effect, Mary was able to appeal to villagers of all community groups and has been very skilful in encouraging them to air their views. She has also been very successful in promoting income-earning schemes for resource poor households (including a tassar silk worm rearing scheme) and encouraging villagers to work together on community-based projects. One of the most active of these is the village youth organisation which played a pivotal role (with Mary's help) in uniting villagers to re-establish a village-wide forest protection initiative. Taking advantage of KRIBHCO's earlier efforts to distribute free tree seedlings to villagers, the youth group took over the village's tree nursery, enlarged it and has since taken up tree planting on private land. Its members also organised meetings to discuss forest protection issues and proposed a ban on fuel wood selling and the cutting of green wood. These meetings proved very successful in re-kindling villagers enthusiasm for JFM and promoting sufficient unity for them to call in the Forest Department to hold a 'general body' meeting attended by over 50 per cent of right-holders.

Ambatoli's VFPMC was actually formalised in 1998 and when I last visited in March 1999, it appeared to be working well. Committee meetings are held weekly and general body members are expected to donate a 'sub' of 2 rupees which is placed in a community bank to fund village development initiatives such as soil conservation measures and the purchase of *Terminalia arjuna* trees for silk worm rearing. Fuel wood selling and the felling of green trees has been banned and villagers now rely on the collection of dry branches, leaves and weed bushes for their fuel. Cattle grazing has also been restricted to wasteland areas or paddy fields and during the ploughing season, most are stall fed on rice straw, household leftovers and the leaves and twigs of *piyar, asan, bahera, putri* (*Combretum decandrum*) and *footkal* trees.

Although regeneration in Ambatoli forest is clearly visible and villagers are enthusiastic about the wider village development benefits of JFM, the scheme's impact on fuel wood gatherers has been significant. Instead of spending a couple of hours in the forest cutting easily available saplings and *sal* coppice shoots, the ban on felling standing trees forces fuel collectors to spend much longer searching for dry branches or weed species such as *putus* or *tetair*. It has also had a major impact on the poorest households who previously used to make an income from selling fuel wood at market and who now have to search for alternative employment or undertake seasonal migration.

Conclusion

Although much populist discourse about forest-dependent communities fails to recognise the difficulties they must face in overcoming intra-village tensions and accepting the time, effort and institutional costs of community-based resource management, it has been important in drawing attention to the tremendous practical and socio-cultural importance of forests. Despite Ambatoli's income inequalities, lack of homogeneity and problems in maintaining unity, for example, villagers' concern about the implications of forest decline for local livelihoods was sufficient to stimulate the desire (if not the reality) of village-wide autonomous forest protection. The centrality of forests to *Sarna* religion and villagers' belief in forest-dwelling *bhuts*, meanwhile, helped to provide a solid enough foundation for the establishment of incentive-led forest management.

Indeed, Ambatoli's success in adopting JFM demonstrates that with sufficient motivation (stimulated, in this case, by Mary Tete and the village youth group), it is sometimes possible to establish participatory forest management in a relatively unfavourable 'area type three' village (Shepherd, 1993a; 1993b). Jamtoli's experience, meanwhile, shows that even villages with apparently little potential for being 'rolled back' to 'stage two' can, with the right leadership, decision-making institutions and high levels of intra-village co-operation, sometimes establish successful autonomous forest management systems. To ignore such villages because they don't display a range of broad criteria identified in the common property resource management literature would seem to be a significant oversight when one considers the number of resource-poor villagers whose livelihoods are at stake.

Nevertheless, responsibility for the success of JFM lies less heavily with local villagers than it does with the Forest Department's ability to supply the necessary 'helpful state intervention' (Shepherd, 1993a; 1993b). Initially, it seemed as though the Bihar JFM programme was going to be a great success as villager's enthusiasm about its provisions for returning 80 per cent of the benefits of forest protection to local people meant that forest officers were swamped by requests to formalise VFPMCs. Instead of taking advantage of local communities' positive response to the demonstration effect (and declining areas of forest available to non-protecting villagers) associated with JFM, however, the Bihar Forest Department has largely failed to capitalise on the programme's excellent start.

One of the main problems appears to be a general loss of interest in JFM 'at the top' (discussion with Sanjay Kumar, 2000); possibly due to the uncertainty prior to and following the formation of a separate Jharkhand State. Whatever the reason, the effect has been to dampen many forest officers' initial enthusiasm for securing villagers' participation by creating employment and income-generating opportunities. Consequently, many of the plans to reduce the scheme's financial lead time (through the cultivation of fast-growing trees like eucalyptus and acacia and (potentially) valuable grasses) seem to have fallen through. VFPMCs have also been disappointed with the Forest Department's failure to launch promised village development programmes (discussion with Sanjay Kumar, 2000).

The programme's initial sensitivity to local situations and its willingness to adopt a fairly bottom-up and flexible approach when superimposing JFM on to existing forest management systems, meanwhile, seem to have been downgraded into a less imaginative and essentially blueprint-based 'management package' (Menon, 1994). As a result, villages like Jamtoli, which have shown themselves eager to take on board the responsibilities demanded by the programme, have been left high and dry by the Forest Department's subsequent neglect with regard to formalising their VFPMC.

Although many people in the research area are still very interested in JFM, the rate of VFPMC formation has slowed down dramatically. Indeed, the Forest Department's village selection procedure seems to have given way to a situation where VFPMCs are formalised only when villagers who have already 'voluntarily protected a forest patch come forward to meet (and pester) the Range or Divisional Forest Officer' (discussion with Sanjay Kumar, 2000).

One of the major difficulties with JFM is that it demands an important leap of faith by local people plus a series of significant short and medium-term sacrifices in return for (possible) long-term financial gain. Ambatoli's villagers, for example, succeeded in initiating JFM despite their deeply entrenched opposition to the Forest Department and their suspicion that local forest officers were uninterested in formalising their VFPMC. In restricting their own use of Ambatoli forest and denying themselves income-earning opportunities from it, they have made much greater sacrifices than the Forest Department to ensure that JFM works.

But as intra-village co-operation and enthusiasm for the scheme cannot be taken for granted in resource poor villages where many households go hungry in the pre-harvest period, the Forest Department's failure to support existing VFPMCs and help them to develop income-generating schemes may have a dramatic effect. While Jamtoli, with its long history of forest protection, high degree of food security and wider village co-operation, will probably continue to maintain some form of forest protection, less unified villages may, in the absence of alternative financial options, be unable to resist the temptation to fell their growing tree crop.

For the sake of villagers' livelihoods, forest sustainability and the maintenance of co-operation between local people, therefore, it is imperative that Bihar's JFM programme is not allowed to 'fizzle out' due to a lack of interest in it by the upper echelons of the Forest Department. Even if no more VFPMCs are formalised, efforts must be made to maintain support for existing JFM committees if the Forest Department is to avoid alienating local communities and wasting the financial resources that it has already invested in JFM.

In some cases, individual forest officers have used their influence to maximise the scheme's potential by promoting more flexible and locally appropriate JFM initiatives. This was especially so in the early days of the programme when Forest Department staff were keen to encourage key target villages like Jamtoli to adopt JFM. Another good example is the formalisation of the Maheshpur VFPMC where the maximum of five female executive committee members was waived to accommodate the pre-existing all-female autonomous forest protection committee. More recently, a Conservator of Forests, Mr.

Dhirendra Kumar, has made significant efforts to promote and sanction JFM forest micro plans throughout the North Chotanagpur Commisionary area. In the village of Nagri near Ranchi, meanwhile, forest officers have made important steps in maintaining villagers interest in JFM by sharing the benefits of an early *sal* crop thinning with VFPMC members (Kumar and Rangan, 1996).

Unless the new Jharkhand Forest Department undertakes further initiatives of this type in the near future, however, local communities themselves may well lose interest in JFM. And if that happens, its potential for promoting sustainable forest management, community empowerment and wider village development will be lost, along with many of the gains that have been made with the recent shift towards participatory forest management. Perhaps it is time for a thorough review of JFM by India's forest intellectuals, but with a view to providing a set of practical, locally-oriented and socio-politically sensitive guidelines rather than a solely theoretical supply side populist alternative.

Notes:

[1] As Shepherd's first model assumes an absence of external intervention 'in the form of state level land tenure rules, forest reserves, land-use regulations etc.' (Shepherd, 1993a), it is more appropriate to engage with the second model (Shepherd 1993a; 1993b) which is meant as a guide to the area types in which participatory forest management is most likely to be successful.

[2] Significantly, Tiwary's comparative study of 'ecological institutions' in Bihar and West Bengal indicates that forest management institutions are more successful when they take on board wider village development issues (Tiwary, 2001).

[3] Simon told me that some months after the establishment of the forest protection committee, a compromise was agreed with the Divisional Forest Officer whereby the committee would continue to protect the forest and co-operate with legal coupe cutting (in the 409 acres of non-protected forest) if villagers could claim 50 per cent of the timber from any trees that they planted in this area.

[4] Most villagers believed there to be little likelihood that eucalyptus grown as a cash crop would displace food crops as it has done elsewhere in India (Shiva et al, 1983; Foley and Bernard, 1985), because even if there was a good market for eucalyptus wood, few local people have enough food security to risk planting any of their good quality land to trees. Nor would eucalyptus replace local fruit trees or multipurpose trees such as *sal*, as its timber is not as strong and *sal* is 'much more useful to local people than eucalyptus' (interview with Simon Oraon, 1995). *Sal* is also faster growing, more 'gregarious' (it grows at a density of 4000 stems per acre compared to the norm of 1000 stems in plantations) and more profitable than eucalyptus, with ten year old poles fetching around 300 rupees each (interview with Sanjay Kumar, 1993).

12 Gender, Silvicultural Knowledge and Forest Management

Introduction

As mentioned in Chapter Four, forest use and management have come to be regarded in the eco-feminist/WED literature as predominantly female tasks as it is usually women who are responsible for fuel wood collection in the developing world (Dankelman and Davidson, 1988; Momsen and Townshend, 1987; Henshall Momsen, 1991; Sontheimer, 1991, World Resources Institute, 1994; Sittirak, 1998). In particular, WED accounts of women as environmental custodians with well developed environmental knowledge systems have had much appeal for donor agencies working on gender- and environmentally-oriented development projects. Recent interest in participatory initiatives has also helped to encourage women's involvement in forest and other environmental management schemes.

Although eco-feminist/WED discourse has been tremendously valuable in challenging the more traditional male-oriented nature of development projects, it has also attracted criticism for its overly-simplistic, undifferentiated approach (Agarwal, 1992; 1994a; 1994b; Jackson, 1993a; 1993b; Locke, 1999). In particular, there has been concern about its failure to distinguish between environmental use and management and its tendency to gloss over the wider social relations that influence women's access to and control over natural resources (Braidotti et al, 1994; Jackson, 1993a; 1993b; Agarwal, 1994a; 1994b; Joekes et al, 1994). To date, however, relatively few in-depth empirical studies have been carried out on the socio-political embeddedness of gender relations and how this affects the ability of men and women from different socio-economic groups to develop and articulate environmental knowledge.[1]

In many ways, Jharkhand, with its relatively high levels of gender equality, is an ideal place to test the validity of eco-feminist/WED claims about women's role as 'primary environmental carers'. Indeed, as a result of women's greater autonomy in hill and tribal areas where they 'still maintain a reciprocal link with nature's resources ... that stems from a given organization of production, reproduction, and distribution, including a given gender division of labour' (Agarwal, 1992, p. 150), one might expect Jharkhandi women to have particularly detailed environmental knowledges. One might also predict that they would have developed careful strategies for managing the environmental resources that they have primary responsibility for using.

According to Kelkar and Nathan (1991), however, forest use in Jharkhand is shared by men and women to a much greater extent than certain (notably Vandana Shiva's) radical eco-feminist/WED accounts allow for. They also argue that silvicultural knowledges are less 'female dominated' than Shiva's (1988) descriptions of forest management in northern India, and maintain that in Jharkhand, forest management has traditionally been much more of a male than a female activity (Kelkar and Nathan, 1991). In an attempt to clarify this situation with reference to local level data, the following four sections will examine the ways in which silvicultural knowledges, forest use and forest management vary according to gender, community, education, class, 'economic value' and seniority in the fieldwork villages.

Gender Divisions of Forest Labour

As in Jharkhandi agriculture, there are some fairly well-defined gender divisions of forest labour amongst *adivasis* in the research area (Jewitt, 2000a; 2000b; Jewitt and Kumar, 2000). At first glance, these appear to confirm WED accounts of women undertaking fuel wood collection and other forest gathering work while men have primary responsibility for the more economically valuable forest resources; notably timber. Certainly, there is a generally held perception within Jharkhand that the collection of fuel wood and NTFPs is women's work (Centre for Women's Development Studies, 1991; Kakat, 1993). Most gathering is carried out in all-female groups (never alone for fear of *otangas*) and women often help one another to lift and balance their headload bundles. During fuel wood collection trips, women frequently bring back a few *sal* leaves, 'tooth brush' twigs (*datun*), fruit, seeds or mushrooms: items typically associated with female gathering work (see plate 12.1). Many women also collect *ranu* (herbs used to ferment rice beer) and those who are knowledgeable about herbal medicine sometimes bring back plants such as *chiraita* (*Swertia chirayita*) or *sataur* (*Asparagus recemosus*) which are used to cure fever and stomach upsets.

The cutting and collection of large timber, meanwhile, are tasks that are generally carried out by men although this appears to be for strength-related rather than socio-cultural or economic reasons as most of the timber cut is used domestically. Moreover, there is nothing to stop women from helping out with such work if they so choose (or if labour is short). The main exception to this is when the timber is to be used to make the stage (*madwa*) where marriage rituals are performed and must, for socio-cultural reasons, be cut by men. Hunting is another activity which is taboo to women although the existence of the ritual *adivasi jani sikar* (women's hunt) every twelve years is thought to be a

Plate 12.1   The collection of *sal* leaves and twigs

remnant from an earlier time when gender equality was greater and women participated freely in hunting activities (Roy, 1915; Kelkar and Nathan, 1991).

Men from all community groups spend quite significant amounts of time in forests trying to shoot birds or rabbits with bows and arrows. During these trips, many men also collect medicinal herbs and *ranu* (see plate 12.2). Reflecting the tendency for work involving sustained periods away from the home to be carried out primarily by men, cattle grazing is a fairly male-dominated activity although young (pre-teenage) girls often share such tasks with older male relatives. The collection of animal fodder, meanwhile, tends to be carried out by men when large branches are being sawn off a tree and by women if leaves or small branches are being gathered.

In spite of these broad gender specialisations, it is important to recognise the substantial flexibility (and interesting anomalies) that exist in forest-based task allocation as well as to acknowledge the ways in which existing divisions of labour are challenged and re-negotiated as intra-household power relations change. In contrast to the cultural constraints affecting women's participation in key areas of agricultural production, there are surprisingly few restrictions on women's forest-related work. With the exception of hunting, the collection of *madwa* timber, and the carrying of forest produce on the shoulder (which is permitted only for men), there are no real taboos restricting forest-related activities to either men or women.

In addition to this flexibility, it is common to find examples of men undertaking so-called women's jobs and women doing men's jobs, 'even if the modalities by which they do so may differ and reflect other social patterns of disadvantage to women (e.g. they tend to have access to animal traction, bicycles, or equipment such as saws, denied to women)' (Joekes et al, 1994, p. 139). A good example of this from Ambatoli is the tendency for men to go in groups to the forest to cut fuel wood and use bullock carts to bring it home. Men also tend to use bicycles to take fuel wood to market whereas women go by foot, headloading their heavy bundles of wood.

Sometimes, anomalies to widely understood gender divisions of labour reflect specific factors such as environmental degradation. Where deforestation is a major problem, for example, it is quite common for men to help women with fuel wood collection. In Jamtoli, fuel wood collection is carried out equally by *adivasi* men and women although it is the women who are primarily responsible for finding additional fuel if the wood taken from the coupe doesn't last all year (discussion with Simon Oraon, 1993). Socio-economic status may also influence task allocation. Amongst poorer families, for example, it is common for household members (especially men) to undertake seasonal migration: a situation that forces the rest of the family to share out additional duties. Wealthier households, by contrast, often encourage women to work close to home where possible and rely on paid labourers to undertake their share of outside work. In many other cases, however, divisions of

Plate 12.2   Men collecting medicinal plants

labour simply reflect intra-household bargaining strategies working within wider structural factors of class, community and religion.

Good examples of this are the relatively wealthy Moslem households in the neighbouring village of Patratoli. In an attempt to practice a loose form of female seclusion, several of these families discourage their womenfolk from undertaking forest-based gathering work: a trend that has encouraged other well-off village women to bargain for help with such tasks. As noted in Chapter Nine, there is a similar tendency amongst the wealthier households in Ambatoli to relieve their womenfolk of jobs that cannot be completed within or near to the homestead. This is particularly pronounced amongst the better-off Backward Caste families although women from some of the wealthiest *adivasi* families sometimes manage to 'buy themselves out' of the most tiresome forest-based labour (notably fuel wood collection) by getting their husbands to purchase cart loads of fire wood or hiring labourers to collect it for them.

As is the case for agricultural work, the extent to which such gender bargaining succeeds depends strongly on intra-household power relations and how these respond to both external events (such as deforestation) and internal changes (such as a new household member). In addition to the assertiveness of individual family members, age or seniority can be an important influence on forest-based task allocation as older married women often manage to avoid the most unpopular jobs. As a result, the burden of drudgery (and of environmental degradation) tends to fall most heavily upon young married women in the poorest households, although young men in either *purdah*-observing or casual labourer households also suffer the effects.

Indeed, it appears that in the research area, forest-based work is shared between men and women to a significantly greater extent than Shiva's (1988) description of gender-environment relations in northern India: a situation that may reflect factors such as forest loss and changing social aspirations amongst the wealthiest villagers. Group interviews in Ambatoli and Patratoli hamlets indicated that although *adivasi* women spend more of their time collecting subsistence-related forest produce than *adivasi* men, the total number of hours spent in the forest by men and women is roughly equal. This is because men use forests for other activities such as hunting, cattle grazing and the collection of large timber, *jaree bootee* or *ranu*. Amongst the wealthier Moslem and Backward Caste families, meanwhile, forest-based fuel wood collection and gathering is dominated by men as women are discouraged, where possible, from undertaking such activities. Indeed, the only contribution that most women from these households make to forest gathering is the collection of leaves and twigs from around the homestead for use as tinder.

As is the case with fuel wood, the collection of tree-borne food, seeds and other NTFPs is more evenly divided between men and women than many eco-feminist accounts suggest. Although Kelkar and Nathan's scenario of *adivasi* men shaking the trees and their wives and children collecting the produce is much more common for homestead than it is for forest trees in the study villages, it is certainly true that *adivasi* men do their fair share of forest-based gathering work. In Moslem and the better off Backward Caste households, meanwhile, men have an even greater responsibility for such tasks.[2]

Generally speaking, the collection of forest-based fruit, seeds, mushrooms and *ranu* is carried out by all-female (*adivasi* and Scheduled Caste) and all-male groups, or by children who take cattle into the jungle to graze. The most common practice is for one person to climb the tree and either shake, cut or throw down the produce which is then collected and shared out amongst the rest of the group. This does not cause any particular problems for the all-female groups as most *adivasi* women are very competent at climbing trees. Indeed, according to Roy, the inability to climb trees was traditionally an accepted reason for an *adivasi* man to divorce his wife (Roy, 1928). Fruit and seeds in very tall trees that cannot easily be climbed are often poked down with long sticks or knocked off the branch by carefully aimed stones.

An interesting reflection of the socio-economic discrimination that exists against women, even in areas of relatively high gender equality, is the fact that the NTFPs most frequently collected by Jharkhandi men are economically valuable tree oil seeds such as *sal*, *mahua* and *piyar*. Men also market these items whereas the items that women are responsible for selling tend to be the less valuable fruits, *sal* leaves, wild vegetables, mushrooms and tooth brush twigs. In most villages, the income obtained from the sale of forest products is kept by the person selling them: a situation that tends to favour men who market the most valuable commodities.

Overall, therefore, gender divisions of forest labour in the research area show only limited support for the eco-feminist/WED idea of women as the main gatherers. The fact that forest gathering is undertaken primarily by men in many of Patratoli's Moslem and some of Ambatoli's better off Backward Caste (and even *adivasi*) households indicates a high degree of variation in forest-based task allocation. To rule out Jharkhand's gender equality as an explanation for this high degree of task sharing, however, it is important to investigate whether Jharkhandi women are able to make use of their greater autonomy and put their silvicultural knowledge into practice.

Gender Differences in Silvicultural Knowledge

As noted in Chapter Nine, another important aspect of eco-feminist/WED discourse is the belief that women have better-developed environmental knowledges than men. Speaking of India, for example, Shiva claims that 'it is women's work that protects and conserves nature's life in forestry' (Shiva, 1988, p. 65) and argues that women have 'a holistic and ecological knowledge of nature's processes' (Shiva, 1988, p. 24). Evidence from Jharkhand, however, fails to support this view and highlights the dangers of 'invisibilising' men's silvicultural expertise and ignoring intra-gender variations in environmental knowledge possession.

But before turning to the local level data, it is important to acknowledge that accessing local knowledge systems can be problematic and is frequently complicated by the constraints that many people (especially women) face in articulating it (Jackson, 1993b). In addition, knowledge possession tends to be strongly 'politically determined' (Scoones and Thompson, 1993, p. 16). For women, in particular, the development and vocalisation of environmental knowledge is often influenced by factors such as class, community, education, religion, age, individual circumstances and personalities.

One of the most significant gender-related differences in silvicultural knowledge in Ambatoli and Jamtoli stems from a basic familiarity with forest areas. Almost all of Ambatoli's and Jamtoli's men have spent most of their lives in these villages and are well accustomed to local forests as they have been using them since childhood for playing in, hunting or grazing cattle. Most newly-married women, by contrast, first come into contact with Ambatoli's and Jamtoli's forests when they move to the villages after they marry. Consequently, they have to learn from scratch where the forest boundaries are situated and where to find the best fuel wood, fruit, mushrooms, wild vegetables etc. This is a hard enough task for those women who grew up in forested areas, but those moving from town-based households have to learn quickly how to undertake forest-based gathering work. In some cases, women from relatively well-off (mostly non-*adivasi*) town-based families who marry into rural households find themselves doing jobs that would have been regarded as socially unacceptable in their natal places.

Of the eighteen Ambatoli women with whom I conducted in-depth interviews, eight said that they were concerned when they first saw Ambatoli forest as it is much bigger than the forests they had grown up near and they were frightened of being chased by wild animals or *bhuts*. Two of these eight (Niramala and Rina Sahu) come from fairly well-off Backward Caste households and have always been discouraged from undertaking forest-based work. Speaking of when she first moved to Ambatoli, Niramala said: 'I didn't like living in a mud-built house so close to the forest and I was frightened of the wild animals that live there' (discussion with Niramala Sahu, 1993). Similarly, her sister-in-law, Rina, commented 'I was really shocked when I first came here and saw what a *junglee* [wild, forested] area it is. I've never been into the forest, but if I did, I'd be frightened of being attacked by bears or a *bhut*. I have to work much harder here than I did in my parents' place. Although my husband collects fuel wood from the forest, I still have to cultivate the vegetable garden, do the cooking and keep the house clean. My mother-in-law taught me how to do these things' (discussion with Rina Sahu, 1993).

Such concerns were not exclusive to women from better-off households. One of my town-born *adivasi* interviewees said: 'When I first went to the jungle, I was surprised at how big it was and was frightened that I might see tigers' (discussion with Rahil Orain, 1993). Another commented: 'When I first married, I missed all my childhood friends, I didn't like the local people and wondered how I would ever manage to live here. I was frightened of going to the forest in case I was attacked by snakes and wild animals' (discussion with Sattan Lohrain, 1993).

A further four *adivasi* women told me that they had never collected fuel wood from the forest before they came to Ambatoli and had to be shown by their mothers-in-law or husbands how to recognise the different forest flora and how to cut and carry fuel wood bundles. Sila Lohrain said: 'When I first came to the village from the outskirts of Ranchi, I didn't know anything about fuel wood collection. I was frightened when I first saw the forest as it is so big and I thought I might be attacked by wild animals or evil spirits. People used to laugh at me because I was so unaccustomed to the work that I kept on dropping down asleep. I was worried about going to the forest as I didn't know which trees were which, so I had to learn these things from my mother-in-law. She also taught me which wild forest vegetables can be eaten, where to find them and how to cook them' (discussion with Sila Lohrain, 1993).

Etwari Orain, meanwhile, commented: 'I learned about the different forest trees and wild vegetables from my mother-in-law and my sister-in-law showed me where to collect fuel wood from. My mother-in-law taught me how to collect jungle *jaree bootee* for making medicine, but when I'd learned these things, she left most of the forest-based work to me and my sister-in-law' (discussion with Etwari Orain, 1993). In a similar vein, Sattan Lohrain said: 'When I first came here, I didn't know how to do anything. My mother-in-law taught me everything about cooking and housework and I learned about fuel wood collection from my sister-in-law as my mother-in-law hardly goes to the forest' (discussion with Sattan Lohrain, 1993).

Even women who had grown up in forested villages said that it took them some time to learn where to find the best fuel wood, fruit, mushrooms and *ranu*. Jitan Manjhi, a Scheduled Caste woman who was born in a village with a bigger and more dense forest than Ambatoli's said: 'I had to re-learn everything to my mother-in-law's satisfaction. She taught me where the forest boundary was, where I could collect fuel wood from and where the best wild vegetables and *jaree bootee* grow' (discussion with Jitan Manjhi, 1993). In Neera Lohrain's natal village, meanwhile, most villagers use dry leaves and branches as fuel, so she had had very little experience of cutting trees before moving to Ambatoli. She commented: 'I used an axe for the first time when I came here. It was really hard work until I got used to it' (discussion with Neera Lohrain, 1993).

Although most married women quickly become familiar with the ecology of their village forest, very few appear to develop more than a basic knowledge about the medicinal properties of forest plants and none are considered to be 'experts' in treating 'natural' illnesses. The treatment of witchcraft- and *bhut*-related illnesses, on the other hand, are regarded as strictly male preserves. According to Kelkar and Nathan, this is because healing is 'a male monopoly'; the only medical knowledges specifically associated with women being related to contraception and abortion (Kelkar and Nathan, 1991). This is not quite so in Ambatoli and Jamtoli where a number of women know where to obtain and how to use basic medicinal plants like *lal dawai* and *chiraita*. As for contraception and abortion, none of the methods mentioned to me had anything to do with herbal medicine. Several village men, meanwhile, mentioned that they sometimes collect an energy-giving plant called *parsawanti* which is also used to help women to get back their strength after childbirth.

Following the logic of eco-feminist/WED ideology, however, the task in which women should possess the most detailed silvicultural knowledge is fuel wood collection which (despite men's frequent contribution) is typically regarded as women's work. The early WED literature, for example, makes much of the fact that women prefer to collect dead branches for fuel rather than cutting living trees, citing this as evidence for their environmental concern (Fortmann, 1986). As Jackson (1993b) points out, however, there are good practical reasons for this practice; notably that it is easier to collect, carry and burn dead wood. Moreover, most women resort to cutting green wood once their supplies of dry branches run out.

Indeed, before the establishment of Ambatoli's JFM committee, women's fuel wood cutting activities were often carried out so carelessly that far from 'managing and renewing the diversity of the forest' (Shiva, 1988, p. 60) they inhibited forest regeneration as well as causing forest decline. Although most fuel wood collectors preferred to gather dry wood 'because it is much lighter and does not have to be dried before use' (discussion with Hiramuni Orain, 1993), supplies tended to last only for a few weeks at the beginning of the fuel collection season. Thereafter, villagers used to cut small trees and (when male assistance was available) large branches from mature trees. A number of the poorest villagers also used to cut additional trees to sell at market as a means of meeting shortfalls in their subsistence foodstuffs.

This is not to say that such activities were permitted without censure. Indeed, many of the older village women tried hard to slow the rate of forest decline, fearing that there would be no forest left for their grandchildren. Some were quite severe in scolding their daughters and daughters-in-law for cutting young trees; encouraging them instead to cut weed species like *putus* or to search harder for dry branches. Others were more pragmatic, saying 'they won't start thinking about the future of the forest until they have children of their own and start having to consider their futures' (discussion with Sumari Orain, 1993).

Certainly, it was quite common in 1993 to see Ambatoli's young (often unmarried) women cutting saplings without appearing to think (or care) whether they would regenerate from coppice growth. During one fuel wood collection trip with two teenage *adivasi* girls who were paid to collect fuel wood for a fairly well-off Oraon family, I tried to establish how they selected the saplings that they would cut. Thinking that they would try to avoid felling good fruit or timber species, I was surprised when one of the girls replied: 'We don't really think about what sort of trees we are cutting. We just cut any tree that is the right size' (discussion with Maino Kumari, 1993). As I watched, she felled a young *kend* tree, trimmed its lateral branches and added it to her head load bundle.

Although young women like Maino were well aware that their activities were regarded as careless by many of the older female villagers, they perceived these women to be unjustified in their scolding given that they managed to delegate the lion's share of fuel wood collection to their daughters and daughters-in-law. One young woman also pointed out that 'the forest has declined a lot since these women were young. They don't understand how much time it takes to collect fuel wood nowadays because they rarely go into the forest except to collect *jaree bootee* and *mahua* flowers' (discussion with Nera Lohrain, 1993). In a similar vein, several disagreed with the idea that they would change their fuel collection methods after having children because child care, combined with agricultural work, housework and (for many) wage labour, would make them even busier than they were. They also emphasised the risk of being caught by the forest guard and charged for (real or otherwise) illegal tree felling if they were to spend additional time in the forest searching for dry branches and weed species.

An even more significant departure from eco-feminist/WED views about women's role as environmental custodians, however, is the fact that prior to the establishment of village-wide forest protection, few female fuel wood collectors felled trees in ways that were likely to stimulate coppice regeneration. When I questioned women about coppicing techniques during a series of all-female group discussions in 1993, a typical response was 'we don't really think about whether the tree will sprout again after it has been cut' (Kaparia all-female group discussion, 1993). When I tried to find out whether any efforts were made to cut trees at a slanting angle to allow water to drain from the cut stump (thereby discouraging rotting), most women replied 'we cut straight across. We didn't know that cutting it an angle would help the tree to survive' (Kaparia all-female group discussion, 1993).

In part, this failure to encourage coppice regeneration may have been linked to the fact that the tools women used to fell trees (usually a small axe or sickle shaped *daoli*) have short handles which prevented them from cutting a tree at an angle close to the base. More significant, however, seemed to be gender differences in silvicultural knowledges and attitudes towards forest protection. Indeed, throughout the research area women appeared to be rather in the dark about preparing tree stumps in ways that would enhance coppice growth: a situation that rather makes a mockery of eco-feminist claims that women have better environmental knowledge than men because of their innate link with nature.

Many of Ambatoli's older village men, by contrast, displayed considerable expertise in coppicing and pollarding trees. Some of their skill may have been linked to their use of heavier, long-handled axes that enabled them to cut the base of the tree at a slanting angle. A more convincing explanation is the fact that between 1965 and 1972, forest contractors frequently employed local men to undertake forest felling and thinning operations and trained them how to coppice trees scientifically.[3] Prior to the establishment of JFM, however, their skills did little to slow forest decline as most men preferred to coppice and pollard their own trees rather than those in Ambatoli forest. The reason for this was that they went into Ambatoli forest 'to steal, so why waste time cutting a tree properly when the forest guard might come and catch us?' (discussion with Tila Oraon, 1993).

In contrast to Shiva's claim that it is 'primarily women who use and manage the produce of forests and trees' (Shiva, 1988, p. 60) it seems to be Jharkhandi men who possess the greatest environmental knowledges. Indeed, even in villages with relatively high levels of gender equality and minor socio-economic inequalities, there are often households in which women play virtually no part in forest gathering and, as a result, have few opportunities to obtain silvicultural expertise. More crucially, it is usually men who have the political power within the village to implement their ecological knowledge in a coherent forest management strategy.

So, although eco-feminist/WED discourses have been very useful for stimulating research into gender-environment relations, they seem to have little bearing on the empirical reality of forest use and management in a predominantly tribal area. The possibility that they are more representative of gender-environment relations elsewhere is also open to doubt as areas with greater economic and gender inequalities tend to place even more emphasis on restricting women's work either to household jobs or (in the case of well educated women) well paid white collar employment. In such situations, men's role in forest management is likely to be even more pronounced than in the research area.

Gender and Autonomous Forest Management in Ambatoli and Jamtoli

In most Jharkhandi villages, women have traditionally been caught in a kind of catch 22 situation in that they undertake more fuel wood collection than (most) men but have less well developed silvicultural knowledges and little input into local environmental decision-making bodies. Prior to the establishment of JFM in Ambatoli, for example, gender conflict was a frequent consequence of the lack of public communication between men and women over environmental management issues. Indeed, men have long blamed women for forest decline, but have also discouraged them from being members of autonomous forest protection committees because they perceived this to be a male role. In the words of one Ambatoli man: 'Who would do the housework if women were allowed to be members of a forest protection committee?' (Ambatoli group interview, 1993). Women, on the other hand, have often pointed out that 'men always scold and blame women for forest decline, but if they don't collect fuel wood, how will the meals get cooked?' (discussion with Rahil Orain, 1993). Many women also argued that men make their own contribution to forest decline as they cut big trees that will not re-grow.

In spite of such comments, many women came to feel rather guilty about their contribution to forest decline and frustrated by their lack of confidence in their own abilities to address it. By 1993, a number of Ambatoli women had become so concerned about deforestation and the seeming inability of their menfolk to organise a successful forest protection initiative that they talked about forming a women's forest protection committee. Many saw forest decline as one of their 'biggest problems' that would not be solved by men who 'just get drunk and forget about forest protection'. A women's group, they believed, would 'co-operate and listen to one another' (discussion with Gandri Orain, 1993).

When pressed, however, none of the women who had mentioned the female forest protection committee was willing to try to organise it. They also contradicted what was said about a women's committee having fewer disputes by adding that an 'all-female group would be prone to gossip which would result in a lot of arguments and conflict' (interviews with Sila Lohrain and Gandri Orain, 1993): a comment that hints at their uneasiness about intervening in the male-dominated domain of resource management and also helps to explain why all of Ambatoli's attempts at autonomous forest protection have been led by men.

Interestingly and in spite of the well documented subsistence-related and socio-cultural relationship between *adivasis* and forests, the most successful hamlet-based forest protection committees in Ambatoli were set up by non-tribal men (Indru Swansi, Ram Prasad Sahu and Jageshwar Ram Manjhi). In addition, Bineswar Thakur who worked with Jageshwar Ram on the XISS education programme tried hard to re-establish a forest protection system in Bartoli (discussion with Bineswar Thakur, 1993). The interest of these men in forest protection seemed to stem primarily from the fact that they came from relatively well-off (or well-educated) households and, not wishing their wives to be involved in forest gathering work, had to undertake (or pay for) much of the family's fuel wood collection themselves. Reflecting this, the *tola*-based committees that they established sought primarily to address fuel wood shortages: a situation that contrasts with most *adivasi* forest protection initiatives which tend to be established only when shortages of large timber (for house-building and plough-making) become apparent (Mehrotra and Kishore, 1990; Kant et al, 1991; Agarwal, 1997b).

In Ambatoli, Deepatoli and Kaparia hamlets where *adivasi* men had a more central role in setting up *tola*-based forest protection initiatives, the committees all broke up within a short time. Some non-tribal villagers attributed this to a relative lack of interest in forest protection by Ambatoli's *adivasi* men coupled with a tendency for alcohol-fuelled disputes to break out amongst committee members. With the exception of female-headed households, women were not allowed to become committee members: a situation that helped to encourage the male-dominated committee meetings to develop into drinking sessions.

Even in Jamtoli where women have better-developed silvicultural knowledges and have become quite skilled in felling, coppicing and replanting as a result of their participation in the annual coupe, their role in forest management was, until the formation of the VFPMC, very limited. When Simon first established his forest protection system to address the village's shortage of good timber, membership was set up on the basis of one adult male per household. Although women were expected to guard the forest, almost all decisions regarding forest protection, the punishment of offenders, the distribution of timber and fuel wood amongst households and the species to be cut and replanted were made by men.

Despite their different degrees of success in sustaining autonomous forest management, therefore, both villages conformed to the wider Jharkhand trend for resource management to be carried out primarily by men; usually during meetings of male-dominated institutions such as *Parhas* or Panchayats (Kelkar and Nathan, 1991). They also displayed a rather unfair tendency to expect women to collect forest produce and participate in the guarding of the forest whilst simultaneously blaming them for forest decline and discouraging them from playing a role in forest management.

In spite of such inequalities, however, few women in the study villages appeared to be worried about their lack of participation in autonomous forest protection. One Oraon woman from Jamtoli commented that she was glad not to have been obliged to attend forest committee meetings as she already had 'too much work to do' (discussion with Allo Bhagat, 1993). Others said that they could always make their feelings known to their household committee member, or to Simon Oraon himself, who was 'always happy to listen' (Kaxitoli group discussion, 1993). In Ambatoli, too, most women were glad not to have had to attend *tola*-based committee meetings, reasoning that they would get to hear if anything of importance had been discussed.

Interestingly, this lack of women's participation in autonomous forest protection systems did not seem to prevent them from meeting their requirements from forests; as has sometimes been the case elsewhere (Sarin, 1993). Most Jamtoli women said that the annual coupe was 'a much easier way of obtaining fuel wood than head loading it daily from the forest' (Kaxitoli group discussion, 1993), and even when the forest protection system was first set up, they 'didn't face too many difficulties in collecting fuel as deforestation in surrounding areas was not so great then and dead wood was more plentiful' (Kaxitoli group discussion, 1993). In Ambatoli, meanwhile, most of the hamlet-based initiatives focused on preventing pilfering by outsiders rather than restricting villagers' own use of the protected area. In addition, many of the most active male committee members were fuel wood collectors themselves and were well aware of the difficulties that fire wood shortages caused.

Gender and Joint Forest Management

When JFM was introduced, it presented quite a radical challenge to traditional environmental management systems with its requirements for women to become members of the VFPMC. Moreover, it provided women with an almost unprecedented opportunity to have a say in the strongly male-dominated domain of forest management. Reflecting the reality of many Panchayati Raj Institutions (Mukhopadhyay, 1996; Singh, 1998), however, women tend, in the absence of outside support to increase their self-confidence, to be regarded as token VFPMC members with the majority of forest management decisions continuing to be made by men.

There are, of course, a few notable exceptions. In the *adivasi*-dominated (Oraon) villages of Barwe in Ormanjhi Block and Jahanabaj in Bero Block, women were so encouraged by the JFM programme that they succeeded in establishing female-dominated forest protection committees in 1992 and 1994 respectively.[4] A more famous example is Maheshpur village in Anghara Block, where an all-women forest protection committee took over from the former male-dominated committee when disputes arose in 1991 (Roy and Mukherjee, 1991).[5] Highlighting the danger of taking an undifferentiated view of gender relations, however, the Maheshpur committee suffered from conflict between Moslem and *adivasi* VFPMCs members over how the forest was to be used. Many of the village's better-off Moslem women preferred to use kerosene rather than wood as a cooking fuel and try to minimise the amount of forest-based work that they undertook. Consequently, they relied little on NTFPs as a source of income and saw the forest primarily in terms of its timber revenue. Most of Maheshpur's *adivasi* women, by contrast, depended heavily on the forest for their day-to-day subsistence and favoured a management system that maximised NTFP and fuel wood yields. The failure of these different groups to reach a compromise caused the committee to break up.

Reinforcing a more general tendency for JFM to benefit men more than women, however, is the Bihar programme's stipulation that the financial benefits of JFM must be shared amongst the 'general body' committee members (of which there is only one per household): a situation which privileges men as they are much more likely than women to become the designated committee member. Extensive interviews conducted by Sanjay Kumar in Ranchi District suggested that in many JFM villages, this is causing resentment amongst women who feel both that committee membership should be made more accessible and that women should receive a separate share of the benefits accruing from JFM. Some bolstered their argument by pointing out that in addition to having the primary responsibility for household subsistence, they were suffering the most from the closure of forests by JFM committees as it affected their fuel wood collecting activities.

To a large degree, however, their pleas have fallen on deaf ears as despite the central government's recent efforts to make JFM more gender sensitive (Government of India, 2000), no alternative fora have yet been created in which women can be encouraged to air their views. More often than not, therefore, few female executive committee members actually attend VFPMC meetings, let alone contribute to decision-making. In Jamtoli, for example, the female VFPMC members initially turned up at the weekly meetings only when specifically invited (discussion with Anand Jha, 1993) and nowadays rarely attend at all (discussion with Sanjay Kumar, 2000). Indeed in 1993, one executive body member was happy to announce that 'women are only on the VFPMC because they are one of the main causes of forest decline' (discussion with Sukra Bhagat, 1993).

Nevertheless, other committee members were more positive about the women's contribution, saying that when they do attend committee meetings, they make 'good suggestions' and they also contribute to the guarding of the forest (interviews with Budhwa Oraon and Ram Prasad Lohra, 1997). Simon Oraon is also positive about the inclusion of women on the VFPMC saying that they have helped to make forest protection even more effective than before. In spite of his encouragement, however, the women committee members themselves feel able to make only minimal contributions to VFPMC meetings and prefer to leave major forest-related decisions to Simon and other male committee members.

In Ambatoli, women's involvement in JFM has been strongly encouraged by Mary Tete who played an important role in pushing the villagers to get their VFPMC formalised by the Forest Department. By making herself approachable and actively creating opportunities for women and men to air their views (separately and together) about forest-related and other issues, she has managed to increase communication and reduce gender tension. As a result, she has been able to maintain a good sense of the benefits and costs of the JFM system for both sexes.

By and large, the Ambatoli villagers are currently quite happy with the JFM system as their unity has remained strong and forest regeneration from protected root-stock has been rapid. The main difficulty created by the new system is a shortage of fuel wood resulting from the ban on cutting green trees in the protected area. Instead, villagers have to spend significantly longer searching for dry branches and weed bushes such as *putus* and *tetair*. This is a particularly onerous pursuit for young women from poorer households who have neither the economic resources to buy themselves out of such work nor the seniority to delegate it to other household members. It has also increased the work of male fuel collectors from the better-off Backward Caste and Moslem households although some try to minimise the drudgery by going to the forest in groups and taking a bullock cart with them to transport the wood that they find.

Although Mary is well aware of these difficulties, few objections appear to have been made to the forest protection committee. When I questioned the chairman about the problems fuel wood collectors face as a result of forest protection, he replied that 'certainly fuel wood collection takes more time and effort nowadays, but the women don't mind because they are committed to forest protection' (interviews with Basant Kumar Pradhan, 1998). Although both female and male fuel wood collectors express rather different views in private, they tend to keep quiet in public as they don't want to be heard complaining about a system that is visibly assisting forest regeneration. Indeed, despite Mary's encouragement, only a handful of women regularly attend committee meetings and even fewer feel able to make suggestions to the executive committee. In many cases, this stems from a fear of 'making fools of themselves' or seeming to 'question the committee's decision' (Nehalu all-female group discussion, 1998).

Broadly speaking, this concern about speaking out in public seems to be rather less pronounced amongst *adivasi* and Scheduled Caste women but is significant amongst those Backward Caste and Moslem women who undertake little or no forest-based gathering work. Indeed, the latter usually stay away from public meetings or keep silent if they do attend. Not unexpectedly, Backward Caste and Moslem men from the same households are often quite active in committee meetings as they are more aware of the fuel collection problem than *adivasi* men who can delegate such tasks to their womenfolk.

In addition to these gender and community-related factors, however, it is clear that age, education, marital status (for women) and socio-economic standing as well as individual personalities play an important role in the willingness of villagers to participate in group discussions. Although mothers-in-law are usually keen to delegate fuel wood collection work to their daughters-in-law, they are often quite forthcoming with their views in committee meetings. At the other extreme, however, there are a number of mature and otherwise outspoken *adivasi* women who feel that there is no point in speaking out as their suggestions would not be taken seriously. Sila Lohrain, for example, commented that: 'My husband doesn't listen to me when I make suggestions, so why should the committee take notice?' (discussion with Sila Lohrain, 1998).

To alleviate such difficulties, Mary Tete has tried to encourage Ambatoli's women to establish an informal forest protection committee to work alongside the official one and provide women with a forum in which they can air their views more freely. It would also give women an opportunity to put forward and discuss their opinions about benefit sharing well in advance of a decision needing to be made. Unfortunately, she has not been successful to date; ostensibly because the women have been too busy for committee meetings, but more realistically because they are concerned that they might be perceived to be challenging the authority of the official committee. And without assistance in the form of more female Forest Department staff, gender-progressive NGOs and all-women groups to bolster women's self confidence and help them to increase their bargaining power within forest management institutions, the possibility of women making the transition from minimal to interactive (empowering) participation (Agarwal, 2001) in JFM is likely to remain limited.

Conclusion

Contrary to eco-feminist/WED emphases on women's role as environmental custodians, therefore, evidence from the research area suggests that the 'job of [forest] management is essentially a male one' (Kelkar and Nathan, 1991, p. 116) as it is men who have the most comprehensive silvicultural knowledge systems and local-level political power to make use of them. The wider Jharkhand tendency to consider resource management and village-based decision-making as male-dominated activities also supports Agarwal's suggestion that the failure of many women to put forward their views 'is part of a larger issue of the cultural construction of gender, and social perceptions about women's capabilities and place in society, from which even hill and tribal communities are not immune' (Agarwal, 1997b, p. 33).

Although opportunities for women's participation in JFM are significantly greater than in most autonomous forest management systems, many women will remain uneasy about exercising this new power until gender progressive initiatives from the state, donors and NGOs coupled with wider structural change start to erode existing gender inequalities and increase women's inter- and intra-community bargaining power (Agarwal, 2001). Also, the fact that quite significant sums of money may be at stake makes the whole issue of forest management potentially more contentious. As a result, many women will continue to suffer in silence rather than risk being accused of challenging committee decisions: a situation that may well be reinforced by the constant threat of witchcraft in the region. So although the Bihar/Jharkhand maximum of five women on the VFPMC is even less likely to encourage women's active participation than JFM programmes with no such limit, it is also less likely to upset traditional socio-political structures and gender relations. While this is undoubtedly a bad thing in terms of women's long-term empowerment, it may not be such an inappropriate short-term intervention for schemes that are unable or unwilling to investigate and work around the embedded micro-political ecologies of inter- and intra-gender relations and social norms.

Notes:

[1] Much of the data used in this section was first published in Jewitt, S. (2000a) 'Mothering Earth? Gender and Environmental Protection in the Jharkhand, India', *The Journal of Peasant Studies* 27 (2) pp. 94-131.

[2] Indeed, the only gathering activity in which women from all communities frequently participate is the collection of produce from homestead trees.

[3] This consists of manipulating the stump's cambium layer to reduce the number of coppice shoots produced (as well as their competition for nutrients). Saplings felled by female fuel wood collectors, by contrast, tend to sprout too many shoots which compete with one another for nutrients and quickly die off.

[4] One explanation for the existence of more female-led forest protection committees in villages dominated by Oraons rather than by other tribal groups is that Oraons have strong village-based affiliations and are therefore more aware of village and forest boundaries than tribes which are more hamlet- or clan-oriented. In addition, the traditionally strong participation of Oraons in agricultural production gives them a good sense of village and forest boundaries.

[5] The committee meets weekly and fines people for taking wood without permission. It also operates a bank to which members contribute one rupee per week and from which they can, if required, take loans (Roy and Mukherjee, 1991).

13 Conclusions

The Advantages of Populist Approaches

Some of the greatest benefits that populist and eco-feminist/WED ideas have brought to the wider development literature are a greater awareness of the socio-economic and environmental problems generated by top-down, male-dominated development projects (Esteva, 1987; Rahnema, 1988; Shiva, 1998; Parajuli, 1991; Escobar, 1992; 1995; Alvarez, 1994; Peet and Watts, 1996a; Sittirak, 1998) and a corresponding interest in indigenous knowledge systems as a potential foundation for more appropriate development practices (Brokensha et al, 1980; Chambers, 1983; Chambers, Pacey and Thrupp, 1989; Verhelst, 1990; Scoones and Thompson, 1993; 1994a; Agrawal, 1995; Warren et al, 1995a; Gujit and Kaul Shah, 1998a; Agrawal and Sivaramakrishnan, 2000a). Another extremely valuable contribution has come from the broader yet strongly populist anti-development literature which has drawn attention to the 'dominating knowledges' (Apffel Marglin and Marglin, 1990) and neo-colonial/orientalist pretensions inherent in much so-called development practice. In combination with certain eco-feminist influences (Shiva, 1988; 1991a; Mies and Shiva, 1993), this literature has been particularly valuable in emphasising the role of women in environmental management and in highlighting the potential displacement and de-legitimisation of local agro-ecological knowledges by Western 'dominating knowledges' and modern farming practices.

A less obvious but very valuable contribution of radical populist, eco-feminist and anti-developmentist ideology has been the increased dialogue that they have helped to promote between traditional and alternative schools of development thinking (Watts and Peet, 1996). In addition to encouraging the 'development pendulum' (Drinkwater, 1994a) to swing towards more bottom-up approaches, the resulting discourses have helped to generate interest in a more appropriate and sustainable route to development. Indeed, some would argue that populist thinking has been responsible for stimulating a 'paradigm shift' away from the idea of transferring Western technology to (supposedly) stupid and lazy Third World peasants towards more participatory and locally empowering development initiatives (Vivian, 1992; Scoones and Thompson, 1994a; Agrawal, 1995; World Bank, 1996; 2001; DfID, 1997; 2001; FAO, 1999; 2001; Agrawal and Sivaramakrishnan, 2000a).

Certainly, a willingness to learn from farmers' agro-ecological knowledges and experiments (Brokensha et al, 1980; Chambers 1983; Richards, 1985; 1994; Chambers, Pacey and Thrupp, 1989) and the possibility of combining local and scientific knowledges (Ostrom et al,

1993; Millar, 1994; Scoones and Thompson, 1993; 1994a) have been apparent in the research area with the Birsa Agricultural University's remarkably successful attempts to learn from, develop and improve local environmental knowledges. Similarly, in the early 1990s, the Bihar Forest Department made an effort to move away from the dominating discourses of scientific forestry towards a more 'people-oriented' and participatory JFM programme.

Populist Critiques

In spite of the substantial benefits that populist philosophy and eco-feminist/WED discourse has brought to development thinking, there are a number of issues that these approaches must address if they are to offer serious practical 'development alternatives' or 'alternatives to development' (Watts, 1993; Peet and Watts, 1996a). First of all, there would seem to be little benefit in exchanging a top-down, inappropriate, male-dominated development paradigm for an equally narrow, conceptually simplistic and romanticised populist or eco-feminist one. As Agrawal and Sivaramakrishnan point out, agrarian environments are home not to harmonious and self-sufficient communities (the 'exoticised indigene' or 'essentialised natural woman') but to complex and politically fragmented social formations 'that reflect the diversity and flux of their landscapes' (Agrawal and Sivaramakrishnan, 2000b, p. 10). Likewise, the heterogeneity of interests that characterise most local communities undermines simplistic assumptions about their supposedly innate effectiveness as natural resource managers and promoters of development (Guha, 2000; Jackson and Chattopadhyay, 2000; Springer, 2000; Zook, 2000). Attention surely needs to be placed, therefore, on local actors working in local situations within the constraints of wider social structures and power relations. In terms of solving practical development problems, attempts to promote local participation and empowerment will achieve little if they ignore the power structures within which local decisions are made and fail to acknowledge that non-dominant groups, especially women, are often unable to vocalise alternative viewpoints.

Gender

In contrast to romantic eco-feminist/WED claims about a special women-environment link, empirical data from Jharkhand indicates that cultural restrictions on women's abilities to accumulate, vocalize and use agro-ecological knowledge exist even in an area characterised by relatively high female autonomy. It also highlights the importance of 'material realities' (Agarwal, 1992) and taboos in determining gender divisions of labour and restricting women's mobility. Indeed, the failure of eco-feminist/WED approaches to acknowledge, or better still, challenge existing socio-political structures which constrain women to 'invisible work' with nature (Nanda, 1991, p. 51) urgently needs to be addressed. This is especially important as the practical implications of WED discourse, when adopted by development policy-makers, are rather different for the educated women who promote such ideas compared to those rural women who actually carry out the lion's share of environmental work. As Meera Nanda points out, 'Third World feminists need to be very cautious with 'difference theories' since the political context in which most Third World women live their lives needs to stress gender equality if any real gains are to be made for the majority' (Nanda, 1991, p. 36).

One of the major problems with eco-feminist/WED accounts of gender-environment relations is that they often fail to differentiate between environmental use and environmental concern. While evidence from the research area supports the view that women have a major (although not exclusive) role in forest-based gathering work, it also highlights the environmental damage that can result from the careless cutting of young saplings. Moreover, the socio-political constraints that many women face in obtaining and making use of silvicultural knowledge plus the strong tradition of male-dominance in the region's resource management systems highlight the dangers of 'invisibilising' men and their environmental knowledge systems (Jackson, 1993a; 1993b; Leach, Joekes and Green, 1995; Kaul Shah, 1998; Sarin et al, 1998; Locke, 1999; Singh, 1999). As for task allocation, empirical data from both Ambatoli and Jamtoli emphasise the complexity, fluidity and frequent re-negotiation of gender divisions of labour in response to changes in intra-household power relations and bargaining strategies (Jackson, 1993a; 1992b; 1995).

In practical terms, therefore, it is rather worrying that romanticised and undifferentiated WED discourse has gained such a foothold in development policy-making. Despite its obvious appeal, it presents a dangerously simplistic image of gender-environment relations and provides a very shaky foundation on which to build environmentally-oriented development projects. Potentially even more serious, however, is the possibility that policy-makers' emphasis on women's role as environmental custodians may enlarge their work loads without enabling them to effectively increase their control over the resources they are assumed to want to protect.

As Agarwal points out, India's prevailing traditions of patrilinearity and patrilocality often discourage women from having the same long-term interests in land as men and 'women's concerns, even if pressing, do not automatically translate into environmental action by women themselves or by the community' (Agarwal, 1997b, p. 37). A related problem is that development planners often 'neglect the importance of disaggregating women, fail to understand the role that resource use plays in gendered livelihood strategies and conceal aspects of access, power and control that influence different women's resource use needs' (Locke, 1999, p. 275).

With respect to JFM,. for example, most programmes recognise women's role as fuel wood collectors and many have made attempts on paper to allow women to participate in forest management, but few actually recognize the constraints (and wider issues of gender ideology) that women face in taking up such opportunities in practice (Sarin, 1995; 1998b; Kaul Shah and Shah, 1995; Agarwal, 1997b; Locke, 1999). As a result, they frequently leave many women (and some men) worse off than before in terms of the time taken to collect fuel wood (Sarin, 1993; 1995; Kaul Shah and Shah, 1995; Agarwal, 1997a; 1997b; 1998; 2001; Kaul Shah, 1998; Saxena, 1998). There is also concern that a shift from citizenship to membership as the main resource use criterion may threaten women's traditional usufruct rights to forests (Agarwal, 1997b).

By challenging the traditionally male-dominated nature of environmental management, participatory programmes also run the risk of upsetting accepted power relations and creating gender tension (Crawley, 1998; Gujit and Kaul Shah, 1998). Certainly the populist assumption that the adoption of participatory or learning approaches can easily empower local people has received a lot of criticism for its political naiveté (Long and Long, 1992; Long and Villareal, 1994; Scoones and Thompson, 1994a). According to Jackson, empowerment 'does not necessarily follow from participation - particularly for women - because participation of women does not necessarily express their gender interests' (Jackson, 1994, p. 120). The concept of empowerment is also very simplistic in its assumption that it can break down the barriers that women face in obtaining and articulating agro-ecological knowledge simply by creating greater opportunities for their participation in development projects. As Crawley points out, 'empowerment is a relative concept, it cannot take place without the relative disempowerment of another group, and power is usually not given up voluntarily, rather it must be taken' (Crawley, 1998, p. 31).

Populism

One of the biggest problems with broader populist approaches is that they often seek to replace traditional development blueprints with a supply side populism that often reflects the orientalist and paternalistic visions of Western(ised) intellectuals rather than the desires of local people. The romanticised image of an 'original affluent' society (Sahlins, 1972) or peasant moral economy is particularly dangerous as it has potential for condemning so-called traditional societies to lifestyles that are unable to provide basic calorific requirements and health care (Kelkar and Nathan, 1991): lifestyles that may not be what these societies want. In situations where peasant farmers must 'intensify or die' (Bebbington, 1996, p. 92), the promotion of indigenous technology is an inappropriate intervention if the root causes of poverty remain unaddressed.

A related problem is the populist tendency to represent indigenous knowledges as static yet holistic and somehow more wholesome than modern scientific knowledge. But this fails to take account of the knowledge imperfections inherent within all world views as well as the knowledge adjustments and changes in aspirations that have occurred (and will continue to occur) in response to factors such as market integration and interaction with other knowledge systems (Blaikie, 1985; Bebbington, 1990a; 1990b; 1992; 1994a; 1996; Bandyopadhyay, 1992; Ghai and Vivian, 1992b; Guha; 1993; Marsden, 1994; Agrawal, 1995; Horta, 2000; Rangan, 2000a; 2000b). Consequently, there is a danger that uncritical populist emphases on indigenous knowledges may not only overrate their capability to solve development problems but could also discourage the use of conventional scientific practices that may be more effective (Nanda, 2001a).

To take agricultural development as an example, it is unclear what the future holds for Jharkhand or Asia as a whole; especially with current declining trends in staples yield growth. In order to feed the 4000 million people who will depend on rice as their major staple by 2015, global paddy production will have to increase by 50 per cent to maintain current nutritional standards or by 70 per cent to achieve satisfactory nutritional levels (Greenland, 1997). Given populist fears about and criticism of science, biotechnology and GM foods (not to mention patenting issues - Shiva, 1991b; 1993), public sector research in this direction has been severely hampered. Private sector research, meanwhile, caters to wealthy farmers' demands for traits such as prolonged shelf life rather than poor people's requirements for plants that can raise yields, resist moisture stress, fix nitrogen and improve the health of millions (Lipton, 1999). Indeed, Conway (1997) argues that GM crops have potential for increasing agricultural productivity in just those unfavourable, rainfed environments that benefited little from the green revolution. According to Lipton, however, the tendency of the anti GM lobby to exaggerate the risk from GM crops 'belittles the far greater risk of worsening nutrition in their absence' (Lipton, 1999, p. 3). Moreover, 'if the duck we are shooting is the third great breakthrough of the century in tropical staples - after maize hybrids and wheat and rice semidwarfs - then GM plants quack plausibly, and are the only duck on the block' (Lipton, 1999, p. 20).

Given a harsh choice between the ability to feed themselves and the risk of future environmental problems (or even increased agrarian class differentiation - Byres, 1981; 1983; Brass, 1990; 1994), there is little doubt that most Jharkhandi farmers would choose the former. But as professional risk minimisers, they would probably find ways, in conjunction with local knowledge and indigenous practices, of avoiding or dealing with environmental problems when they arose. In the past, farmers in India's advanced agricultural areas have overwhelmingly voted with their pockets for modern farming methods; despite their practical experience (as well as Marxist, populist and anti-developmentist critiques) of the green revolution's side effects. In relatively backward, rainfed, rice-based ecosystems with less scope for economic polarisation than Punjab's wheat-based systems, meanwhile, most farmers are keen to continue growing MVs (especially vegetables) and using agro-chemicals because they have found them to be useful for increasing food security.

Another critical question is the extent to which populist environmental discourse is part of an orientalist vision that appeals to Western 'green thinking' and a more long-standing romantic fascination with the idea of self-sufficient, environmentally-sensitive communities. Surely much of the appeal of JFM and similar environmental rehabilitation programmes reflects these romantic images (plus the West's success in placing the burden of environmental protection on developing countries - Moore, Chaloner and Stott, 1996). But this does not mean that they represent what local people want and need. As Shepherd's models show, the need to increase food production or to cope with population pressure and land fragmentation can mean that local development priorities can be far removed from populist visions of self-sufficiency and environmental stewardship. Consequently, in areas where the centrality of forests to local subsistence is little more than a populist myth (Kelkar and Nathan, 1991), the romantic appeal of participatory forest management is obviously misguided, resource wasteful and potentially harmful for local subsistence strategies.

Another criticism that applies particularly to supply side populist and anti-developmentist writings is that given the many examples of local resistance to state-imposed modernisation (Chambers, 1983; Scott, 1985; Guha, 1989; 1992; Kelkar and Nathan, 1991; Long and Long, 1992; Redclift, 1992; Scoones and Thompson, 1993; 1994a; Escobar, 1995; Sivaramakrishnan, 1999) the assumption that local people cannot determine or articulate what they want from development is rather patronising. In the fieldwork area, local people have a very clear idea of their development priorities and are perfectly capable of articulating them. Equally, they have proved themselves able to resist the kind of development that they don't want, as has been the case with their long history of opposition to commercial forest management.

Indeed, far from being helplessly duped (as Vandana Shiva seems to suggest) into adopting ecologically unsound modern seeds and farming techniques, most local people are making hard decisions about how best to meet their annual household food requirements. The fact that so many villagers have wholeheartedly adopted modern farming methods reflects the ability of these techniques to increase food security, reduce the need for unpopular household survival strategies (like seasonal migration) and help local people to meet their aspirations for a better material standard of living.

Local people in Jharkhand have also made efforts to ensure that these opportunities remain available by establishing a kind of demand side (but not necessarily populist) development that seeks to retain local input into and control over the development process. At the local level,

Simon Oraon has been very successful in obtaining state funding for locally desired and appropriate development projects that meet local people's basic needs and cultural aspirations. He has also established a forest protection system which enables Jamtoli villagers to meet their annual subsistence requirements for fuel wood and timber as well as helping to maintain a strong local sense of 'forest culture'.

At the regional level, local people's votes supported the Jharkhand movement in its quest to regain customary rights to land and forest resources and obtain the political power necessary to gain greater control over the development process. At the same time, the movement built upon and promoted a broad sense of Jharkhandi cultural identity whilst attempting to retain an accurate understanding of the type of development that local people wanted. As a demand side development movement, it has also (arguably) suffered less from the sociological and political naiveté that supply side populism has often attracted criticism for (Long and Long, 1992; Scoones and Thompson, 1993; 1994a; 1994b; Bebbington, 1994a; Agrawal, 1995).

Before demand side development can be proposed as a serious alternative to top-down or supply side populist models, however, it is important to bear in mind the wider structural limitations on its expansion. In most cases, demand side populist (and new social) movements develop as a means of either resisting specific state-imposed initiatives or achieving greater control over local development (Schuurman, 1993). But not all communities are capable of organising themselves collectively and it usually takes a particular incident or set of circumstances to act as a catalyst around which a community (or local leader(s)) can mobilise support. The heavy reliance of demand side initiatives on actors and place therefore often restricts their scope to fairly small scale, locally-oriented, non-blueprint solutions to specific development problems.

The Jharkhand movement, for example, succeeded in causing the Indian state to take notice of its demands by holding periodic economic blockades that threatened 'the accumulation possibilities of the dominant classes' (Blaikie, 1985, p. 147) by preventing resources from leaving the region. It also managed to play its 'indigenous' and 'environmental' cards in ways that attracted the support of national and international NGOs, thereby forcing the Indian state to stand by its environmental and tribal development rhetoric (Vivian 1992; Omvedt, 1994; Karlsson, 2000). On the other hand, the central government kept a watchful eye on the activities of known Jharkhandi activists. Major leaders such as Ram Dayal Munda had their mail routinely censored and were expected to seek police permission if they wished to travel beyond a certain distance from their residences. The state's interest in Jharkhand reflected, in large part, its role as a major source of raw materials for Indian industry and its ability to hold large parts of the country to ransom with its economic blockades. In addition, the financial turnover of Jharkhand's mining activities created huge potential for corruption in both state and private sectors. Indeed, it is well known that a number of people would have liked to 'dispose' of those Jharkhand leaders who could not be bribed into backhanding large slices of the mine sector's profits (discussion with Ram Dayal Munda 1995).

Now that a separate Jharkhand State exists, the movement's role in challenging the autonomy of the Bihar and central governments has effectively been brought to a close. The extent to which the leaders of the new State succeed in maintaining their accountability to local people (as well as a sense of their changing development priorities) will be an important indication of the long-term success of similar movements elsewhere (Vivian, 1992; Colchester, 1994; Peet and Watts, 1996a; Bebbington, 1996; Kaufman, 1997a; Rangan, 2000a).

Beyond the Populist Impasse

Given that neither supply side nor demand side populism is likely to act as a panacea for development problems or even as a practical alternative to traditional development blueprints, in which direction should development thinking and practice head next? Obviously, there are no simple answers to this question, but there are some important lessons about the promotion of more appropriate forms of development that can be learned from both populist critiques of development and critiques of populist development. Some of the most sensible pointers can be found in beyond farmer first approaches and the growing field of political ecology which both seek to inject a more politically, sociologically and environmentally differentiated view of development into the naive populism of approaches like 'farmer first'. Taking these ideas forward in the gender-environment field, meanwhile, 'feminist environmentalism' (Agarwal, 1992) and gender analysis perspectives have done much to stimulate investigations into 'conjugal contracts', gender bargaining and the wider structural characteristics of gender-environment relations. Also of great value for linking agency and structure at the micro and macro level are the actor oriented approaches advocated by Norman and Ann Long (1992).

At the macro level, actor oriented and political ecology approaches can be particularly helpful in highlighting the ways in which development discourses can be manipulated by various actors to serve different political and economic goals. To understand the Indian state's recent interest in participatory forest management, for example, it is important to consider the political agendas hidden beneath its new populist rhetoric. Given the suspicion that social forestry was little more than a veiled attempt to grow wood for the pulp and paper industry (Guha 1983; Alvarez, 1994), one cannot help but wonder whether India's forest intellectuals have helped to stimulate a genuine shift towards a more 'people oriented' system of forestry or whether JFM is simply a convenient means for the state to get local people (where it has failed) to use their resources to regenerate and conserve forests as future sources of timber for industrial markets.

At the local level too, difference sensitive and politically astute actor oriented, beyond farmer first and gender analysis approaches have great potential for grounding supply side populist development initiatives; assisting them to investigate spatially and temporally specific development problems and generate realistic and locally acceptable solutions. By highlighting the dynamics of different actors' ongoing struggles over resources such approaches can help to explain why a problem that exists in one place may be easily overcome in another. They may also reveal why a solution that works in one area may not easily be transplanted elsewhere.

With respect to the causes of forest decline in Ambatoli, for example, the simple binary tensions over access to forest produce that existed between local people and the Forest Department were only part of the story. More significant were wider tensions over food security coupled with inter- and intra-village, community, wealth, gender and actor-related conflicts over both forest felling and perceived favouritism from the forest guard; all of which inhibited the development of a wider sense of village unity. Jamtoli, by contrast, is a village with greater community homogeneity which was united and encouraged to prosper by an energetic local leader (Simon Oraon) who was trusted for his honesty and respected for his expertise in agriculture, forest management and wider village development.

Similarly with regard to gender-environmental relations, actor oriented and gender analysis approaches are very helpful in explaining inequalities in environmental use and agro-ecological knowledge possession. In Ambatoli, for example, women's natal to marital village transfers of agro-ecological knowledge were very actor oriented in that they reflected the personalities and experiences of the women concerned and the extent to which they wanted or felt able to discuss their agricultural knowledges within the household. Superimposed on these differences were socio-cultural restrictions (mediated by wealth) on the participation of women in agriculture and the extent to which they could use or articulate the agro-ecological knowledge that they did possess. A similar situation was found with respect to gender differences in the possession and control over silvicultural knowledges.

What seems to be needed, therefore, is the replacement of instrumental 'add women and stir' (Braidotti et al, 1994) and supply side populist development initiatives with more locally-oriented systems of analysis that can investigate the institutional ecologies, socio-economic relations and environmental contexts of particular places and work around underlying inequalities in power, wealth, property and resource allocation rights. Particularly important is the need to ground general theories on resource management, gender-environment relations and indigenous knowledge with a flexible and locally sensitive analytical framework that can investigate socio-economic, regional, gender, age-related and other differences between local people and the impacts that these can have on wider socio-political relations.

On the gender-environment front, there is some evidence to suggest that donor agencies are starting to take these ideas on board. The Organisation for Economic Co-operation and Development's Development Assistance Committee's statement on gender equality, for example, accepts the need for 'further development of gender analysis' (Development Assistance Committee, 1995, p. 3), arguing that '[s]pecific efforts to enhance the role of women in development remain as necessary as ever but the focus must widen to encompass both women's and men's roles, their responsibilities, needs, access to resources and decision-making as well as the social relations between women and men' (Development Assistance Committee, 1995, p. 1). Similarly, DfID's White Paper on International Development acknowledges gender inequalities in access to and control over resources as well as in participation in public life (DfID, 1997). As Razavi and Miller (1995) point out, however, 'the challenge remains in ensuring that paper commitments are translated into concrete action' (Razavi and Miller, 1995: Executive Summary, Section IV). Many forest projects, for example, have failed in practice to assess the success of their 'gender sensitive' policies (Jahan, 1998).

But even more important than the adoption of gender analysis, actor oriented and beyond farmer first approaches into the rhetoric of development policy documents is the need to 'ground' their theoretical views with the help of site-specific empirical research at the agriculture-forest interface. In spite of the now long-standing interest in indigenous knowledges, for example, the embedded and often highly interlinked nature of local technical, socio-cultural and magico-religious knowledge systems is still largely ignored by development policy-makers and practitioners.

Data from Jharkhand, for example, demonstrates the importance of detailed locally-based fieldwork (by insiders such as Mary Tete as well as outsiders) for understanding what local people want from development and what different actors are likely to consider to be culturally insensitive, irrelevant or inappropriate. A knowledge of local socio-cultural and magico-religious beliefs (and the political ecologies that lie behind them), for example, can be crucial for understanding gender variations in agro-ecological knowledge possession or integrating traditional with modern scientific knowledges without de-legitimising the former.

An understanding of local aspirations can also be very important for distinguishing between populist myths and local realities. As Bebbington's work has shown, local people may prefer to discard any romantic attachment that they do have to their traditional agro-ecological knowledges in favour of modern agricultural practices which enable them to meet their household subsistence requirements without having to migrate (Bebbington, 1990a; 1990b; 1992; 1994a). Similarly, many rural people may not fit the populist stereotype of environmentally sensitive Noble Savages, needing instead to convert forests to agricultural land from which they can feed their families (Pathak, 1994).

For participatory initiatives to be successful, therefore, it is extremely important that they address a strongly felt development need and are funded and planned to give long-term support to their target communities. In the Mathwad Forest Range, Madhya Pradesh, where Pathak did his field research, for example, JFM would be an inappropriate solution to local problems as villagers there have much greater requirements for agricultural land than for forest produce (Pathak, 1994). In the research area, by contrast, JFM fits many villagers' requirements better as agricultural intensification is possible and shortages of fuel wood and timber tend to be seen by most villagers as a greater problem than shortages of land. Indeed, many successful forest protection committees help to maintain their support bases by diversifying into issues of wider village development (Tiwary, 2001). Simon Oraon's forest protection strategy, for example, has been based around the intensification of agriculture as a means of addressing the poverty that lies at the root of much local forest exploitation and succeeds because it links through to wider village development priorities. An additional factor in favour of JFM is that many Jharkhandis have a strong sense of socio-cultural attachment to forests: a situation that has been enhanced by the Jharkhand movement's emphasis on the centrality of forests to Jharkhandi culture. If the programme is to succeed, however, the new Jharkhand State needs to ensure that it offers the same level of commitment and flexibility that it expects from local communities.

Obviously, for many donor agencies working within strict time and financial constraints, it is unrealistic to undertake the long-term research necessary to identify local people's development aspirations and understand their wider socio-economic, political and cultural environments. On the other hand, there is some doubt as to whether the short-term participatory techniques employed by many NGOs and donor agencies are capable of accurately identifying local development priorities and constraints (Blackburn and Holland, 1998b; Gujit and Kaul Shah, 1998a; Sarin, 1998b; Karlsson, 2000). Rapid Rural and Participatory Appraisal techniques have often been found to over-represent dominant views and give only a superficial impression of wider village politics (Mosse, 1994; Drinkwater, 1994b; Mayoux, 1995; Crawley, 1998; Pangare, 1998; Locke, 1999). When coupled with instrumental populist approaches such as WED that focus on undifferentiated and tokenistic categories, such methodologies are likely to miss the complexity of local power and authority hierarchies which actively disadvantage women and other marginal groups (Sen and Grown, 1987; Moser, 1993; Braidotti et al, 1994; Mosse, 1994; Cornwall, 1998; Kaul Shah, 1998; Sarin, 1998b; Singh, 1999).

As projects which fail to benefit their target groups cannot be regarded as successful, therefore, a compromise needs to be found. In practical terms, an increase in locally-recruited development practitioners and project staff can help to maximise awareness of local socio-economic, cultural, political and environmental conditions and enable programmes to work around these. To encourage and maintain women's input, meanwhile, culturally-sensitive participatory approaches implemented by female forest officers, NGO staff and development workers can be effective in both creating fora in which local women feel happy to air their views and finding appropriate ways of increasing women's opportunities to develop their knowledge bases. The success of Mary Tete's efforts in Ambatoli is a case in point.

Without more fundamental structural change, however, such measures may struggle to promote more than superficial empowerment for women and marginal groups given the entrenched nature of inequalities in access to decision-making institutions and control over land and other resources. Although India's positive discrimination policies have helped Scheduled Castes and Tribes to obtain college places and public sector jobs and the Constitution (Seventy-Third Amendment) Act of 1992 has created theoretical opportunities for women's political participation, the obstacles are many and change is likely to be slow.

Nevertheless, radical populist critiques have been important in stimulating a shift away from top-down approaches towards a new 'learning paradigm' (Pretty and Chambers, 1993; 1994) and eco-feminist/WED discourse has had a significant impact on donor attitudes towards gender-environment relations. So now the ball has started rolling, it is important to maximise its potential for local level change. To promote realistic rather than romanticised impressions of local people's development priorities and constraints, it is crucial to encourage flexible, locally-oriented development projects that can investigate and work around existing political ecologies, power structures and socio-cultural norms to create real empowerment opportunities for women and resource poor villagers.

Indeed, a more sensitive and appropriate supply side populism of this type could work in tandem with demand side development initiatives and new social movements. In India, at least, successful opposition to the 1981 Draft Forest Bill and widespread support for the Chipko movement have both made it easier for indigenous and environmental groups to hold the Indian state to its populist rhetoric on environmental conservation and social justice (Bandyopadhyay, 1992; Ghai and Vivian, 1992b; Guha, 1993; Gadgil and Guha, 1994; Omvedt, 1994). The granting of a separate Jharkhand State also reflects this power to a large extent.

Whilst the strength of these groups *vis-à-vis* the state should not be overestimated (Vivian, 1992), an increase in global environmental activism has certainly helped to reduce the 'Achilles heel of localism' (Esteva and Prakash, 1992) that has often constrained them in the past. In addition, when clashes do occur, this often provides a wider forum for debate which, although it is a painfully slow process, does have potential for swinging the pendulum more firmly towards appropriate development approaches.

Ironically, given increasing trends towards globalisation, one of the most important lessons about development that still needs be learned is that geography does matter and development programmes are unlikely to assist the most needy unless they take account of this. Instead of grand development theories and blueprints or technical (and even participatory) fixes (Scoones and Thompson, 1994c), therefore, a spatially oriented 'new professionalism' (Pretty and Chambers, 1993; 1994) is required that can respond sensitively to different local needs and situations.

Bibliography

Adams, W.M. (1990), *Green Development: Environment and Sustainability in the Third World*, London, Routledge.

Agarwal, B. (1986a), *Cold Hearths and Barren Slopes: The Woodfuel Crisis in the Third World*, New Delhi, Allied.

Agarwal, B. (1986b), 'Women, Poverty and Agricultural Growth in India', *The Journal of Peasant Studies*, vol. 13 (4), pp. 29-45.

Agarwal, B. (1992), 'The Gender and Environment Debate: Lessons from India', *Feminist Studies*, vol. 18 (1), pp. 119-158.

Agarwal, B. (1994a), 'Gender, Resistance and Land: Interlinked Struggles over Resources and Meanings in South Asia', *The Journal of Peasant Studies*, vol. 22 (1), pp. 81-125.

Agarwal, B. (1994b), *A Field of One's Own. Gender and Land Rights in South Asia*, Cambridge, Cambridge University Press.

Agarwal, B. (1997a), 'Gender, Environment, and Poverty Interlinks: Regional Variations and Temporal Shifts in Rural India', *World Development*, vol. 25 (1), pp. 23-52.

Agarwal, B. (1997b), 'Environmental Action, Gender Equity and Women's Participation', *Development and Change*, vol. 28, pp. 1-44.

Agarwal, B. (1998), 'Environmental Management, Equity and Ecofeminism: Debating India's Experience', *The Journal of Peasant Studies*, vol. 25 (4), pp. 55-95.

Agarwal, B. (2001), 'Participatory Exclusions, Community Forestry, and Gender: An Analysis for South Asian and a Conceptual Framework', *World Development*, vol. 29 (10), pp. 1623-1648.

Agarwal, C. and Singh, K. (1995), 'Forest Co-Operative Societies in Kangra (H.P): A Case-Study', *Wastelands News*, XI (2), pp. 38-48.

Agarwala, V.P. (1990), *Forests in India.* (second edition), New Delhi, Oxford and IBH Publishing Company.

Agrawal, A. (1995), 'Dismantling the Divide Between Indigenous and Scientific Knowledge', *Development and Change*, vol. 26, pp. 413-439.

Agrawal, A. and Sivaramakrishnan, K. (eds) (2000a), *Agrarian Environments. Resources, Representation, and Rule in India*, Durham and London, Duke University Press.

Agrawal, A. and Sivaramakrishnan, K. (2000b), 'Introduction: Agrarian Environments', in A. Agrawal and K. Sivaramakrishnan (eds) (2000), *Agrarian Environments. Resources, Representation, and Rule in India*, Durham and London, Duke University Press, pp. 1-11.

Altieri, M.A. and Yurjevic, A. (1995), 'The Latin American Consortium on Agroecology and Development (CLADES) - Fostering Rural Development Based on Indigenous Knowledge', in D.M. Warren, L.J. Slikkerveer, and D. Brokensha (eds), *The Cultural Dimension of*

Development. Indigenous Knowledge Systems, London, Intermediate Technology Publications, pp. 458-463.

Altieri, M.A. with contributions by Norgaard, R.B., Hecht, S.B., Farell, J.G. and Liebman, M. (1987), *Agroecology. The Scientific Basis of Alternative Agriculture*, Boulder, Westview Press and London, Intermediate Technology Publications.

Alvarez, C. (1994), *Science, Development and Violence. The Revolt Against Modernity*, Delhi, Oxford University Press.

Anderson, R.S. and Huber, W. (1988), *The Hour of the Fox. Tropical Forests, the World Bank and Indigenous People of Central India*, Seattle, University of Washington Press.

Apffel Marglin, F. and Marglin, S.A. (eds) (1990), *Dominating Knowledge. Development, Culture and Resistance*, Oxford, Clarendon Press.

Appadurai, A. (1990), 'Technology and the Reproduction of Values in Rural Western India', in F. Apffel Marglin and S.A. Marglin (eds), *Dominating Knowledge. Development, Culture and Resistance*, Oxford, Clarendon Press, pp. 185-216.

Arnold, J.E.M. (1990), *Social Forestry and Communal Management in India*, ODI Social Forestry Network. Paper 11b. London, ODI.

Arnold, J.E.M. (1991), *Tree Products in Agroecosystems: Economic and Policy Issues*, International Institute for Environment and Development, Sustainable Agriculture Series, Gatekeeper Series No 28.

Arnold, J.E.M. and Campbell, J.G. (1985), 'Collective Management of Hill Forests in Nepal: The Community Forestry Development Project', in the Panel on Common Property Resource Management, Board on Science and Technology for International Development, Office of International Affairs and National Research Council, 1985. *Proceedings of the Conference on Common Property Resource Management. April 21-26, 1985*, Washington, D.C., National Academy Press, pp. 425-454.

Arora, D. (1997), 'From State Regulation to People's Participation: The Case of Forest Management in India', *Economic and Political Weekly*, vol. 19 (27), pp. 2551-2565.

Arul, N.J. and Poffenberger, M. (n.d.)., 'FPC Case Studies. Type 1 Case Study: Gamtalao Khurd Village - Mandvi South Range', in R.S. Pathan, N.J. Arul and M. Poffenberger, *Forest Protection Committees in Gujarat - Joint Management Initiative*. Sustainable Forest Management Working Paper Series, Working Paper No. 7, New Delhi, Ford Foundation, pp. 14-18.

Bahaguna, V.K. (1991), 'Changing Dimensions of Forest Management in India: Options and Roles for Peoples' Participation', in R. Singh (ed), *Managing the Village Commons. (Proceedings of the National Workshop on Managing Common Lands for Sustainable Development of Our Villages: A Search for Participatory Management Models)*, December 15-16, 1991 Bhubaneshwar. Bhopal, Indian Institute of Forest Management, pp. 46-52.

Bahaguna, V.K. (1994), 'Collective Forest Management in India', *Ambio*, vol. 23 (4-5), pp. 269-273.

Baines, G.B.K. (1989), 'Traditional Resource Management in the Melanesian South Pacific: A Development Dilemma', in F. Berkes (ed), *Common Property Resources. Ecology and Community-Based Sustainable Development*, London, Belhaven Press, pp. 273-295.

Baker, J.M. (2000), 'Colonial Influences on Property, Community, and Land Use in Kangra, Himachal Pradesh', in A. Agrawal and K. Sivaramakrishnan, (eds) (2000), *Agrarian Environments. Resources, Representation, and Rule in India*, Durham and London, Duke University Press, pp. 47-67.

Baland, J.M. and Platteau, J.P. (1996), *Halting Degradation of Natural Resources: Is there a Role for Rural Communities?*, Oxford: Clarendon Press.

Bandyopadhyay, J. (1992), 'From Environmental Conflicts to Sustainable Mountain Transformation: Ecological Action in the Garhwal Himalaya', in D. Ghai and J.M. Vivian (eds), *Grassroots Environmental Action. People's Participation in Sustainable Development*, London, Routledge, pp. 259-278.

Banuri, T. (1990), 'Development and the Politics of Knowledge: A Critical Interpretation of the Social Role of Modernization', in F. Apffel Marglin and S.A. Marglin (eds), *Dominating Knowledge. Development, Culture and Resistance*. Oxford, Clarendon Press, pp. 29-72.

Banuri, T. and Apffel Marglin, F. (eds) (1993a), *Who Will Save the Forests? Knowledge, Power and Environmental Destruction*, London, Zed Books.

Banuri, T. and Apffel Marglin, F. (1993b), 'A Systems-of-Knowledge Analysis of Deforestation', in T. Banuri and F. Apffel Marglin (eds), *Who Will Save the Forests? Knowledge, Power and Environmental Destruction*, London, Zed Books, pp. 1-23.

Banuri, T. and Apffel Marglin, F. (1993c), 'The Environmental Crisis and the Space for Alternatives: India, Finland and Maine', in T. Banuri and F. Apffel Marglin (eds), *Who Will Save the Forests? Knowledge, Power and Environmental Destruction*, London, Zed Books, pp. 24-52.

Bardhan, P. (1984), *The Political Economy of Development in India*, Oxford, Basil Blackwell.

Barker, F., Hulme, P., Iversen, M., and Loxley, D. (eds) (1986), *Europe and its Others. Volume 1*, Colchester, University of Essex.

Barlett, P. F. (1980), 'Introduction: Development Issues and Economic Anthropology', in P.F. Barlett (ed), *Agricultural Decision Making. Anthropological Contributions to Rural Development*, New York, Academic Press, pp. 1-16.

Bates, R.H. (1988), *Towards a Political Economy of Development*, Berkeley, University of California Press.

Baudet, H. (1965), *Paradise on Earth - Some Thoughts on European Images of Non-European Man*, Newhaven, Yale University Press.

Bayliss-Smith, T.P. (1982), *The Ecology of Agricultural Systems*, Cambridge, Cambridge University Press.

Bebbington, A.J. (1990a), 'Indigenous Agriculture in the Central Ecuadorian Andes. The Culture, Ecology and Institutional Conditions of its Construction and its Change', Ph.D. Dissertation, Graduate School of Geography, Clark University, Worcester.

Bebbington, A.J. (1990b), 'Local Conceptions of Indigenous Agricultural Revolution: Peasant Organisations, NGOs and the Search for Indigenous Models of Rural Development in Central Ecuador', Paper Presented for the Workshop on 'Relevance, Realism and Choice in Social Development Research', University of Hull, 10-12 January, 1990.

Bebbington, A.J. (1992), *Searching for an 'Indigenous' Agricultural Development: Indian Organisations and NGOs in the Central Andes of Ecuador*, Centre of Latin American Studies, University of Cambridge, Working Paper No: 45.

Bebbington, A.J. (1994a), 'Composing Rural Livelihoods: From Farming Systems to Food Systems', in I. Scoones and J. Thompson (eds), *Beyond Farmer First. Rural People's Knowledge, Agricultural Research and Extension Practice* London, Intermediate Technology Publications, pp. 88-93.

Bebbington, A.J. (1994b), 'Federations and Food Systems: Organisations for Enhancing Rural Livelihoods', in I. Scoones and J. Thompson (eds), *Beyond Farmer First. Rural People's Knowledge, Agricultural Research and Extension Practice*, London, Intermediate Technology Publications, pp. 220-224.

Bebbington, A.J. (1996), Movements, Modernizations, and Markets. Indigenous Organizations and Agrarian Strategies in Ecuador', in R. Peet and M. Watts (eds) *Liberation Ecologies: Development, Environment, Social Movements*, London, Routledge, pp. 86-109.

Bennett, J.W. (1980), 'Management Style: A Concept and a Method for the Analysis of Family-Operated Agricultural Enterprise', in P.F. Barlett (ed), *Agricultural Decision Making. Anthropological Contributions to Rural Development*, New York, Academic Press, pp. 203-237.

Berkes, F. (ed) (1989a), *Common Property Resources. Ecology and Community-Based Sustainable Development*, London, Belhaven Press.

Berkes, F. (1989b), 'Cooperation from the Perspective of Human Ecology', in F. Berkes (ed), *Common Property Resources. Ecology and Community-Based Sustainable Development*, London, Belhaven Press, pp. 70-88.

Berkes, F. and Taghi Farvar, M. (1989), 'Introduction and Overview', in F. Berkes (ed), *Common Property Resources. Ecology and Community-Based Sustainable Development*, London, Belhaven Press, pp. 1-17.

Bernstein, H. (1978), 'Notes on Capital and Peasantry', *Review of African Political Economy*, vol. 10, pp. 60-73.

Berry, S.A. (1980), 'Decision Making and Policymaking in Rural Development', in P.F. Barlett (ed), *Agricultural Decision Making. Anthropological Contributions to Rural Development*, New York, Academic Press. pp. 321-336.

Bhattacharaya, F. (1998), 'Forest and Forest Dwellers in Modern Bengali Fiction' in Jeffery, R. (ed) (1998), *The Social Construction of Indian Forests*, Delhi, Manohar, pp. 25-40.

Birkeland, J. (1993), 'Ecofeminism: Linking Theory and Practice', in G. Gaard (ed), *Ecofeminism: Women, Animals, Nature*, Philadelphia, PA, Temple University Press.

Birsa Agricultural University (1992), *Annual Report 1991-1992*, Ranchi, Birsa
    Agricultural University.
Birsa Agricultural University (1995), *Annual Report 1994-1995*, Ranchi, Birsa
    Agricultural University.
Birsa Agricultural University (1996), *Indian Council of Agricultural Research.
    National Agricultural Research Project for Zone IV, V and VI in Bihar.
    'Impact'*, Ranchi, Birsa Agricultural University.
Blackburn, J. with Holland, J. (eds) (1998a), *Who Changes? Institutionalizing
    Participation in Development*, London, Intermediate Technology
    Publications.
Blackburn, J. and Holland, J. (1998b) 'General Introduction', in Blackburn, J.
    with J. Holland (eds) (1998), *Who Changes? Institutionalizing
    Participation in Development*, London, Intermediate Technology
    Publications, pp. 1-8.
Blaikie, P.M. (1975), *Family Planning in India: Diffusion and Policy*, London,
    Edward Arnold.
Blaikie, P.M. (1985), *The Political Economy of Soil Erosion in Developing
    Countries*, New York, Longman Scientific and Technical.
Blaikie, P.M. and Brookfield, H. (1987a), *Land Degradation and Society*,
    London, Methuen.
Blaikie, P.M. and Brookfield, H. (1987b), 'Defining and Debating the Problem',
    in P.M. Blaikie and H. Brookfield, *Land Degradation and Society*,
    London, Methuen, pp. 1-26.
Blaikie, P.M. and Brookfield, H. (1987c), 'Approaches to the Study of Land
    Degradation' in P.M. Blaikie and H. Brookfield, *Land Degradation and
    Society*, London, Methuen, pp. 27-48.
Blaikie, P.M., Harriss, J.C. and Pain, A.N. (1986), 'The Management and Use of
    Common Property Resources in Tamil Nadu, India' in Panel on
    Common Property Resource Management, Board on Science and
    Technology for International Development, Office of International Affairs
    and National Research Council, *Proceedings of the Conference on
    Common Property Resource Management, April 21-26, 1985*,
    Washington, D.C., National Academy Press, pp. 481-504.
Blauert, J. (1988), 'Autochthonous Development and Environmental Knowledge
    in Oaxaca, Mexico', in P. Blaikie and T. Unwin, (eds), *Environmental
    Crises in Developing Countries*. Monograph No. 5, Developing Areas
    Research Group, Institute of British Geographers, pp. 33-54.
Blauert, J. and Guidi, M. (1992), 'Strategies for Autochthonous Development:
    Two Initiatives in Rural Oaxaca, Mexico', in D. Ghai and J.M. Vivian
    (eds), *Grassroots Environmental Action. People's Participation in
    Sustainable Development*, London, Routledge, pp. 188-220.
Booth, D. (1985), 'Marxism and Development Sociology: Interpreting the
    Impasse', *World Development*, vol. 13 (7), pp. 761-87.
Booth, D. (ed) (1994), *Rethinking Social Development: Theory, Research and
    Practice*, Harlow, Longman Scientific and Technical.
Bose, S.R. and Ghosh, P.P. (1976), *Agro Economic Survey of Bihar. A Pilot
    Study*, Patna, B.K. Enterprise.

Boserup, E. (1970), *Women's Role in Economic Development*, New York, St Martin's Press.

Boster, J. (1984), 'Inferring Decision Making from Preferences and Behavior: An Analysis of Aguaruna Jivaro Manioc Selection', *Human Ecology*, vol. 12 (4), pp. 343-358.

Box, L. (1989), 'Virgilio's Theorem: a Method for Adaptive Agricultural Research', in R. Chambers, A. Pacey, and L.A. Thrupp (eds), *Farmer First. Farmer Innovation and Agricultural Research*, London, Intermediate Technology Publications, pp. 61-67.

Boyer, P.S. and Nissenbaum, S. (1974), *Salem Possessed: the Social Origins of Witchcraft*, Cambridge, Harvard University Press.

Bradnock, R.W. and Saunders, P.L., (2000), 'Sea-level Rise, Subsidence and Submergence. The Political Ecology of Environmental Change in the Bengal Delta', in Stott, P. and Sullivan, S., (2000), *Political Ecology. Science, Myth and Power*, London, Arnold, pp. 67-90.

Braidotti, R., Charkiewicz, E., Hausler, S. and Wieringa, S. (1994), *Women, the Environment and Sustainable Development: Towards a Theoretical Synthesis*, London, Zed Books.

Brass, T. (1990), 'Class Struggle and the De-proletarianisation of Agricultural Labour in Haryana (India)', *The Journal of Peasant Studies*, vol. 18 (1).

Brass, T. (1994), 'Some Observations on Unfree Labor, Capitalist Restructuring, and De-proletarianisation', *International Review of Social History*, vol. 39 (2).

Brokensha, D.W. and Riley, B.W. (1980), 'Mbeere Knowledge of their Vegetation and its Relevance for Development (Kenya)', in D.W. Brokensha, D.M. Warren and O. Werner (eds), *Indigenous Knowledge Systems and Development*, Lanham, MD, University Press of America, pp.113-132.

Brokensha, D.W., Warren, D.M. and Werner, O. (eds) (1980), *Indigenous Knowledge Systems and Development*, Lanham, MD, University Press of America.

Brush, S. (1980), 'Potato Taxonomies in Andean Agriculture', in D.W. Brokensha, D.M. Warren and O. Werner (eds), *Indigenous Knowledge Systems and Development*, Lanham, MD, University Press of America, pp. 37-48.

Bryant, R.L. (1992), 'Political Ecology: an Emerging Research Agenda in Third-World Studies', *Political Geography*, vol. 11, pp. 12-36.

Bryant, R.L. (1997), 'Beyond the Impasse: the Power of Political Ecology in Third World Environmental Research' *Area*, vol. 29, pp. 1-15.

Bryant, R.L. and Bailey, S. (1997), *Third World Political Ecology*, London, Routledge.

Byres, T.J. (1979), 'Of Neo-Populist Pipe Dreams', *The Journal of Peasant Studies*, vol. 6, pp. 210-44.

Byres, T.J. (1981), 'Agrarian Structure, the New Technology and Class Action in India', *The Journal of Peasant Studies*, vol. 8, pp. 322-42.

Byres, T.J. (1983), *Green Revolution in India*, Milton Keynes, Open University Press.

Campbell, A. (1994), 'Community First: Landcare in Australia', in I. Scoones and J. Thompson (eds), *Beyond Farmer First. Rural People's Knowledge,*

Agricultural Research and Extension Practice, London, Intermediate
     Technology Publications, pp. 252-258.

Cancian, F. (1972), *Change and Uncertainty in a Peasant Economy*, Stanford,
     Stanford University Press.

Cancian, F. (1980), 'Risk and Uncertainty in Agricultural Decision Making', in
     P.F. Barlett (ed), *Agricultural Decision Making. Anthropological
     Contributions to Rural Development*, New York, Academic Press, pp.
     161-176.

Carrin-Bouez, M. (1990), *Inner Frontiers: Santhal Responses to Acculturation*,
     Working Paper. DERAP Development Research and Action Programme,
     Chr. Michelsen Institute, Department of Social Science and
     Development.

Carson, R. (1964), *The Silent Spring*, New York, Fawcett.

Casley, D.J. and Lury, D.A. (1987), *Data Collection in Developing Countries*,
     Oxford, Clarendon Press.

Centre for Womens' Development Studies (1991), *Tribal Women: Saviours of the
     Forest*, Report of a Workshop (3rd to 5th April, 1990, Ranchi, Bihar),
     Ranchi, Centre for Womens' Development Studies.

Chambers, R. (1978), 'Towards Rural Futures', *IDS Discussion paper* No. 134,
     London, Institute of Development Studies/International Institute of
     Envionment and Development.

Chambers, R. (1979), 'Rural Development: Whose Knowledge Counts?', Special
     Issue, vol. 10 (2), *IDS Bulletin*, Institute of Development Studies,
     University of Sussex.

Chambers, R. (1981) 'Rapid Rural Approach: Rationale and Repertoire', *Public
     Administration and Development*, vol. 1, pp. 95-106.

Chambers, R. (1983), *Rural Development: Putting the Last First*, New York,
     Longman.

Chambers, R. (1987), 'Sustainable livelihoods', *IDS Discussion paper* No. 240,
     London, Institute of Development Studies/International Institute of
     Envionment and Development.

Chambers, R. (1989), 'The State and Rural Development', *IDS Discussion paper*
     No. 269, London, Institute of Development Studies/International
     Institute of Envionment and Development.

Chambers, R. (1992), 'Rural Appraisal: Rapid, Relaxed and Participatory', *IDS
     Discussion paper* No. 311, London, Institute of Development
     Studies/International Institute of Envionment and Development.

Chambers, R. (1994), 'Paradigm Shifts and the Practice of Participatory
     Development', *IDS Working Paper* No. 2, London, Institute of
     Development Studies/International Institute of Envionment and
     Development.

Chambers, R. (1995), 'Poverty and Livelihoods: Whose Reality Counts?', *IDS
     Discussion paper* No. 347, London, Institute of Development
     tudies/International Institute of Envionment and Development.

Chambers, R. (1998), 'Foreword' in Gujit, I. and Kaul Shah, M. (eds) (1998), *The
     Myth of Community. Gender Issues in Participatory Development*,
     London, Intermediate Technology Publications, pp. xvii-xx.

Chambers, R. and Conway, G. (1992), 'Sustainable Rural Livelihoods: Practical Concepts for the 21st Century', *IDS Discussion Paper* No. 296, Brighton: IDS.

Chambers, R. and Ghildyal, B.P. (1985), 'Agricultural Research for Resource-Poor Farmers: The Farmer-First-and-Last Model', *Agricultural Administration*, vol. 20, pp. 1-30.

Chambers, R. and Leach, M. (1990), 'Trees as Savings and Security for the Rural Poor', *Unasylva*, vol. 41, p. 161.

Chambers, R. and Longhurst, R. (1986), 'Trees, Seasons and the Poor', *IDS Bulletin*, vol. 17 (3), pp. 44-50.

Chambers, R., Pacey, A. and Thrupp, L.A. (eds) (1989), *Farmer First. Farmer Innovation and Agricultural Research*, London, Intermediate Technology Publications.

Chandrakanth, M.G. Gilless, J.K. Gowramma, V. and Nagaraja, M.G. (1990), 'Temple Forests in India's Forest Development', *Agroforestry Systems*, vol. 11, pp. 199-211.

Chapman, G.P. (1983), 'The Folklore of the Perceived Environment in Bihar', *Environment and Planning A*, vol. 15, pp. 945-968.

Chaturvedi, A.N. (1996), 'Forestry in India: Tasks Ahead', *Wastelands News*, August-October 1996, pp. 12-14.

Chatwin, B. (1988), *The Songlines*, London, Picador.

Chaudhuri, N. (1941), 'The Sun as a Folk God', *Man in India*, 21 (1), pp. 8-16.

Chibnik, M. (1980), 'The Statistical Behaviour Approach: The Choice between Wage Labor and Cash Cropping in Rural Belize', in P.F. Barlett (ed), *Agricultural Decision Making. Anthropological Contributions to Rural Development*, New York, Academic Press, pp. 87-114.

Clifford, J. and Marcus, G. (eds) (1986), *Writing Culture. The Poetics and the Politics of Ethnography*, Berkeley and Los Angeles, University of California Press.

Cohen, P. (1967), 'Economic Analysis and Economic Man', in R. Firth (ed), *Themes in Economic Anthropology*, London, Tavistock, pp. 91-118.

Colchester, M. (1994), 'Sustaining the Forests: The Community-Based Approach in South and South-East Asia', *Development and Change*, vol. 25, pp. 69-100.

Collins, G.F.S. (1921), 'Extension of Minor Forests for the Purposes of Forest Privileges in Ankola, Kumta and Honavar Talukas and Bhaktal Petha' in *Modifications in the Forest Settlements Kanara Coastal Tract*, Part 1 No. F.O.R., S.R.VI. 7/1920, Karwar, Mahomedan Press.

Conway, G., Chambers, R., McCracken, J. and Pretty, J. (1988), *RRA Notes*, Number 1, London, International Institute for Environment and Development.

Conway, G.R. (1998), *The Doubly Green Revolution: Food for All in the 21st Century*, Cornell University Press, Ithaca.

Conway, G.R. and Barbier, E.B. (1990), *After the Green Revolution. Sustainable Agriculture for Development*, London. Earthscan.

Cooke, R. (1999) 'On Sustainable JFM', *Wastelands News* XIV (2), pp. 69-71.

Corbridge, S.E. (1986), 'State Tribe and Religion: Policy and Politics in India's Jharkhand. 1900-1980', Ph.D. Dissertation, University of Cambridge.

Corbridge, S.E. (1990), 'Post-Marxism and Development Studies. Beyond the Impasse', *World Development*, vol. 18 (5), pp. 623-39.

Corbridge, S.E. (1991a), 'Ousting *Singbonga*: the Struggle for India's Jharkhand', in C. Dixon and M.J. Heffernan (eds), *Colonialism and Development in the Contemporary World*, London, Mansell, pp. 153-182.

Corbridge, S.E. (1991b), 'The Poverty of Planning or Planning for Poverty?: An Eye to Economic Liberalisation in India', *Progress in Human Geography*, vol. 15 (4), pp. 467-476.

Corbridge, S.E. (1993), 'Marxisms, Modernities, and Moralities: Development Praxis and the Claims of Distant Strangers', *Environment and Planning D: Society and Space*, vol. 11, pp. 449-472.

Corbridge, S.E. (ed) (1995a), *Development Studies: A Reader*, London, Edward Arnold.

Corbridge, S.E. (1995b), 'Thinking About Development. Editor's Introduction', in Corbridge, S.E. (ed), *Development Studies: A Reader*, London, Edward Arnold, pp. 1-16.

Corbridge, S.E. (1995c), 'Forest Struggles and Forest Protection in Tribal Bihar, India', Public Lecture: Department of Geography, University of Colorado, Boulder, April, 1995.

Corbridge, S.E. and Jewitt, S. (1997), 'From Forest Struggles to Forest Citizens? Joint Forest Management in the Unquiet Woods of India's Jharkhand', *Environment and Planning A*, vol. 29 (12), pp. 2145-2164.

Cornwall, A. (1998), 'Gender, Participation and the Politics of Difference', in Gujit, I. and Kaul Shah, M. (eds) (1998), *The Myth of Community. Gender Issues in Participatory Development*, London, Intermediate Technology Publications, pp. 46-57.

Cornwall, A., Gujit, I. and Welbourn, A. (1994), 'Acknowledging Process: Challenges for Agricultural Research and Extension Methodology', in I. Scoones and J. Thompson (eds), *Beyond Farmer First. Rural People's Knowledge, Agricultural Research and Extension Practice*, London, Intermediate Technology Publications, pp. 98-117.

Correa, M. (1997), 'Gender and Joint Forest Planning and Management: a Research Study in Uttar Kannada District, Karnataka', New Delhi: India Development Services, Bangalore and Society for Promotion of Wasteland Development.

Crawley, H. (1998), 'Living up to the Empowerment Claim? The Potential of PRA', in Gujit, I. and Kaul Shah, M. (eds) (1998), *The Myth of Community. Gender Issues in Participatory Development*, London, Intermediate Technology Publications, pp. 24-34.

D'Abreo, D. (1982), *People and Forests. The Forest Bill and a New Forest Policy*, New Delhi, Indian Social Institute.

Daly, M. (1978), *Gyn/ecology, the Metaphysics of Radical Feminism*, Boston, Beacon Press.

Dankelman, I. and Davidson, J. (1988), *Women and Environment in the Third World. Alliance for the Future*, London, Earthscan.

Dasgupta, S. (1985), 'Adivasi Politics in Midnapur, 1760-1824', in Guha, R., *Subaltern Studies IV*, New Delhi, Oxford University Press, pp. 101-135.

de Beauvoir, S. (1988), *The Second Sex*, London, Picador.

Devalle, S.B.C. (1992), *Discourses of Ethnicity. Culture and Protest in Jharkhand*, New Delhi, Sage.

Devavaram, J., Arunothayam, E., Prasad R. and Pretty, J. (1999), 'Watershed and Community in Tamil Nadu, India', in F. Hinchcliffe, J. Thompson, J. Pretty, I. Gujit and P. Shah (eds) (1999), *Fertile Ground. The Impacts of Participatory Watershed Management*, London, Intermediate Technology Publications, pp. 324-331.

Development Assistance Committee (DAC) of the Organisation for Economic Co-operation and Development (1995), 'Gender Equality: Moving Towards Sustainable, People-Centred Development', http://www.oecd.org/dac/htm.

Development Studies Unit, Department of Social Anthropology, Stockholm University (1996), 'Gender and Forestry - Important Issues', *Wastelands News*, vol. XI, (3), pp.27-41.

DfID (1997), 'Eliminating Poverty: A Challenge for the 21st Century', White Paper on International Development, London, HMSO.

DfID (1999), *Sustainable Livelihoods Guidance Sheets*, London, DfID.

DfID (2001), *Departmental Report 2001*, London, The Stationary Office, http://www.dfid.gov.uk.

Dhar, T.N. (1999), 'The Tragedy of the Commons Must be Averted' *Wastelands News* XIV (4) 1999, pp. 55-63.

Dixon, C. (1990), *Rural Development in the Third World*, London, Routledge.

Dogra, B. (1983), *People and Forests (A Report on the Himalayas)*, New Delhi, Indian Social Institute.

Drinkwater, M. (1994a), 'Knowledge, Consciousness and Prejudice: Adaptive Agricultural Research in Zambia', in I. Scoones and J. Thompson (eds), *Beyond Farmer First. Rural People's Knowledge, Agricultural Research and Extension Practice*, London, Intermediate Technology Publications, pp. 32-41.

Drinkwater, M. (1994b), 'Developing Interaction and Understanding: RRA and Farmer Research Groups in Zambia', in I. Scoones and J. Thompson (eds), *Beyond Farmer First. Rural People's Knowledge, Agricultural Research and Extension Practice*, London, Intermediate Technology Publications, pp. 133-139.

Dyson, T. and M. Moore (1983), 'Kinship Structure, Female Autonomy and Demographic Behaviour', *Population and Development Review*, vol. 9 (1), pp. 35-60.

Eckholm, E. (1984), *Fuelwood: The Energy Crisis that Won't Go Away*, London, Earthscan.

Ehrlich, P. (1966), *The Population Bomb*, New York, Ballantine.

Elson, D. (1988), 'Dominance and Dependency in the World Economy', in B. Crow and M. Thorpe (eds), *Survival and Change in the Third World*, Cambridge, Polity/Open University Press, pp. 264-87.

Elwin, H.V.H. (1936), *Leaves from the Jungle. Life in a Gond Village*, London, John Murray.

Elwin, H.V.H. (1939), *The Baiga*, London, John Murray.

Elwin, H.V.H. (1964), *The Tribal World of Verrier Elwin. An Autobiography*, London, Oxford University Press.

Escobar, A. (1992), 'Reflections on 'Development'', *Futures*, vol. 24, pp. 411-36.

Escobar, A. (1995), *Encountering Development. The Making and Unmaking of The Third World*, Princeton, Princeton University Press.

Escobar, A. (1996), 'Constructing Nature. Elements for a Poststructural Political Ecology', in R. Peet and M. Watts (eds) *Liberation Ecologies: Development, Environment, Social Movements*, London, Routledge, pp. 46-68.

Esteva, G. (1987), 'Regenerating People's Space', *Alternatives*, vol. XII, pp. 125-52.

Esteva, G. (1992), 'Development', in Sachs, W. 1992. (ed), *The Development Dictionary*, London, Zed Books, pp. 6-25.

Esteva, G. and Prakash, M.S. (1992), 'Grassroots Resistance to Sustainable Development: Lessons from the Banks of the Narmada', *The Ecologist*, vol. 22, pp. 45-51.

Fairhead, J. (n.d)., 'Indigenous Technical Knowledge and Natural Resource Management in Sub-Saharan Africa: a Critical Overview', Unpublished draft.

Fairhead, J. and Leach, M. (1994), 'Declarations of Difference', in I. Scoones and J. Thompson (eds), *Beyond Farmer First. Rural People's Knowledge, Agricultural Research and Extension Practice*, London, Intermediate Technology Publications, pp. 75-79.

Falconer, J. (1987), *Forestry Extension: A Review of the Key Issues*, ODI Social Forestry Network, Paper 4e. London, ODI.

Fanon, F. (1967), *The Wretched of the Earth*, Harmondsworth, Penguin Books.

FAO (1985), *Tree Growing by Rural People*, FAO Forestry Paper 64, Rome, FAO.

FAO (1986), *Forestry Extension Organisation*, FAO Forestry Paper 66, Rome, FAO.

FAO (1986b), *Monitoring and Evaluating Social Forestry in India - an Operational Guide*, FAO Forestry Paper 75, Rome, FAO.

FAO, Committee on Agriculture (1999), *Report on Follow-up to Agenda 21. Highlights of FAO Programmes*, 15th Session Rome, 25-29 January 1999, Rome: FAO.

FAO (2001), 'Participation', www.fao.org/participation/ourvision.html.

Farmer, A. and Bates, R. (1996), 'Community Versus Market', *Comparative Political Studies*, vol. 29 (4), pp. 379-400.

Farrington, J. (1988), Farmer Participatory Research, Editorial Introduction, *Experimental Agriculture*, vol. 24, pp. 269-279.

Fernandes, W. (1983), 'Towards a New Forest Policy: An Introduction', in W. Fernandes and S. Kulkarni (eds), *Towards a New Forest Policy. Peoples' Rights and Environmental Needs*, New Delhi, Indian Social Institute, pp. 1-24.

Fernandes, W. (ed) (1998), *Drafting a People's Forest Bill: The Forest Dweller-Social Activist Alternative*, A Sponsored Publication of Indian Social Institute, New Delhi No. 60.69. New Delhi, Indian Social Institute.

Fernandes, W. and Kulkarni, S. (eds) (1983), *Towards a New Forest Policy. Peoples' Rights and Environmental Needs*, New Delhi, Indian Social Institute.

Fernandez, A.P. (1999), 'Equity, Local groups and Credit: lessons from MYRADA's Work in South India', in F. Hinchcliffe, J. Thompson, J. Pretty, I. Gujit and P. Shah (eds) (1999) *Fertile Ground. The Impacts of Participatory Watershed Management*, London, Intermediate Technology Publications, pp. 295-308.

Foley, G. and Barnard, G. (1985), *Farm and Community Forestry*, ODI Social Forestry Network, Paper 1b, London, ODI.

Fortmann, L. (1986), 'Women in Subsistence Forestry', *Journal of Forestry*, vol. 84.

Foucault, M. (1990), *The Archaeology of Knowledge*, London, Routledge.

Freeman, R. (1998), 'Folk-Models of the Forest Environment in Highland Malabar', in Jeffery, R. (ed) (1998), *The Social Construction of Indian Forests*, Delhi, Manohar, pp. 55-78.

Gadgil, M. (1983), 'Forestry with a Social Purpose', in W. Fernandes and S. Kulkarni (eds), *Towards a New Forest Policy. Peoples' Rights and Environmental Needs*, New Delhi, Indian Social Institute, pp. 111-134.

Gadgil, M. and Guha, R. (1992), *This Fissured Land. An Ecological History of India*, New Delhi, Oxford University Press.

Gadgil, M. and Guha, R. (1994), 'Ecological Conflicts and the Environmental Movement in India', *Development and Change*, vol. 25, pp. 101-136.

Gadgil, M. and Guha, R. (1995), *Ecology and Equity. The Use and Abuse of Nature in Contemporary India*, London, Routledge.

Gadgil, M. and Iyer, P. (1989), 'On the Diversification of Common Property Resource Use by Indian Society', in F. Berkes (ed), *Common Property Resources. Ecology and Community-Based Sustainable Development*, London, Belhaven Press, pp. 240-255.

Gadgil, M. and Subash Chandran, M.D. (1992), 'Sacred Groves', in G. Sen (ed), *Indigenous Vision. Peoples of India. Attitudes to the Environment*, New Delhi, Sage Publications, pp. 183-188.

Ghai, D. (1994), 'Environment, Livelihood and Empowerment', *Development and Change*, vol. 25 pp. 1-11.

Ghai, D. and Vivian, J.M. (eds) (1992a), *Grassroots Environmental Action. People's Participation in Sustainable Development*, London, Routledge.

Ghai, D. and Vivian, J.M. (1992b), 'Introduction', in D. Ghai and J.M. Vivian (eds), *Grassroots Environmental Action. People's Participation in Sustainable Development*, London, Routledge, pp. 1-19.

Ghate, R. (1998), 'A Modern 'Sacred Grove'' *Wastelands News* XIII (3) 1998, pp. 33-35.

Ghate, R.S. (1992), *Forest Policy and Tribal Development. A Study of Maharashtra*, New Delhi, Concept Publishing Company.

Gibbs, J.N. and Bromley, D.W. (1989), 'Institutional Arrangements for Management of Rural Resources: Common Property Regimes', in F. Berkes (ed), *Common Property Resources. Ecology and Community-Based Sustainable Development*, London, Belhaven Press, pp. 22-32.

Giddens, A. (1990), *The Consequences of Modernity*, Cambridge, Polity Press.

Gleissman, S.R., Garcia, R. and Amador, A. (1981), 'The Ecological Basis for the Application of Traditional Agricultural Technology in the Management of Tropical Agro-Ecosystems', *Agro-Ecosystems*, vol. 7, pp. 173-185.

Goldsmith, E. (1992), 'Gaia is the Source of all Benefits', in G. Sen (ed), *Indigenous Vision. Peoples of India. Attitudes to the Environment*, New Delhi, Sage Publications, pp. 13-24.

Goldsmith, E.R.D., Allen, R., Allaby, M., Davoll, J. and Lawrence, S. (1972), 'A Blueprint for Survival', *The Ecologist*, vol. 2 (1), pp. 1-43.

Government of Bihar (1937), *Annual Progress Report of Forest Administration in the Province of Bihar for 1936-7*, Gulzarbagh, Government of Bihar.

Government of Bihar (1938), *Annual Progress Report of Forest Administration in the Province of Bihar for 1937-8*, Gulzarbagh, Government of Bihar.

Government of Bihar (1939), *Annual Progress Report of Forest Administration in the Province of Bihar for 1938-9*, Gulzarbagh, Government of Bihar.

Government of Bihar (1940), *Annual Progress Report of Forest Administration in the Province of Bihar for 1939-40*, Gulzarbagh, Government of Bihar.

Government of Bihar (1941), *Annual Progress Report of Forest Administration in the Province of Bihar for 1930-41*, Gulzarbagh, Government of Bihar.

Government of Bihar (1947), *Annual Progress Report of Forest Administration of the Province of Bihar for 1946-7*, Gulzarbagh, Government of Bihar.

Government of Bihar (1948), *Annual Progress Report of Forest Administration in the State of Bihar for 1947-8*, Gulzarbagh, Government of Bihar.

Government of Bihar (1949), *Annual Progress Report of Forest Administration in the Province of Bihar for 1948-9*, Gulzarbagh, Government of Bihar.

Government of Bihar (1950), *Annual Progress Report of Forest Administration in the Province of Bihar for 1949-50*, Gulzarbagh, Government of Bihar.

Government of Bihar (1951), *Annual Progress Report of Forest Administration in the State of Bihar for 1950-1*, Gulzarbagh, Government of Bihar.

Government of Bihar (1952), *Annual Progress Report of Forest Administration in the State of Bihar for 1951-2*, Gulzarbagh, Government of Bihar.

Government of Bihar (1953), *Annual Progress Report of Forest Administration in the State of Bihar for 1952-3*, Gulzarbagh, Government of Bihar.

Government of Bihar (1954), *Annual Progress Report of Forest Administration in the State of Bihar for 1953-4*, Gulzarbagh, Government of Bihar.

Government of Bihar (1955), *Annual Progress Report of Forest Administration in the State of Bihar for 1954-5*, Gulzarbagh, Government of Bihar.

Government of Bihar (1956), *Annual Progress Report of Forest Administration in the State of Bihar for 1955-6*, Gulzarbagh, Government of Bihar.

Government of India (1881), *Census of India, 1881*, New Delhi, Government of India.

Government of India (1901), *Census of India, 1901*, New Delhi, Government of India.

Government of India (1907), *Judicial and Administrative Statitstics of British India for 1905-1906 and Preceding Years*, Calcutta, Superintendent, Government Printing.

Government of India (1911a), *Census of India, 1911. Volume 1*, Calcutta, Superintendent, Government Printing.

Government of India (1911b), *Census of India, 1911. Volume V. Bengal, Bihar and Orissa and Sikkim. Part 1. Report*, Calcutta, Superintendent, Government Printing.

Government of India (1921), *Census of India, 1921*, New Delhi, Government of India.

Government of India (1931), *Census of India, 1931*, New Delhi, Government of India.

Government of India (1941), *Census of India, 1941*, New Delhi, Government of India.

Government of India (1951), *Census of India, 1951*, New Delhi, Government of India.

Government of India (1961), *Census of India, 1961*, New Delhi, Government of India.

Government of India (1971), *Census of India, 1971*, New Delhi, Government of India.

Government of India (1981), *Census of India, 1981*, New Delhi, Government of India.

Government of India (1990), *Resolution (1st June) on 'Involvement of Village Communities and Voluntary Agencies for Regeneration of Degraded Forest Land'*, New Delhi, Government of India.

Government of India (1991), *Census of India, 1991*, New Delhi, Government of India.

Government of India, Ministry of Agriculture and Irrigation (1976), *Report of the National Commission on Agriculture, Part IX: Forestry*, New Delhi, Government of India.

Government of India, Ministry of Environment and Forests (1997), *Annual Report 1996-7*, New Delhi, Government of India.

Government of India, Ministry of Environment and Forests (Forest Protection Division) (2000), Resolution No.22-8/2000-JFM (FPD).

Government of Orissa (1990), *Resolution (11th December 1990) on 'Protection of Reserved Forests and Protected forests and Enjoyment of Certain Usufructs by the Community'*, Bhubaneshwar, Government of Orissa.

Government of Orissa (1993), 'Notification No. 16700-10F (Pron) - 20/93-F&E dated 3-7-1993', *Protection of Reserved Forest and Protected Forest Areas by the Community and Enjoyment of Certain Usufructs by the Community*, Bhubaneshwar, Government of Orissa.

Granovetter, M. (1985), 'Economic Action and Social Structure: the Problem of Embeddedness', *American Journal of Sociology*, 91 (3), 481-510.

Greenland, D.J. (1997), *The Sustainability of Rice Farming*, Wallingford, CAB International, IRRI.

Gubbels, P. (1994), 'Populist Pipedream or Practical Paradigm?: Farmer-Driven Research and the Project Agro-Forestier in Burkina Faso', in I. Scoones and J. Thompson (eds), *Beyond Farmer First. Rural People's Knowledge, Agricultural Research and Extension Practice*, London, Intermediate Technology Publications, pp. 238-244.

Guha, R. (1983), 'Forestry in British and Post British India. A Historical Analysis', *Economic and Political Weekly*, October 1983, 1882-1897 and November 1983, 1940-1947.

Guha, R. (1989), *The Unquiet Woods. Ecological Change and Peasant Resistance in the Himalaya*, Delhi, Oxford University Press.

Guha, R. (1992), 'Prehistory of Indian Environmentalism. Intellectual Traditons', *Economic and Political Weekly*, January 4-11, 57-64.

Guha, R. (1993), 'The Malign Encounter: The Chipko Movement and Competing Visions of Nature', in T. Banuri and F. Apffel Marglin (eds), *Who Will Save the Forests? Knowledge, Power and Environmental Destruction*, London, Zed Books, pp. 80-113.

Guha, S. (2000), 'Economic Rents and Natural Resources: Commons and Conflicts in Premodern India', in A. Agrawal and K. Sivaramakrishnan, (eds) (2000), *Agrarian Environments. Resources, Representation, and Rule in India*, Durham and London, Duke University Press, pp. 132-146.

Gujit, I. and Kaul Shah, M. (eds) (1998a), *The Myth of Community. Gender Issues in Participatory Development*, London, Intermediate Technology Publications.

Gujit, I. and Kaul Shah, M. (1998b), 'Waking up to Power, Conflict and Process' in Gujit, I. and Kaul Shah, M. (eds) (1998), *The Myth of Community. Gender Issues in Participatory Development*, London, Intermediate Technology Publications, pp. 1-23.

Gupta, A.K. (1986), 'Socioecology of Stress: Why do Common Property Resource Management Projects Fail?', in Panel on Common Property Resource Management, Board on Science and Technology for International Development, Office of International Affairs and National Research Council, *Proceedings of the Conference on Common Property Resource Management. April 21-26, 1985*, Washington, D.C., National Academy Press, pp. 305-321.

Gupta, A.K. (1989), 'Scientists' Views of Farmers' Practices in India: Barriers to Effective Interaction', in R. Chambers, A. Pacey, and L.A. Thrupp (eds), *Farmer First. Farmer Innovation and Agricultural Research*. London, Intermediate Technology Publications, pp. 24-30.

Gupta, A.K. (1992), 'Sustainability Through Biodiversity: Designing a Crucible of Culture, Creativity and Conscience', Indian Institute of Management, Ahmedabad, Working Paper No. 1005.

Gupta, A.K. (1996), 'Roots of Creativity and Innovation in Indian Society: A Honey Bee Perspective', *Wastelands News*, August-October 1996, pp. 37-68.

Gupta, T.D. (1992a), in *Economic and Political Weekly*, June 27th, 1992.

Gupta, T.D. (1992b), in *Economic and Political Weekly*, February 29th, 1992.

Gururani, S. (2000), 'Regimes of Control, Strategies of Access: Politics of Forest Use in the Uttarakhand Himalaya, India', in A. Agrawal and K. Sivaramakrishnan (eds) (2000), *Agrarian Environments. Resources, Representation, and Rule in India*, Durham and London, Duke University Press, pp. 170-190.

Hadly, M. and Schreckenberg, K. (1995) 'Traditional Ecological Knowledge and UNESCO's Man in the Biosphere (MAB) Programme', in D.M. Warren,

L.J. Slikkerveer, and D. Brokensha (eds), *The Cultural Dimension of Development. Indigenous Knowledge Systems*, London, Intermediate Technology Publications, pp. 464-474.

Haggett, P. (1983), *Geography: a Modern Synthesis, 3rd Edition*, New York, Harper & Row.

Hallett, M.G. (1917), *Bihar and Orissa Gazetteers. Ranchi*, Patna, Superintendent, Government Printing, Bihar and Orissa.

Hardiman, D. (1987), *The Coming of the Devi. Adivasi Assertion in Western India*, New Delhi, Oxford University Press.

Hardin, G. (1968), 'The Tragedy of the Commons', *Science*, vol. 162, pp. 1243-1248.

Hardin, G. (1974), 'Living on a lifeboat', *Bioscience* 24, 561-8.

Harriss, B. and Watson, E. (1987), 'The Sex Ratio in South Asia', in J. Momsen, and J. Townshend, (eds), *The Geography of Gender in the Third World*, London, Hutchinson, 1987, pp. 85-115.

Hecht, S.B. (1985), 'Environment, Development and Politics: Capital Accumulation and the Livestock Sector in Eastern Amazonia', *World Development*, vol. 13 (6), pp. 663-684.

Hecht, S.B. (1987), 'The Evolution of Agroecological Thought', in M.A. Altieri with contributions by R.B. Norgaard, S.B. Hecht, J.G. Farell and M. Liebman, *Agroecology. The Scientific Basis of Alternative Agriculture*, Boulder, Westview Press, London, Intermediate Technology Publications, pp. 1-20.

Hecht, S.B. and Cockburn, A. (1989), *The Fate of the Forest: Developers, Destroyers and Defenders of the Amazon*, London, Verso.

Heinz, H.J. and Maguire, B. (n.d.), 'The Ethno-Biology of the Ko! Bushmen: Their Ethno-Botanical Knowledge and Plant Lore', Occasional Paper No. 1. Gaborone, Botswana Society.

Henshall Momsen, J. (ed) (1991), *Women and Development in the Third World*, London, Routledge.

Hinchcliffe, F., Thompson, J., Pretty, J., Gujit, I. and Shah, P. (eds) (1999) *Fertile Ground. The Impacts of Participatory Watershed Management*, London, Intermediate Technology Publications.

Hindustan Fertiliser Corporation, (1990), 'Towards Sustainable Agriculture', *Prayas Rainfed Farming Project News Letter*, vol. 1 (3), December, 1990.

Hobley, M. (1996), *Participatory Forestry: the Process of Change in India and Nepal*, ODI Rural Development Forestry Study Guide 3, London, ODI.

Hoffman, J.S.J. (1961), 'Principles of Succession and Inheritance among the Mundas', *Man in India*, 41 (4), pp. 324-338.

Holmgren, P., Juma Masakha, E. and Sjoholm, H. (1994), 'Not all African Land is being Degraded: A Recent Survey of Trees on Farms in Kenya Reveals Rapidly Increasing Forest Resources', *Ambio* 23 (7), pp. 390-95.

Horta, C. (2000), 'Biodiversity Conservation and the Political Economy of International Financial Institutions, in Stott, P. and Sullivan, S., (2000), *Political Ecology. Science, Myth and Power*, London, Arnold, pp. 179-202.

Howes, M. (1980), 'The Uses of Indigenous Technical Knowledge in Development', in D.W. Brokensha, D.M. Warren and O. Werner (eds), *Indigenous Knowledge Systems and Development*, Lanham, MD, University Press of America, pp. 341-358.

Howes, M. and Chambers, R. (1980), 'Indigenous Technical Knowledge: Analysis, Implications, and Issues', in D.W. Brokensha, D.M. Warren and O. Werner (eds), *Indigenous Knowledge Systems and Development*, Lanham, MD, University Press of America, pp. 329-340.

Humble, M. (1998), 'Assessing PRA for Implementing Gender and Development', in Gujit, I. and Kaul Shah, M. (eds) (1998), *The Myth of Community. Gender Issues in Participatory Development*, London, Intermediate Technology Publications, pp. 35-45.

IDS Workshop (1989a), 'Farmers' Knowledge, Innovations and Relation to Science', in R. Chambers, A. Pacey, and L.A. Thrupp (eds), *Farmer First. Farmer Innovation and Agricultural Research*, London, Intermediate Technology Publications, pp. 31-38.

IDS Workshop (1989b), 'Interactions for Local Innovation', in R. Chambers, A. Pacey, and L.A. Thrupp (eds), *Farmer First. Farmer Innovation and Agricultural Research*, London, Intermediate Technology Publications, Publications, pp. 43-50.

IDS Workshop (1989c), 'Interactive Research', in R. Chambers, A. Pacey, and L.A. Thrupp (eds), *Farmer First. Farmer Innovation and Agricultural Research*, London, Intermediate Technology Publications, pp. 122-126.

IDS Workshop (1989d), 'Farmers Groups and Workshops', in R. Chambers, A. Pacey, and L.A. Thrupp (eds), *Farmer First. Farmer Innovation and Agricultural Research*, London, Intermediate Technology Publications, pp. 100-105.

Imam, B. (1990), 'Ecologial Background of Chota Nagpur - its Imbalance and Tribal Communities', *Man in India*, vol. 70 (3), pp. 267-277.

Inden, R. (1986), 'Orientalist Constructions of India', *Modern Asian Studies*, vol. 20 (3), pp. 401-446.

Indurkar, P. (1992), *Forestry, Environment and Economic Development*, New Delhi, Ashish Publishing House.

International Rice Research Institute (1983), *Women in Rice Farming*, Proceedings of a Conference on Women in Rice Farming Systems, Aldershot, Gower, IRRI.

International Rice Research Institute (1994), *Integrated Pest Management: The IRRI Perspective*, IRRI Information Series No. 3, Los Banos, IRRI.

Isbell, B.J. (1978), *To Defend Ourselves: Ecology and Ritual in an Andean Village*, Prospect Heights, Waveland Press.

Jackson, C. (1993a), 'Women/Nature or Gender/History? A Critique of Ecofeminist 'Development'', *The Journal of Peasant Studies*, vol. 20 (3), pp. 389-419.

Jackson, C. (1993b), 'Doing what Comes Naturally? Women and Environment in Development', *World Development*, vol. 21 (12), pp. 1947-1963.

Jackson, C. (1994), 'Gender Analysis and Environmentalisms' in M. Redclift and T. Benton (eds), *Social Theory and Global Environment*, London, Routledge, pp. 113-149.

Jackson, C. (1995), 'From Conjugal Contracts to Environmental Relations: Some Thoughts on Labour and Technology', *IDS Bulletin*, 26 (1), pp. 33-46.

Jackson, C. and Chattopadhyay, M. (2000), 'Identities and Livelihoods: Gender, Ethnicity, and Nature in a South Bihar Village', in A. Agrawal and K. Sivaramakrishnan (eds) (2000), *Agrarian Environments. Resources, Representation, and Rule in India*, Durham and London, Duke University Press, pp. 146-169.

Jahan, R. (1998), 'Gender and International Institutions' in World Bank *Proceedings of the Gender and Development Workshop*, Washington D.C., World Bank.

Janzen, D.H. (1973), 'Tropical Agroecosystems', *Science* 182, pp. 1212-1219.

Jeffery, R. (ed) (1998a), *The Social Construction of Indian Forests*, Delhi, Manohar.

Jeffery, R. (1998b), 'Introduction' in Jeffery, R. (ed) (1998), *The Social Construction of Indian Forests*, Delhi, Manohar, pp.1-16.

Jewitt, S. (1995a), 'Voluntary and 'Official' Forest Protection Committees in Bihar: Solutions to India's Deforestation?', *Journal of Biogeography*, 22 (6), pp. 1003-1021.

Jewitt, S. (1995b), 'Europe's 'Others'? Forestry Policy and Practices in Colonial and Post-Colonial India', *Environment and Planning D: Society and Space*, 13, pp. 67-90.

Jewitt, S. (1996), 'Agro-ecological Knowledges and Forest Management in the Jharkhand, India: Tribal development or Populist Impasse?', Ph.D. Dissertation, Cambridge University.

Jewitt, S. (1998), 'Autonomous and Joint Forest Management in India's Jharkhand: Lessons for the Future?', in Jeffery, R. (ed) (1998), *The Social Construction of Indian Forests*, Delhi, Manohar, 121-143, pp. 145-168.

Jewitt, S. (2000a), 'Mothering Earth? Gender and Environmental Protection in the Jharkhand, India', *The Journal of Peasant Studies*, vol. 27 (2), pp 94-131.

Jewitt, S. (2000b), 'Unequal Knowledges in Jharkhand, India: Deromanticizing Women's Agroecological Expertise', *Development and Change,* vol. 31 (5) pp. 961-985.

Jewitt, S. and Kumar, S. (2000), 'A Political Ecology of Forest Management: Gender and Silvicultural Knowledge in the Jharkhand, India', in Stott, P. and Sullivan, S. (2000), *Political Ecology. Science, Myth and Power*, London, Arnold, pp. 91-113.

Jodha, N. S. (1986), 'Common Property Resources and the Rural Poor', *Economic and Political Weekly*, vol. 21 (27), pp. 1169-1181.

Joeckes, S. with Heyzer, N. Oniang'o, R and Salles, V. (1994), 'Gender, Environment and Population', *Development and Change*, vol. 25, pp. 137-165.

Johnston, B. and Kilby, P. (1975), *Agriculture and Structural Transformation*, London, Hutchinson.

Joshi, G. (1983), 'Forest Policy and Tribal Development: Problems of Implementation, Ecology and Exploitation', in W. Fernandes and S. Kulkarni (eds), *Towards a New Forest Policy. Peoples' Rights and Environmental Needs*, New Delhi, Indian Social Institute, pp. 25-47.

Kakat, S. (1993), 'Where are the Women?', *Hindustan Times*, 7th February, 1993, Calcutta, Hindustan Times.

Kalam, M.A. (1998), 'Sacred Groves in Coorg, Karnataka', in Jeffery, R. (ed) (1998), *The Social Construction of Indian Forests*, Delhi, Manohar, pp. 41-54.

Kandiyoti, D. (1988), 'Bargaining with Patriarchy', *Gender and Society*, vol. 2 (3) pp. 274-90.

Kant, S., Singh, N.M. and Singh, K.K. (1991), *Community Based Forest Management Systems (Case Studies from Orissa)*, New Delhi, Indian Institute of Forest Management, Bhopal, Swedish International Development Agency, New Delhi and ISO/Swedforest, New Delhi.

Kapadia, K. (1995), 'Where Angels Fear to Tread?: 'Third World Women' and 'Development'', Review Article, *The Journal of Peasant Studies*, vol. 22 (2), pp. 356-368.

Karlsson, B. G. (2000) *Contested Belonging. An Indigenous People's Struggle for Forest and Identity in Sub-Himalayan Bengal*, Richmond, Curzon.

Kaufman, M. (1997a), 'Community Power, Grassroots Democracy, and the Transformation of Social Life' in M. Kaufman and H.D. Alfonso, (eds) (1997), *Community Power and Grassroots Democracy. The Transformation of Social Life*, London, Zed Books, pp. 1-24.

Kaufman, M. (1997b), 'Differential Participation: Men, Women and Popular Power' in M. Kaufman and H.D. Alfonso, (eds) (1997), *Community Power and Grassroots Democracy. The Transformation of Social Life*, London, Zed Books, pp. 151-169.

Kaul Shah, M. (1998), '"Salt and Spices": Gender issues in Participatory Programme Implementation in AKRSP, India', in Gujit, I. and Kaul Shah, M. (eds) (1998), *The Myth of Community. Gender Issues in Participatory Development*, London, Intermediate Technology Publications, pp. 243-253.

Kaul Shah, M. and Shah, P. (1995), 'Gender, Environment and Livelihood Security: An Alternative Viewpoint from India', *IDS Bulletin*, vol. 26 (1), pp. 75-82.

Kean, S.A. (1988), 'Developing a Partnership between Farmers and Scientists: the Example of Zambia's Adaptive Research Planning Team', *Experimental Agriculture*, vol. 24, pp. 289-299.

Kearney, R. (1984), *Dialogues with Contemporary Continental Thinkers*, Manchester, Manchester University Press.

Keil, R., Bell, D.V.J., Penz, P. and Fawcett, L. (eds) (1998), *Political Ecology. Global and Local*, London, Routledge.

Kelkar, G. and Nathan, D. (1991), *Gender and Tribe. Women, Land and Forests in Jharkhand*, New Delhi, Kali for Women.

Khiewtam, Ramesh, S. and Ramakrishnan, P.S. (1989), Socio-cultural Studies of the Sacred Groves at Cherrapunji and Adjoining Areas in North-Eastern India, *Man in India*, 69 (1), pp. 64-71.

Khush, G.S. (1995), 'Breaking the Yield Frontier of Rice', *Geojournal*, vol. 35, pp. 329-332.

Kiernan, V.G. (1969), *The Lords of Humankind. Black Man, Yellow Man and White Man in an Age of Empire*, Hammondsworth, Penguin.

Kishwan, J. (1996), 'Forestry Beyond Forests', *Wastelands News*, August-October 1996, pp. 5-9.

Kitching, G. (1982), *Development and Underdevelopment in Historical Perspective. Populism, Nationalism and Industrialisation (Revised Edition)*, London, Routledge.

Kittredge, G.L. (1972), *Witchcraft in Old and New England*, New York, Atheneum.

Korton, D.C. (1980), 'Community Organisation and Rural Development - A Learning Process Approach', *Public Administration Review*, vol. 40 (5), pp. 480-511.

Korton, D.C. (ed) (1986), *Community Management. Asian Experience and Perspecitves*, New Delhi, Kumarian Press.

Krishnakumar, A. (1999) 'Terminator of Food Security', *Wastelands News*, XIV (2) pp. 63-66.

Kulkarni, S. (1983), 'The Forest Policy and Forest Bill. A Critique and Suggestions for Change', in W. Fernandes and S. Kulkarni (eds), *Towards a New Forest Policy. Peoples' Rights and Environmental Needs*, New Delhi, Indian Social Institute, pp. 84-102.

Kulkarni, S. (1987), 'Forest Legislation and Tribals: Comments on Forest Policy Resolution', *Economic and Political Weekly*, December 12, 1987, pp. 143-148.

Kulkarni, S. (1992), 'Encroachment on Forests. Government Versus People', *Economic and Political Weekly*, vol. 17 (3), pp. 55-59.

Kumar, N. (1970), *Bihar District Gazetteers. Ranchi*, Patna, Government of Bihar, Gazetteers Branch.

Kumar, P.J.D. (1991), 'Between Common Property and State Monopoly: The Institutional Context of Wasteland Regeneration in India', in Singh, R. (ed), *Managing the Village Commons, (Proceedings of the National Workshop on Managing Common Lands for Sustainable Development of Our Villages: A Search for Participatory Management Models)*, December 15-16, 1991 Bhubaneshwar, Bhopal, Indian Institute of Forest Management, pp. 53-74.

Kumar, S. (1998), 'Food, Forest and the People', *Wastelands News* XIII (4) pp. 29-33.

Kumar, S. and Rangan, R. (1996), 'Feel of First Fruits of JFM in Bihar', *Wastelands News*, August-October, pp. 23-24.

Kumar, S. (2001), 'Indigenous Communities' Knowledge of Local Ecological Services', *Economic and Political Weekly*, vol. 36 (30), pp. 2859-2869.

Lal, M. (1983), *The Munda Elites. Recruitment, Network, Attitudes, Perception and Role in Social Transformation*, New Delhi, Amar Prakashan.

Lal, R. (1998), 'JFM under World Food Programme in Rajasthan', *Wastelands News*, XIII (3) pp. 56-57.

Leach, M. (1991), 'Engendered Environments: Understanding Natural Resource Management in the West African Forest Zone', *IDS Working Paper* No. 16.

Leach, M., Joekes, S. and Green, C. (1995), 'Editorial: Gender Relations and Environmental Change', *IDS Bulletin*, vol. 26, (1), pp. 1-8.

Lele, L. (1991), 'Sustainable Development: a Critical Review', *World Development*, vol. 19, pp. 607-621.

Lightfoot, C. De Guia, J.R. and Ocado, F. (1988), 'A Participatory Method for Systems-Problem Research: Rehabilitating Marginal Uplands in the Philippines', *Experimental Agriculture*, vol. 24, pp. 301-309.

Linkenbach, A. (1994), 'Ecological Movements and the Critique of Development', *Thesis Eleven*, vol. 39, pp. 63-85.

Linkenbach, A. (1998), 'Forests in Garhwal and the Construction of Space', in Jeffery, R. (ed) (1998), *The Social Construction of Indian Forests*, Delhi, Manohar, pp. 79-105.

Lipton, M. (1977), *Why Poor People Stay Poor: A Study of Urban Bias in World Development*, London, Temple Smith.

Lipton, M. (1999), 'Reviving Global Poverty Reduction: What Role for Genetically Modified Plants?', 1999 Sir John Crawford Memorial Lecture, Washington D.C., Consultative Group on International Agricultural Research (CGIAR).

Lipton, M. with Longhurst, R. (1989), *New Seeds and Poor People*, London, Unwin and Hyman.

Locke, C. (1995), 'Planning for the Participation of Vulnerable Groups in Communal Management of Forest Resources: The Case of the Western Ghats Forestry Project', Unpublished Ph.D. Thesis, Centre for Development Studies, Swansea.

Locke, C. (1999), 'Constructing a Gender Policy for Joint Forest Management in India', *Development and Change*, vol. 30, pp. 265-285.

Lohmann, L. (1993) 'Green Orientalism' *The Ecologist*, vol. 23 (6), pp. 34-38.

Long, N. and Long, A. (eds) (1992), *Battlefields of Knowledge. The Interlocking of Theory and Practice in Social Research and Development*, London, Routledge.

Long, N. and Villareal, M. (1994), 'The Interweaving of Knowledge and Power in Development Interfaces', in I. Scoones and J. Thompson (eds), *Beyond Farmer First. Rural People's Knowledge, Agricultural Research and Extension Practice*, London, Intermediate Technology Publications, pp. 41-52.

Mahapatra, S. (1992), 'Invocation: Rites of Propitiation in Tribal Societies', in G. Sen (ed), *Indigenous Vision. Peoples of India. Attitudes to the Environment*, New Delhi, Sage Publications, pp. 63-74.

Marglin, S.A. (1990a), 'Towards the Decolonization of the Mind', in F. Apffel Marglin and S.A. Marglin (eds), *Dominating Knowledge. Development, Culture and Resistance*, Oxford, Clarendon Press, pp. 1-28.

Marglin, S.A. (1990b), 'Losing Touch: the Cultural Conditions of Worker Accommodation and Resistance', in F. Apffel Marglin and S.A. Marglin (eds), *Dominating Knowledge. Development, Culture and Resistance*, Oxford, Clarendon Press, pp. 217-282.

Marsden, D. (1994), 'Indigenous Management and the Management of Indigenous Knowledge', in I. Scoones and J. Thompson (eds), *Beyond Farmer First.*

Rural People's Knowledge, Agricultural Research and Extension Practice, London, Intermediate Technology Publications, pp. 52-57.

Matose, F. and Mukamuri, B. (1994), 'Trees, People and Communities in Zimbabwe's Communal Lands', in I. Scoones and J. Thompson (eds), *Beyond Farmer First. Rural People's Knowledge, Agricultural Research and Extension Practice*, London, Intermediate Technology Publications, pp. 69-75.

Maurya, D.M. (1989), 'The Innovative Approach of Indian Farmers', in R. Chambers, A. Pacey, and L.A. Thrupp (eds), *Farmer First. Farmer Innovation and Agricultural Research*, London, Intermediate Technology Publications, pp. 9-13.

Maurya, D.M., Bottrall, A. and Farrington, J. (1988), 'Improved Livelihoods, Genetic Diversity and Farmer Participation: A Strategy for Rice Breeding in Rainfed Areas of India', *Experimental Agriculture*, vol. 24, pp. 311-320.

Mawdsley, E. (1998), 'After Chipko: From Environment to Region in Uttaranchal', *The Journal of Peasant Studies*, vol. 25 (4), pp. 36-54.

Mayoux, L. (1995), 'Beyond Naivety: Women, Gender Inequality and Participatory Development', *Development and Change*, vol. 26 (2), pp. 235-258.

McCracken, J. (1995), 'International Institute for Environment and Development (IIED) and Rapid Rural Appraisal for Indigenous Sustainable Development', in D.M. Warren, L.J. Slikkerveer, and D. Brokensha (eds), *The Cultural Dimension of Development. Indigenous Knowledge Systems*, London, Intermediate Technology Publications, pp. 451-453.

McNeely, J.A. (1995), 'IUCN and Indigenous Peoples: How to Promote Sustainable Development', in D.M. Warren, L.J. Slikkerveer, and D. Brokensha (eds), *The Cultural Dimension of Development. Indigenous Knowledge Systems*, London, Intermediate Technology Publications, pp. 445-450.

Meadows, D.H. (1972), *Limits to Growth. A Report on the Club of Rome's Project on the Predicament of Mankind*, London, Potomac Associates Book.

Meehan, P. (1980), 'Science, Ethnoscience and Agricultural Knowledge Utilization', in D.W. Brokensha, D.M. Warren and O. Werner (eds), *Indigenous Knowledge Systems and Development*, Lanham, MD, University Press of America, pp. 383-392.

Mehrotra, S. and Kishore, C. (1990), *A Study of Voluntary Forest Protection in Chotanagpur, Bihar*, New Delhi, ISO/Swedforest, Indian Institute of Forest Management, Bhopal.

Menon, A. (1995), 'Constructing the 'Local'. Decentralising Forest Management', *Economic and Political Weekly*, August 26, pp. 2110-2111.

Merchant, C. (1989), *The Death of Nature: Women, Ecology and the Scientific Revolution,* San Francisco, Harper and Row.

Messerschmidt, D.A. (1987), 'Conservation and Society in Nepal: Traditional Forest Management and Innovative Development', in P.D. Little, M.M. Horowitz, with A.E. Nyerges (eds), *Lands at Risk in the Third World. Local Level Perspectives*, Boulder, Westview Press, pp. 373-397.

Messerschmidt, D.A. (1995), 'Local Traditions and Community Forest
    Management: A View from Nepal', in D.M. Warren, L.J. Slikkerveer,
    and D. Brokensha (eds), *The Cultural Dimension of Development.
    Indigenous Knowledge Systems*, London, Intermediate Technology
    Publications, pp. 231-244.
Mies, M. and Shiva, V. (1993), *Ecofeminism*, New Delhi, Kali for Women.
Millar, D. (1994), 'Experimenting Farmers in Northern Gambia', in I. Scoones and
    J. Thompson (eds), *Beyond Farmer First. Rural People's Knowledge,
    Agricultural Research and Extension Practice*, London, Intermediate
    Technology Publications, pp. 160-165.
Mishra, R. (1993), 'People's Zeal Inspires State Participation', *Wastelands News*,
    May-July, pp. 31-33.
Mishra, V.K. (1998) 'Parliament Approves Expert Committee Disproves. Rights
    of Panchayats on Minor Forest Produce', *Wastelands News*, XIII (4) pp.
    8-9.
Mitra, S.C. (1919), 'The Mango Tree in the Marriage Ritual of the Aborigines of
    Chota Nagpur and Santalia', *Journal of the Bihar and Orissa Research
    Society*, vol. 5 (2), pp. 259-271.
Moore, P.D., Chaloner, W.G. and Stott, P.A. (1996), *Global Environmental
    Change*, Oxford, Blackwell Science.
Moser, C. (1993), *Gender, Planning and Development. Theory, Practice and
    Training*, London, Routledge.
Mosse, D. (1994), 'Authority, Gender and Knowledge: Theoretical Reflections on
    the Practice of Participatory Rural Appraisal', *Development and Change*,
    vol. 25, pp. 497-526.
Mueller, A. (1987), 'Power and Naming in the Development Institution. The
    "Discovery" of "Women in Peru."' Paper Presented at the Fourteenth
    Annual Third World Conference, Chicago.
Mukherji, S.D. (1998), 'Is Handing Over Forests to Local Communities a
    Solution to Deforestation?', *Wastelands News*, XIII (4), pp. 22-28.
Mukhopadhyay, A. (1996), 'The Coming of Women into Panchayati Raj',
    Occasional Paper No.2, School of Women's Studies, Jadavpur
    University, Calcutta.
Munda, R.D. (1988), 'The Jharkhand Movement. Retrospect and Prospect', *Social
    Change*, vol. 18 (2), pp. 28-42.
Mundy, P.A. and Compton, J.L. (1995), 'Indigenous Communication and
    Indigenous Knowledge', in D.M. Warren, L.J. Slikkerveer, and D.
    Brokensha (eds), *The Cultural Dimension of Development. Indigenous
    Knowledge Systems*, London, Intermediate Technology Publications, pp.
    112-123.
Murthy, Ranjani K. (1998), 'Learning about Participation from Gender relations of
    Female Infanticide', in Gujit, I. and Kaul Shah, M. (eds) (1998), *The
    Myth of Community. Gender Issues in Participatory Development*,
    London, Intermediate Technology Publications, pp. 78-92.
MYRADA (1998), 'Equity in Watershed Management' *Wastelands News*, XIII (3)
    1998 pp. 15-17.
Nair, K. (1979), *In Defence of the Irrational Peasant*, Chicago, University of
    Chicago Press.

Nanda, M. (1991), 'Is Modern Science a Western Patriarchal Myth? A Critique of the Populist Orthodoxy', *South Asia Bulletin*, vol. XI, pp. 36-61.

Nanda, M. (2001a), 'Breaking the spell of Dharma. A Case for Indian Enlightenment', *Economic and Political Weekly*, vol. 36 (27) pp. 2251-2566.

Nanda, M. (2001b), 'We are All Hybrids Now: The Dangerous Epistemology of Post-Colonial Populism', *The Journal of Peasant Studies*, vol. 28 (2), pp. 162-186.

Nandy, A. and Visvanathan, S. (1990), 'Modern Medicine and its Non-Modern Critics: A Study in Discourse', in F. Apffel Marglin and S.A. Marglin (eds), *Dominating Knowledge. Development, Culture and Resistance*, Oxford, Clarendon Press, pp. 145-184.

Narayan, D. and Rietbergen-McCracken, J., (1998), *Participation and Social Assessment: Tools and Techniques*, Washington D.C., World Bank.

Nathan, D. (1988), 'Factors in the Jharkhand Movement', *Economic and Political Weekly*, January 30, pp. 185-187.

Neefjes, K. (2000), *Environments and Livelihoods: Strategies for Sustainability*, Oxford, Oxfam.

Nesmith, C. (1991), 'Women and Trees: Social Forestry in West Bengal, India', Unpublished Ph.D. Dissertation, University of Cambridge.

Netting, R. (1976), 'What Alpine Peasants have in Common: Observations on Communal Tenure in a Swiss Village', *Human Ecology*, vol. 4, pp. 135-46.

Netting, R. (1997), 'Unequal Commoners and Uncommon Equity: Property and Community among Smallholder Farmers', *The Ecologist*, vol. 27 (1).

Norem, R.H., Yoder, R. and Martin, Y. (1989), 'Indigenous Agricultural Knowledge and Gender Issues in Third World Agricultural Development', in D.M. Warren, L.J. Slikkerveer and S. Oguntunji Titilola (eds), *Indigenous Knowledge Systems: Implications for Agriculture and International Development*, Studies in Technology and Social Change, No. 11, Technology and Social Change Program, Ames, Iowa State University, pp. 91-100.

Norgaard, R. (1992), *Development Reportrayed*, London, Routledge.

Noronha, R. (1981), 'Why is it So Difficult to Grow Fuelwood?', *Unasylva*, vol. 33, p. 131.

Noronha, R. and Spears, J.S. (1988), 'Sociological Variables in Forest Project Design', in M. Cernea (ed), *Putting People First*, Oxford, Oxford University Press, pp. 227-266.

Okali, C., Sumberg, J. and Farrington, J. (1994), *'Farmer Participatory Research: Rhetoric and Reality'*, London, Intermediate Technology Publications.

Omvedt, G. (1994), 'Peasants, Dalits and Women: Democracy and India's New Social Movements', *Journal of Contemporary Asia*, vol. 24 (1), pp. 35-48.

Oommen, T.K. (1981), 'Gandhi and Village: Towards a Critical Appraisal', in S.K. Lal (ed), *Gandhiji and Village*, New Delhi, Agricole Publishing Company, pp. 1-11.

Ortner, S.B. (1974), 'Is Female to Male as Nature is to Culture?', in M.Z. Rosaldo and Lamphere, L. (eds), *Women, Culture and Society*, Stanford, Stanford University Press.

Ostrom, E. (1990), *Governing the Commons. The Evolution of Institutions for Collective Action*, Cambridge, Cambridge University Press.

Ostrom, E. (1999), 'Self-Governance and Forest Resources', Occasional Paper No. 20, Center for International Forestry Research, Bogor, Indonesia. http://www.cgiar.cifor.

Ostrom, E., Schroeder, L. and Wynne, S. (1993), *Institutional Incentives and Sustainable Development. Infrastructure Policies in Perspective*, Boulder, Westview Press.

Ouedraogo, K., Traore, O. and Nebie, K. (1994), 'Forest Management and Farmers Participation. The Case of Bougnounou Village in the Province of Sissili, Burkino Faso', *Forest, Trees and People Newsletter*, vol. 24, pp. 44-47.

Pagden, A. (1982), *The Fall of Natural Man, The American and the Origins of Comparative Ethnology*, Cambridge, Cambridge University Press.

Palit, S. (1993), *The Future of Indian Forest Management: Into the Twenty First Century*, New Delhi: National Support Group for Joint Forest Management, Society for Promotion of Wastelands Development and Ford Foundation. Joint Forest Management Working Paper 14.

Pangare, V.H. (1998), 'Gender Dynamics. Creating Gender Awareness in Watershed Development and Management', *Wastelands News*, XIII (3), pp. 26-27.

Parajuli, P. (1991), Power and Knowledge in Development Discourse, *International Social Science Journal*, vol. 127, pp. 173-90.

Pathak, A. (1994), *Contested Domains. The State, Peasants and Forests in Contemporary India*, New Delhi, Sage.

Pathan, R. S. (1998), 'The Community Action in Development Leads to Sustenance...Let Us Prove It', *Wastelands News*, XIII (4), pp. 34-36.

Pathan, R.S. (n.d)., 'The Evolution of the Forest Protection Committee Program', in R.S. Pathan, N.J. Arul and M. Poffenberger (eds), *Forest Protection Committees in Gujarat - Joint Management Initiative*, Sustainable Forest Management Working Paper Series. Working Paper No. 7, New Delhi, Ford Foundation, pp. 1-13.

Peet, R. and Watts, M. (1996a), 'Liberation Ecology: Development, Sustainability, and Environment in an Age of Market Triumphalism', in R. Peet and M. Watts (eds) *Liberation Ecologies: Development, Environment, Social Movements*, London, Routledge, pp. 1-45.

Peet, R. and Watts, M. (eds) (1996b), *Liberation Ecologies: Development, Environment, Social Movements*, London, Routledge.

Pereira, W. (1992), The Sustainable Lifestyle of the Warlis. Pages 189-204 in G. Sen (ed), *Indigenous Vision. Peoples of India. Attitudes to the Environment*, New Delhi, Sage Publications.

Pieterse, J.N. (1991), 'Dilemmas of Development Discourse: The Crisis of Developmentalism and Comparative Method', *Comparative Change*, vol. 22, pp. 5-29.

Pingali, P.L., Hossain, M. and Gerpacio, R.V. (1997), *Asian Rice Bowls. The Returning Crisis?*, International Rice Research Institute, Manila.

Pinney, C. (1988), 'Representations of India: Normalization and the Other', *Pacific Viewpoint*, vol. 29, pp. 144-163.

Poffenberger, M. (1990), *Joint Forest Management for Forest Lands. Experiences from South Asia*, A Ford Foundation Statement, New Delhi, Ford Foundation.

Poffenberger, M. (1996), 'Valuing the Forests' in Poffenberger, M. and McGean, B. (1996), *Village Voices, Forest Choices. Joint Forest Management in India*, New Delhi, Oxford University Press, pp. 259-286.

Potter, D. (1998), 'NGOs and Forest Management in Karnataka', in Jeffery, R. (ed) (1998), *The Social Construction of Indian Forests*, Delhi, Manohar, pp. 121-143.

Prabhu, P. (1983), 'Social Forestry: an Adivasi Viewpoint', in W. Fernandes and S. Kulkarni (eds), *Towards a New Forest Policy. Peoples' Rights and Environmental Needs*, New Delhi, Indian Social Institute, pp. 134-144.

Pretty, J. and Chambers, R. (1993), 'Towards a Learning Paradigm: New Professionalism and Institutions for Agriculture', *Institute of Development Studies Discussion Paper* 334, London, Institute of Development Studies/International Institute of Environment and Development.

Pretty, J. and Chambers, R. (1994), 'Towards a New Learning Paradigm: New Professionalism and Institutions for Agriculture', in I. Scoones and J. Thompson (eds), *Beyond Farmer First. Rural People's Knowledge, Agricultural Research and Extension Practice*, London, Intermediate Technology Publications, pp. 182-203.

Pretty, J. and Shah, P. (1999), 'Soil and Water Conservation: A Brief History of Coercion and Control', in F. Hinchcliffe, J. Thompson, J. Pretty, I. Gujit and P. Shah, (eds) (1999) *Fertile Ground. The Impacts of Participatory Watershed Management*, London, Intermediate Technology Publications, pp. 1-26.

Pretty, J. and Ward, H. (2001), 'Social Capital and the Environment', *World Development*, vol. 29 (2), pp. 209-227.

Rahman, H.Z. (2001), *Re-thinking Local Governance towards a Livelihood Focus*, PPRC Policy Paper 1/2001, London: Power and Participation Research Centre.

Rahnema, M. (1988), 'On a New Variety of AIDS and its Pathogens: Homo-economicus, Development and Aid', *Alternatives*, vol. 13 (1), pp. 117-136.

Ramakrishnan, P.S. and Patnaik, S. (1992), 'Jhum: Slash and Burn Cultivation', in G. Sen (ed), *Indigenous Vision. Peoples of India. Attitudes to the Environment*, New Delhi, Sage Publications, pp. 215-220.

Ranchi District Gazetteer (1905), *Statistics 1901-02*, Calcutta, The Bengal Secretariat Book Depot.

Ranchi District Gazetteer (1915), *Statistics 1900-1901 to 1910-1911*, Calcutta, Government Press.

Rangan, H. (1996), 'From Chipko to Uttaranchal: Development, Environment and Social Protest in the Garhwal Himalayas, India', in R. Peet and M.

Watts, *Liberation Ecologies: Environment, Development, Social Movements*, London, Routledge, pp. 205-226.

Rangan, H. (2000a), *Of Myths and Movements. Rewriting Chipko into Himalayan History*, London, Verso.

Rangan, H. (2000b), 'State Economic Policies and Changing Regional Landscapes in the Uttarakhand Himalaya', in A. Agrawal and K. Sivaramakrishnan, (eds) (2000), *Agrarian Environments. Resources, Representation, and Rule in India*, Durham and London, Duke University Press, pp. 23-46.

Rastogi, A. (1998) 'Conflict Resolution. A Challenge for Joint Forest Management in India', *Wastelands News*, XIII (4), pp. 17-21.

Razavi, S. and Miller, C. (1995), 'Gender Mainstreaming: A Study of Efforts by the UNDP, the World Bank and the ILO to Institutionalize Gender Issues', Occasional Paper No. 4, UN Fourth World Conference on Women. August 1995.

Reardon, T. and Vosti, S.A. (1995), 'Links Between Rural Poverty and the Environment in Developing Countries: Asset Categories and Investment Poverty', *World Development*, vol. 23 (9), pp. 1495-1506.

Redclift, M. (1984), *Development and the Environmental Crisis: Red or Green Alternatives*, New York, Methuen.

Redclift, M. (1987), *Sustainable Development: Exploring the Contradictions*, London, Methuen.

Redclift, M. (1992), 'Sustainable Development and Popular Participation: A Framework for Analysis', in D. Ghai and J.M. Vivian (eds), *Grassroots Environmental Action. People's Participation in Sustainable Development*, London, Routledge, pp. 23-49.

Reddy, C.R.M. (1999) 'Regeneration Through Watershed Development', *Wastelands News* XIV (2) 1999, pp. 26-27

Regier, H.A., Mason, R.V. and Berkes, F. (1989), 'Reforming the Use of Natural Resources', in F. Berkes (ed), *Common Property Resources. Ecology and Community-Based Sustainable Development*, London, Belhaven Press, pp. 110-126.

Reid, J. (1912), *Final Report on the Survey and Settlement Operations in the District of Ranchi, 1902-1910*, Calcutta, Bengal Secretariat Book Depot.

Rhoades, R.E. (1989), 'The Role of Farmers in the Creation of Agricultural Technology', in R. Chambers, A. Pacey, and L.A. Thrupp (eds), *Farmer First. Farmer Innovation and Agricultural Research*, London, Intermediate Technology Publications, pp. 3-8.

Rhoades, R.E. and Bebbington, A.J. (1988), 'Farmers who Experiment: An Untapped Resource for Agricultural Research and Development', Paper presented at the International Congress on Plant Physiology, New Delhi, 15-20 February.

Rhoades, R.E. and Bebbington, A.J. (1995), 'Farmers Who Experiment: An Untapped Resource for Agricultural Research and Development', in D.M. Warren, L.J. Slikkerveer, and D. Brokensha (eds), *The Cultural Dimension of Development. Indigenous Knowledge Systems*, London, Intermediate Technology Publications, pp. 296-307.

Rhoades, R.E. and Booth, R.H. (1982), 'Farmer-back-to-Farmer: A Model for Generating Acceptable Agricultural Technology' *Agricultural Administration*, vol. 11, pp. 127-137.

Richards, A. (1939), *Land, Labor and Diet in Northern Rhodesia*, London, Routledge and Kegan Paul.

Richards, P. (1980), 'Community Environmental Knowledge in African Rural Development', in D.W. Brokensha, D.M. Warren and O. Werner (eds), *Indigenous Knowledge Systems and Development*, Lanham, MD, University Press of America, pp. 183-196.

Richards, P. (1985), *Indigenous Agricultural Revolution: Ecology and Rice Production in West Africa*, London, Hutchinson.

Richards, P. (1986), *Coping with Hunger: Hazard and Experiment in an African Rice-Farming System*, London, Allen and Unwin.

Richards, P. (1990a), 'Indigenous Approaches to Rural Development: the Agrarian Populist Tradition in West Africa', in M. Altieri and S. Hecht (eds), *Agroecology and Small Farm Development*, New York. CRC Press, pp. 105-111.

Richards, P. (1990b), 'Local Strategies for Coping with Hunger: Northern Nigeria and Central Sierra Leone Compared', *African Affairs*, vol. 89, pp. 265-275.

Richards, P. (1994), 'Local Knowledge Formation and Validation: The Case of Rice Production in Central Sierra Leone', in I. Scoones and J. Thompson (eds), *Beyond Farmer First. Rural People's Knowledge, Agricultural Research and Extension Practice*, London, Intermediate Technology Publications, pp. 165-170.

Rocheleau, D., Thomas-Slayter, B. and Wangari, E. (eds) (1996), *Feminist Political Ecology: Global Issues and Local Experiences*, London, Routledge.

Rocheleau, D., Wachira, K., Malaret, L. and Muchiri Wanjohi, B. (1989), Local Knowledge for Agroforestry and Native Plants, in R. Chambers, A. Pacey, and L.A. Thrupp (eds), *Farmer First. Farmer Innovation and Agricultural Research*, London, Intermediate Technology Publications, pp. 14-23.

Rosner, V. (n.d. (b)), *A Quiver Full of Arrows*, Ranchi, DSS Publications.

Rosner, V. (n.d.(a)), *The Flying Horse of Dharmes*, Ranchi, Satya Bharati Publications.

Roy, S.B. (1991), 'Forest Protection Committees in West Bengal, India: Emerging Policy Issues', in R. Singh (ed), *Managing the Village Commons. (Proceedings of the National Workshop on Managing Common Lands for Sustainable Development of Our Villages: A Search for Participatory Management Models)*, December 15-16, 1991 Bhubaneshwar. Bhopal, Indian Institute of Forest Management, pp. 84-92.

Roy, S.B. and Mukherjee, R. (1991), 'Status of Forest Protection Committees in West Bengal', in R. Singh (ed), *Managing the Village Commons. (Proceedings of the National Workshop on Managing Common Lands for Sustainable Development of Our Villages: A Search for Participatory*

Management Models), December 15-16, 1991 Bhubaneshwar, Bhopal, Indian Institute of Forest Management, pp. 113-116.

Roy, S.C. (1912), *The Mundas and Their Country*, London, Asia Publishing House.

Roy, S.C. (1915), *The Oraons of Chota Nagpur: Their History, Economic Life and Social Organisation*, Ranchi, Man in India Office.

Roy, S.C. (1916), 'The Divine Myths of the Mundas', *Journal of the Bihar and Orissa Research Society*, vol. 2 (2), pp. 201-14.

Roy, S.C. (1919), 'Probable Traces of Totem Worship among the Oraons', *Journal of the Bihar and Orissa Research Society*, vol. 1 (1), pp. 53-56.

Roy, S.C. (1922), 'The Gods of the Oraons', *Man in India*, vol. 2 (3), pp. 137-155.

Roy, S.C. (1926), 'The Oraon Feast of Sal Flowers', *Journal of the Bihar and Orissa Research Society*, vol. 12 (2), pp. 217-19.

Roy, S.C. (1927), 'Marriage Customs of the Oraons', *Journal of the Bihar and Orissa Research Society*, vol. 13 (2), pp. 171-185.

Roy, S.C. (1928), *Oraons Religion and Customs*, New Delhi, Gian Publishing House.

Roy, S.C. (1929), 'Death and its Attendant Ceremonies among the Oraons', *Journal of the Bihar and Orissa Research Society*, vol. 13 (2-3), pp. 278-291.

Roy, S.C. (1931), 'The effects on the Aborigines of Chota Nagpur of their Contact with Western Civilisation', *Man in India*, vol. 62 (1), pp. 65-100.

Roy, S.C. (1932), 'The Study of Folk Lore and Tradition in India', *Man in India*, vol. 63 (1), pp. 85-112.

Roy, S.C. (1946a), 'The Theories of Rent Among the Mundas of Chota Nagpur', *Man in India*, vol. 26 (3), pp. 156-180.

Roy, S.C. (1946b), 'The Aborigines of Chota Nagpur. Their Proper Status in the Reformed Constitution of India', *Man in India*, vol. 26 (2), pp. 120-136.

Roy, S.C. (1981), 'A New Religious Movement Among the Oraons' (Gleanings from S.C. Roy's Note Book), *Man in India*, vol. 61 (4), pp. 379-388.

Rushdie, S. (1983), *Shame*, London, J. Cape.

Saberwal, V.K. (2000), 'Environmental Alarm and Institutionalised Conservation in Himachal Pradesh, 1865-1994', in A. Agrawal and K. Sivaramakrishnan (eds) (2000), *Agrarian Environments. Resources, Representation, and Rule in India*, Durham and London, Duke University Press, pp. 68-85.

Sachidananda (1964), 'Leadership and Culture Change in Kullu', *Man in India*, vol. 44 (2), pp. 116-131.

Sachidananda (1970), *The Changing Munda*, New Delhi, Concept Publishing Company.

Sachs, W. (ed) (1992), *The Development Dictionary*, London, Zed Books.

Sahlins, M. (1972), *Stone Age Economics*, Chicago, Chicago University Press.

Said, E. W. (1978), *Orientalism*, Harmondsworth, Penguin.

Said, E. W. (1986), 'Orientalism Reconsidered', in F. Barker, P. Hulme, M. Iversen and D. Loxley (eds), *Europe and its Others, Volume 1*, Colchester, University of Essex, pp. 12-29.

Said, E. W. (1993), *Culture and Imperialism*, London, Chatto and Windus.

Saigal, S. (1995), 'An Alternative Approach to Regenerate Degraded Forests', *Wastelands News*, vol. XI (2), pp. 70-78.

Salas, M.A. (1994), '"The Technicians Only Believe in Science and Cannot Read the Sky": The Cultural Dimension of the Knowledge Conflict in the Andes', in I. Scoones and J. Thompson (eds), *Beyond Farmer First. Rural People's Knowledge, Agricultural Research and Extension Practice*, London, Intermediate Technology Publications, pp. 57-69.

Sarin, M. (1993), *From Conflict to Collaboration: Local Institutions in Joint Forest Management*, New Delhi: National Support Group for Joint Forest Management, Society for Promotion of Wastelands Development and Ford Foundation, Joint Forest Management Working Paper 14.

Sarin, M. (1995), 'Regenerating India's Forests: Reconciling Gender Equity with Joint Forest Management', *IDS Bulletin*, vol. 26 (1), 83-91.

Sarin, M. (1996), 'Actions of the Voiceless: the Challenges of Addressing Subterranean Conflicts Related to Marginalised Groups and Women in Community Forestry', Theme paper for FAO Conference on 'Addressing Natural Resource Conflicts Through Community Forestry' January-April 1996.

Sarin, M. (1998a), *Who is Gaining? Who is Losing? Gender and Equity Concerns in Joint Forest Management*, New Delhi, Society for the Promotion of Wasteland Development.

Sarin, M. (1998b), 'Community Forest Management: Whose Participation...?', in Gujit, I. and Kaul Shah, M. (eds) (1998), *The Myth of Community. Gender Issues in Participatory Development*, London, Intermediate Technology Publications, pp. 121-130.

Sarin, M. (1998c), 'Gender and Equity concerns in Joint Forest Management', in Kothari, A., Anuradha, R. and Pathak, N. (eds), (1998), *Communities and Conservation: Natural Resource Management in South and Central Asia*, New Delhi, UNESCO and Sage, pp. 323-348.

Sarin, M., Ray, L., Raju, M.S., Chatterjee, M., Banerjee, N. and Hiremath, S. (1998), 'Collaborative System of Forest Management', *Wastelands News*, XIII (4), pp. 43-47.

Saxena, N.C. (1995), 'Forest Policy and Rural Poor in Orissa', *Wastelands News*, vol. XI (2), pp. 9-13.

Saxena, N.C. (1998), 'Extract from *The Saga of Participatory Forest Management in India*, Chapter Three 'Locally Inspired Collective Action', *Wastelands News*, vol. XIII (2), pp. 30-39.

Saxena, N.C., Sarin, M., Singh, R.V. and Shah, T. (1997), 'Western Ghats Forestry Project: Independent Study of Implementation Experience in Kanara Circle', New Delhi, Society for the Promotion of Wasteland Development.

Scheper-Hughes, N. (1992), *Death Without Weeping. The Violence of Everyday Life in Brazil*, Berkeley, University of California Press.

Schultz, T. (1964), *Transforming Traditional Agriculture*, Newhaven, Yale University Press.

Schumacher, R. (1973), *Small is Beautiful*, London, Abacus.

Schuurman, F.J. (1993), 'Modernity, Post-modernity and the New Social Movements', in F.J. Schuurman (ed), *Beyond the Impasse. New Directions in Development Theory*, London, Zed Books.

Scoones, I. and Thompson, J. (1993), 'Challenging the Populist Perspective: Rural People's Knowledge, Agricultural Research and Extension Practice', *Institute of Development Studies Discussion Paper* 332, London, Institute of Development Studies/ International Institute of Environment and Development.

Scoones, I. and Thompson, J. (eds) (1994a), *Beyond Farmer First. Rural People's Knowledge, Agricultural Research and Extension Practice*, London, Intermediate Technology Publications.

Scoones, I. and Thompson, J. (1994b), 'Introduction', in I. Scoones and J. Thompson (eds), *Beyond Farmer First. Rural People's Knowledge, Agricultural Research and Extension Practice*, London, Intermediate Technology Publications, pp. 1-15.

Scoones, I. and Thompson, J. (1994c), 'Knowledge, Power and Agriculture - Towards a Theoretical Understanding', in I. Scoones and J. Thompson (eds), *Beyond Farmer First. Rural People's Knowledge, Agricultural Research and Extension Practice*, London, Intermediate Technology Publications, pp. 16-32.

Scott, A. (1990), *Ideology and the New Social Movements*, London, Unwin Hyman.

Scott, J.C. (1976), *The Moral Economy of the Peasant: Rebellions and Subsistence in South East Asia*, New Haven, Yale University Press.

Scott, J.C. (1985), *Weapons of the Weak. Everyday Forms of Peasant Resistance*, New Haven, Yale University Press.

Sen, G. (ed) (1992), *Indigenous Vision. Peoples of India. Attitudes to the Environment*, New Delhi, Sage Publications.

Sen, G. and C. Grown, (1987), *Development Crises and Alternative Visions. Third World Women's Perspectives*, London, Earthscan Publications.

Sengupta, N. (1991), *Managing Common Property. Irrigation in India and the Philippines*, New Delhi, Sage Publications.

Sengupta, S. (1993), 'Unresolved Issues in Haryana', *Wastelands News*, May-July, 1993.

Shabbeer, S. N. and Tankha, A. (1998) 'The Committee and Commitment', *Wastelands News*, vol. XIII (4), pp. 10-13.

Shah, S.A. (1995), 'Status of Indian Forestry - 1', *Wastelands News*, vol. XI (2), pp. 14-31.

Shenton, R. and Freund, W. (1978), 'The incorporation of Northern Nigeria into the World Capitalist Economy', *Review of African Political Economy*, vol. 13, pp. 8-20.

Shepherd, G. (1985), *Social Forestry in 1985: Lessons to be Learned and Topics to be Addressed*, ODI Social Forestry Network, Paper 1a, London, ODI.

Shepherd, G. (ed) (1992a), *Forest Policies, Forest Politics*, ODI Agricultural Occasional Paper 13, London, ODI.

Shepherd, G. (1992b), 'Forest Policies, Forest Politics', in G. Shepherd (ed), *Forest Policies, Forest Politics*, ODI Agricultural Occasional Paper 13, London, ODI, pp. 5-27.

Shepherd, G. (1992c), 'The reality of the Commons: Answering Hardin from Somalia', in G. Shepherd (ed), *Forest Policies, Forest Politics*, ODI Agricultural Occasional Paper 13, London, ODI, pp. 73-86.

Shepherd, G. (1992d), *Managing Africa's Tropical Dry Forests. A Review of Indigenous Methods*, ODI Agricultural Occasional Paper 14, London, ODI.

Shepherd, G. (1993a), 'Indigenous and Participatory Forest Management and the Role of External Intervention', Paper presented at National Resources Institute workshop on 'Stakeholders and Tradeoffs', March 1993.

Shepherd, G. (1993b), 'Africa: Semi-Arid and Sub-Humid Regions', in D.A. Messerschmidt (ed), *Common Forest Resource Management. Annotated Bibliography of Asia, Africa and Latin America*, Rome, FAO, pp. 99-192.

Shiva, V. (1988), *Staying Alive: Women, Ecology and Survival in India*, New Delhi, Kali for Women.

Shiva, V. (1991a), *The Violence of the Green Revolution. Third World Agriculture, Ecology and Politics*, London, Zed Books.

Shiva, V. (1993), *Monocultures of the Mind. Perspectives on Biodiversity and Biotechnology*, London, Zed Books Ltd.

Shiva, V. (ed) (1991b), *Biodiversity: Social and Ecological Perspectives*, London, Zed Books.

Shiva, V. Saratchandra, H.C. and Bandyopadhyay, J. (1983), 'The Challenge of Social Forestry', in W. Fernandes and S. Kulkarni (eds), *Towards a New Forest Policy. Peoples' Rights and Environmental Needs*, New Delhi, Indian Social Institute, pp. 48-75.

Shiva, V. Saratchandra, H.C. and Bandyopadhyay, J. (1986), 'Social Forestry for Whom?', in D.C. Korton (ed), *Community Management: Asian Experience and Perspectives*, New Delhi, Kumarian Press, pp. 238-46.

Shiva. V. (1992), 'Women's Indigenous Knowledge and Biodiversity Conservation', in G. Sen (ed), *Indigenous Vision. Peoples of India. Attitudes to the Environment*, New Delhi, Sage Publications, pp. 205-214.

Shucking, H. and Anderson, P. (1991), 'Voices Unheeded and Unheard', in V. Shiva (ed), *Biodiversity: Social and Ecological Perspectives*, London, Zed Books, pp. 13-41.

Singh, B.P. (1995), 'People's Forest - The Illambazar Story', *Wastelands News*, vol. XI (2), pp. 82-83.

Singh, C. (1986), *Common Property and Common Poverty. India's Forests, Forest Dwellers and the Law*, Delhi, Oxford University Press.

Singh, H. (1998), 'Baria Ni Kuldi. Women's Empowerment through Thrift/Credit Groups', *Wastelands News*, vol. XIII (3), pp. 13-14.

Singh, I. (1990), *The Great Ascent: The Rural Poor in South Asia*, Baltimore, Johns Hopkins University Press.

Singh, K.S. (1982a), 'Tribal Identity Movements based on Script and Language', *Man in India*, vol. 62 (3), pp. 234-242.

Singh, K.S. (1982b), 'Colonialism, Anthropology and Primitive Society. The Indian Scenario', *Man in India*, vol. 64 (4), pp. 399-413.

Singh, K.S. (ed) (1982c), *Tribal Movements in India Volume I*, New Delhi, Manohar.

Singh, K.S. (ed) (1982d), *Tribal Movements in India Volume II*, New Delhi, Manohar.

Singh, K.S. (1990), 'Environment, Technology and Management in Tribal Areas', *Man in India*, vol. 70 (2), pp. 123-30.

Singh, K.S. (1992), 'The Munda Epic: an Interpretation', in G. Sen (ed), *Indigenous Vision. Peoples of India. Attitudes to the Environment*, New Delhi, Sage Publications, pp. 75-90.

Singh, R. (1991b), 'Managing Common Lands for Sustainable Development of Our Villages: A Search for Participatory Management Models', in R. Singh (ed), *Managing the Village Commons. (Proceedings of the National Workshop on Managing Common Lands for Sustainable Development of Our Villages: A Search for Participatory Management Models)*, December 15-16, 1991 Bhubaneshwar. Bhopal, Indian Institute of Forest Management, pp. 31-36.

Singh, R. (ed) (1991a), *Managing the Village Commons. (Proceedings of the National Workshop on Managing Common Lands for Sustainable Development of Our Villages: A Search for Participatory Management Models)*, December 15-16, 1991 Bhubaneshwar, Bhopal, Indian Institute of Forest Management.

Singh, S. (1999), 'Collective Dilemmas and Collective Pursuits. Community Management of Van Panchayats (Forest Councils) in the UP Hills' *Wastelands News*, vol. XIV (4), pp. 29-45.

Singh, S. and Khare, A. (1993), 'Peoples' Participation in Forest Management', *Wastelands News*, May-July 1993, pp. 34-39.

Sinha, B.B. (1979), *Socio-Economic Life in Chotanagpur*, Delhi, B.R. Publishing Corporation.

Sinha, S.C., Sen, J. and Panchbhai, S. (1969), 'The Concept of Diku amongst the Tribes of Chota Nagpur', *Man in India*, vol. 49 (2), pp. 121-138.

Sinha, V.N.P. (1976), *Chota Nagpur Plateau. A Study in Settlement Geography*, New Delhi, K. B. Publications.

Sittirak, S. (1998), *The Daughters of Development. Women in a Changing Environment*, London, Zed Books.

Sivaramakrishnan, K. (1999), *Statemaking and Environmental Change in Colonial Eastern India*, Stanford: Stanford University Press.

Slater, D. (1985), *New Social Movements and the State in Latin America*, Dordrecht, FORIS Publications Holland.

Slater, D. (1992), 'On the Borders of Social Theory. Learning From Other Regions', *Society and Space*, vol. 10, pp. 307-27.

Society for Promotion of Wastelands Development and Forest and Environment Department, Bihar (1994), *Workshop on Joint Forest Management in Bihar*, Ranchi, Annapurna Press and Process.

Society for Promotion of Wastelands Development (1993), *Joint Forest Management Update 1993*, New Delhi, Society for Promotion of Wastelands Development.

Society for Promotion of Wastelands Development (n.d)., *Working Together: State, People and Forests of Uttara Kannada. Sahyadri Parisara Vardhini in Collaboration with People and Voluntary Organisations of the District*, New Delhi, Society for Promotion of Wastelands Development.

Sontheimer, S. (ed) (1991), *Women and the Environment: A Reader. Crisis and Development in the Third World*, London, Earthscan Publications Ltd.

Spradley, J.P. (1972), 'Adaptive Strategies of Urban Nomads', in J.P. Spradley (ed), *Culture and Cognition: Rules, Maps and Plans*, San Francisco, Chandler.

Springer, J. (2000), 'State Power and Agricultural Transformation in Tamil Nadu', in A. Agrawal and K. Sivaramakrishnan (eds) (2000), *Agrarian Environments. Resources, Representation, and Rule in India*, Durham and London, Duke University Press, pp. 85-106.

Srivastava, J.P.L. and Kaul, R.N. (1994), *Greening of the Common Lands in the Aravallis*, New Delhi, United Printing Press.

Stebbing, E.P. (1922), *The Forests of India. Volume 1*, London, J. Lane.

Stebbing, E.P. (1923), *The Forests of India. Volume 2*, London, J. Lane.

Stevenson, G.G. (1991), *Common Property Economics. A General Theory and Land Use Applications*, Cambridge, Cambridge University Press.

Stolzenbach, A. (1994), 'Learning by Improvisation: Farmer Experimentation in Mali', in I. Scoones and J. Thompson (eds), *Beyond Farmer First. Rural People's Knowledge, Agricultural Research and Extension Practice*, London, Intermediate Technology Publications, pp. 155-59.

Stott, P. and Sullivan, S. (2000a), *Political Ecology. Science, Myth and Power*, London, Arnold.

Stott, P. and Sullivan, S. (2000b), 'Introduction', in Stott, P. and Sullivan, S. (2000), *Political Ecology. Science, Myth and Power*, London, Arnold, pp. 1-11.

Sturtevant, W.C. (1964), 'Studies in Ethnoscience', *American Anthropologist*, vol. 66, pp. 99-131.

Sullivan, S. (2000), 'Getting the Science Right, or Introducing Science in the First Place? Local Facts, Global Discourse - 'Desertification' in North West Namibia', in Stott, P. and Sullivan, S. (2000), *Political Ecology. Science, Myth and Power*, London, Arnold, pp. 15-44.

Sumberg, J. and Okali, C. (1989), 'Farmers, On-Farm Research and New Technology', in R. Chambers, A. Pacey, and L.A. Thrupp (eds), *Farmer First. Farmer Innovation and Agricultural Research*, London, Intermediate Technology Publications, pp. 109-114.

Swaminathan, M.S. (1996), *Sustainable Agriculture. Towards Food Security*, New Delhi, Konark Publishers.

Swedish International Development Agency (1990), 'The S.I.D.A. Supported Bihar Social Forestry Project for Chotanagpur and Santhal Parganas, India', New Delhi, S.I.D.A. and ISO/Swedforest.

Tankha, A. (1998), 'Yes, it is Sambhav Again. "Institutional Intervention"' *Wastelands News*, vol. XIII (3), pp. 64-66.

Thomas, K. (1973), *Religion and the Decline of Magic: Studies in Popular Beliefs in Sixteenth- and Seventeenth-Century England*, Harmondsworth, Penguin.

Thrupp, L.A. (1989), 'Legitemizing Local Knowledge: from Displacement to Empowerment for Third World People', *Agriculture and Human Values*, vol. 3, pp. 13-25.

Tiffen, M., Mortimore, M. and Gichuki, F. (1994), *More People, Less Erosion. Environmental Recovery in Kenya*, Chichester, John Wiley and Sons.

Tilley, C.Y. (1994), *A Phenomenology of Landscape: Places, Paths, and Monuments*, Oxford, Berg.

Tiwari, K.M. (1983), *Social Forestry in India*, Dehra Dun, Natraj Press.

Tiwary, M. (2001), 'Ecological Institutions: Joint Forest Management in Bihar (Jharkhand) and West Bengal, India', Ph.D. Dissertation, Cambridge University.

Tripathi, C.B. (1988), 'Approaches to Tribal Development and Experience: General Policies, Administrative Approach and Personnel Policies and their Implementation. An All India Review,' *Man in India*, vol. 68 (4), pp. 334-355.

Tripathy, R.N. (1989), *Technology, Farm Output and Employment in a Tribal Region*, Delhi, Mittal Publications.

Turton, A.R. (2000), 'Precipitation, People, Pipelines and Power in Southern Africa. Towards a 'Virtual Water'-based Political Ecology Discourse', in Stott, P. and Sullivan, S., (2000), *Political Ecology. Science, Myth and Power*, London, Arnold, pp. 132-153.

United Nations Conference on Environment and Development (UNCED), *Earth Summit '92*, London, The Regency Press Corporation.

Upadhyay, K.P. (1988), *Forests and Food Security in the Densely Populated Upland Ecosystems of the Himalayas*, Rome, FAO Satellite Paper (draft).

Vannucci. M. (1992), 'Tradition and Change', in G. Sen (ed), *Indigenous Vision. Peoples of India. Attitudes to the Environment*, New Delhi, Sage Publications, pp. 25-34.

Vatsyayan, K. (1992), 'Ecology and Indian Myth', in G. Sen (ed), *Indigenous Vision. Peoples of India. Attitudes to the Environment*, New Delhi, Sage Publications, pp. 157-180.

Vedeld, T. (2001), *Participation in Project Preparation: Lessons from World Bank-Assisted Projects in India*, Washington: World Bank.

Verhelst, T.G. (1990), *No Life Without Roots. Culture and Development*, London, Zed Books.

Vivian, J.M. (1992), 'Foundations for Sustainable Development: Participation, Empowerment and Local Resource Management', in D. Ghai and J.M. Vivian (eds), *Grassroots Environmental Action. People's Participation in Sustainable Development*, London, Routledge, pp. 50-77.

Vlassoff, C. (1991), 'Progress and Stagnation: Changes in Fertility and Women's Position in an Indian Village', *Population Studies*, vol. 46, pp. 195-212.

von Furer-Haimendorf, C., with contributions by Yorke, M. and Rao, J. (1989), *Tribes of India. The Struggle for Survival*, Delhi, Oxford University Press.

Wade, R. (1986), 'Common Property Resource Management in South Indian
    Villages', in Panel on Common Property Resource Management, Board
    on Science and Technology for International Development, Office of
    International Affairs and National Research Council, 1985, *Proceedings
    of the Conference on Common Property Resource Management. April 21-
    26, 1985*, Washington, D.C., National Academy Press, pp. 231-257.
Wade, R. (1988), *Village Republics: Economic Conditions for Collective Action
    in South India*, Oakland, ICS Press.
Wade, R. (1990), *Governing the Market: Economic Theory and the Role of
    Government in East Asian Industrialisation*, Princeton, Princeton
    University Press.
WARDA, (1997), 'Salt of the Earth', WARDA Annual Report, 1997, Bouake,
    WARDA.
Warren, D.M. (1974), 'Bono Traditional Healers', *Rural Africana*, vol. 26, pp. 25-
    39.
Warren, D.M. and McKiernan, G. (1995), 'CIKARD: A Global Approach to
    Documenting Indigenous Knowledge for Development' in D.M. Warren,
    L.J. Slikkerveer, and D. Brokensha (eds), *The Cultural Dimension of
    Development. Indigenous Knowledge Systems*, London, Intermediate
    Technology Publications, pp. 426-434.
Warren, D.M., Slikkerveer, L.J. and Oguntunji Titilola, S. (eds) (1989),
    *Indigenous Knowledge Systems: Implications for Agriculture and
    International Development*, Studies in Technology and Social Change,
    No. 11., Technology and Social Change Program, Ames, Iowa State
    University.
Warren, D.M., Slikkerveer, L.J. and Brokensha, D. (1995a), 'Introduction', in
    D.M. Warren, L.J. Slikkerveer, and D. Brokensha (eds), *The Cultural
    Dimension of Development. Indigenous Knowledge Systems*, London,
    Intermediate Technology Publications, pp. xv-xviii.
Warren, D.M., Slikkerveer, L.J. and Brokensha, D. (eds) (1995b), *The Cultural
    Dimension of Development. Indigenous Knowledge Systems*, London,
    Intermediate Technology Publications.
Watts, M. (1982), 'On the Poverty of Theory: Natural Hazards Research in
    Context', in Hewitt, K. (ed), *Interpretations of Calamity from the
    Viewpoint of Human Ecology*, Boston, Allen and Unwin Inc, pp. 231-
    261.
Watts, M. (1983), *Silent Violence: Food Famine and Peasantry in Northern
    Nigeria*, Berkeley, University of California Press.
Watts, M. (1989), 'The Agrarian Crisis in Africa: Debating the Crisis', *Progress
    in Human Geography*, vol. 13 (1), pp. 1-41.
Watts, M. (1993), 'Development 1: Power Knowledge, Discursive Practice',
    Progress in Human Geography, vol. 17 (2), pp. 257-272.
Watts, M. and Peet, R. (1996), 'Conclusion. Towards a Theory of Liberation
    Ecology', in R. Peet and M. Watts (eds) *Liberation Ecologies:
    Development, Environment, Social Movements*, London, Routledge, pp.
    260-269.
Weiner, M. (1978), *Sons of the Soil*, Princeton, Princeton University Press.

Werner, O. and Begishe, K.Y. (1980), 'Ethnoscience and Applied Anthropology', in D.W. Brokensha, D.M. Warren and O. Werner (eds), *Indigenous Knowledge Systems and Development*, Lanham, MD, University Press of America, pp. 151-181.

Werner, O. and Manning, A. (1979), 'Tough Luck Ethnography Versus God's Truth Ethnography in Ethnoscience. Some Thoughts on the Nature of Culture', in B.T. Grindal and D.M. Warren (eds), *Essays in Humanistic Anthropology*, Lanham, MD, University Press of America, pp. 327-374.

Wieringa, S. (1994), 'Women's Interests and Empowerment: Gender Planning Reconsidered', *Development and Change*, vol. 25 (4), pp. 829-848.

World Bank (1981), 'Community Forestry: Notes and Questions from a Field Visit in South Bihar' (draft).

World Bank (1991), *World Development Report, 1991*, Oxford, Oxford Universtiy Press/World Bank.

World Bank (1992a), *World Development Report, 1992*, Oxford, Oxford Universtiy Press/World Bank.

World Bank (1992b), *India Forest Sector Review*, World Bank Report No. 10965 IN1992, Washington D.C., World Bank.

World Bank, (1994), *Enhancing Women's Participation in Economic Development*, Washington D.C., World Bank.

World Bank (1995a), *Towards Gender Equality*, Washington D.C., World Bank.

World Bank (1995b), *Staff Appraisal Report, Madhya Pradesh Forest Project*, Washington D.C., World Bank.

World Bank (1996), *World Bank Participation Sourcebook*, Washington D.C., World Bank.

World Bank (1998), *Implementation Completion Report. West Bengal Forestry Project*, Washington D.C., World Bank.

World Bank (2001), *World Bank Participation Sourcebook*, Washington D.C., World Bank.

World Commission for Environment and Development (1987), *Our Common Future*, Oxford, Oxford University Press.

World Resources Institute (1994), *World Resources 1994-95. People and the Environment*, New York, Oxford University Press.

Xavier Institute of Social Services (1984), *Bero, A Longitudinal Report from 1979-1984*, Ranchi, Xavier Institute of Social Services.

Xaxa, V. (1992), 'Oraons: Religion, Customs and Environment', in G. Sen (ed), *Indigenous Vision. Peoples of India. Attitudes to the Environment*, New Delhi, Sage Publications, pp. 101-110.

Yapa, L. (1996), 'Improved Seeds and Constructed Scarcity' in R. Peet and M. Watts (eds) *Liberation Ecologies: Development, Environment, Social Movements*, London, Routledge, pp. 69-85.

Zimmerer, K.S. (1996), 'Discourses on Soil Loss in Bolivia. Sustainability and the Search for Socioenvironmental "middle ground"', in R. Peet and M. Watts (eds) *Liberation Ecologies: Development, Environment, Social Movements*, London, Routledge, pp. 110-124.

Zook, D.C. (2000), 'Famine in the Landscape: Imagining Hunger in South Asian History, 1860-1990, in A. Agrawal and K. Sivaramakrishnan (eds)

(2000), *Agrarian Environments. Resources, Representation, and Rule in India*, Durham and London, Duke University Press, pp. 107-131.

Appendix I: Latin and Local Tree and Plant Names

Amla=Emblica officinalis
Arhar=Cajanus indicus
Asan=Terminalia tomentosa
Bahera=Terminalia bellirica
Bakain=Melia azedrach
Bamboo=Bambusa arundinacea
Banana=Musa paradisiaca
Bar=Ficus bengalensis
Bel=Aegle marmelos
Ber=Ziziphus mauritiana
Bhelwa=Semecarpus anacardium
Bhuichappa=Kaempferia rotunda
Bodi bean=Vigna catiang
Çhiraita=Swertia chirayita
Dahu=Artocarpus Lakoocha
Dhaunta=Anogeissus latifolia
Dhub grass=Cynodon dactylon
Dumar=Ficus glomerata
Footkal=Ficus retusa
Gamhar=Gmelina arborea
Gondli =Panicum milaire
Gongra=Luffa degyptia
Guava=Psidium guajava
Gulaichi=Plumeria acuminata
Harra=Terminalia chebula
Jackfruit=Artocarpus heterophyllus
Jamun=Prunus cornuta
Jhingi=Luffa acutangula
Jockey=Moringa olifera
Kachnar=Bauhinia purpurea
Karam=Adina cordifolia
Karanj=Pongamia glabra
Karonda=Carissa carandas
Kend=Diospyros melanoxylon
Khayer=Acacia catechu
Koinar=Bauhinia variegata
Koraya=Holarrhena antidysenterica
Kudrum=Ricinus communis
Kujur=Phoenix acaulis
Kurti dal=Dolichos biflorus

344

Kusum=Schleichera oleosa
Lemon=Citrus limon
Mahua=Madhuca indica
Mango=Bassia latifolia
Marua=Eleusine corocana
Neem=Azadirachta indica
Papara=Holoptelea integrifolia
Papaya=Carica papaya
Parsa=Butea monosperma
Peach/Satalu=Prunus persica
Pesar=Cassia fistula
Pipal=Ficus religiosa
Piyar=Buchanania latifolia
Pomegranate=punica granatum
Putri=Combretum decandrum
Putus=Lantana camera
Ramadatoon=Smilex ovalifolia
Rugra=Lycoperdon sp.
Sabai grass=Ichaemum augustifolium
Sagwan=Tectona grandis
Sal=Shorea robusta
Sataur=Asparagus racemosus
Sidha=Lagerstroemia parviflora
Simbal=Bombax ceiba
Sinduar=Vitex negundo
Sisam=Dalbergia sisoo
Surguja=Guizota abyssinica
Tamarind=Tamarindus indica
Tetair=Ipomoea cornea
Tunp=Cedrela toona
Urad=Phaseolus roxburghii

Appendix II: 'Natal village' Women and '*Ghar Jamai*' Men

Natal Village Women

Name and community	Age	Education	Wage work	Distance to natal village	Seed transfers	Children	Years married
Rahil Orain (ST)	20	Primary	No	10 km	No	None	3
Hiramuni Orain (ST)	24	Secondary	Yes	18 km	No	3 girls	6
Nauri Orain (ST)	29	None	No	14 km	Yes	2 boys 1 girl	9
Jatri Orain (ST)	30	None	Yes	30 km	No	3 boys	10
Silamuni Orain (ST)	31	None	No	24 km	Yes	3 boys 1 girl	13
Gandri Orain (ST)	40	None	No	5 km	No	2 boys 2 girls	24
Etwari Orain (ST)	55	None	No	5 km	Yes	1 boy 3 girls	37
Neera Lohrain (ST)	22	None	No	32 km	No	1 boy	6
Sattan Lohrain	32	None	Yes	13 km	Yes	6 girls	14
Sila Lohrain (ST)	37	Primary	Yes	45 km	No	2 boys 3 girls	19
Sanicaria Mahli (ST)	23	None	Yes	9 km	No	2 boys 2 girls	6
Laki Mahli (ST)	32	None	Yes	32 km	No	1 boy 5 girls	14
Pairo Mahli (SC)	33	None	Yes	25 km	No	2 boys 1 girl	2

Natal Village Women (continued)

Name and community	Age	Education	Wage work	Distance to natal village	Seed transfers	Children	Years married
Binita Manjhi (SC)	26	Secondary	Yes	32 km	No	1 girl	5
Jitan Manjhi (SC)	41	None	Yes	24 km	No	1 boy 1 girl	20
Fagni Pradhan (BC)	37	None	No	18 km	Yes	2 boys 3 girls	22
Niramala Sahu (BC)	25	Primary	No	100 km	No	2 boys 1 girl	8
Rina Sahu (BC)	20	Secondary	No	100+ km	No	No	2

Ghar Jamai Men

Name and community	Age	Education	Wage work	Distance to natal village	Seed transfer	Children	Years married
Etwa Mahli (ST)	36	Secondary	Yes	30 km	Yes	2 girls	15
Tila Oraon (ST)	68	Secondary	Retired from Army	5 km	Yes	2 boys	40+
Gandra Oraon (husband of widow Bimla Orain) (ST)	38	Secondary	Yes	13 km	Yes	4 boys 1 girl	18

Appendix III: Song Translations

Between October and December, 1993, I worked very closely with Mukund Nayak, one of the major Ranchi music and dance artists who has worked closely with Jharkhand leaders such as Dr. Ram Dayal Munda and Dr. B.P. Kesari. Mukund also introduced me to other Nagpuri artists such as Madur Mansuri Hasmukh, Manohar Mohanti and Jalpu Lohra. With their help, I was able to record some of the major 'Jharkhand awareness songs' that emphasise how important forests are to both Jharkhandi subsistence and culture. Three of these songs are translated below.

1) *Jagu jawan* (Wake up young people!)

Oh, young people, wake up and pay a little attention to Chota Nagpur.
Young people, wake up and pay a little attention to Chota Nagpur.

Copper, iron, gold, silver. People are taking minerals from this area as if they were jewels, to be sold by outsiders.
Oh, young people, wake up and pay a little attention to Chota Nagpur.
Young people, wake up and pay a little attention to Chota Nagpur.

Others are ploughing our paddy and *bari* land while the village farmers are sitting idle.
Oh, young people, wake up and pay a little attention to Chota Nagpur.
Young people, wake up and pay a little attention to Chota Nagpur.

Contractors are making profits from cutting the forest trees and bushes.
Our fruit groves and forest gardens are vanishing.
Oh, young people, wake up and pay a little attention to Chota Nagpur.
Young people, wake up and pay a little attention to Chota Nagpur.

Lakes and big dams are being constructed to produce electricity. How will this electricity reach your houses?
Oh, young people, wake up and pay a little attention to Chota Nagpur.
Young people, wake up and pay a little attention to Chota Nagpur.

Day after day, thousands more people are being made labourers.
Oh, see how they are being forced to migrate to far-away places.
Oh, young people, wake up and pay a little attention to Chota Nagpur.
Young people, wake up and pay a little attention to Chota Nagpur.

Nagpuri brothers and sisters are getting themselves educated.
But they are facing many difficulties in getting employment.
Oh, young people, wake up and pay a little attention to Chota Nagpur.
Young people, wake up and pay a little attention to Chota Nagpur.

Don't just sit idle. Ask for your rights.
For how long will you tolerate exploitation by others?
Oh, young people, wake up and pay a little attention to Chota Nagpur.
Young people, wake up and pay a little attention to Chota Nagpur.

Little time is left now.
If you get together, you will gain strength.
If necessary, Mukund will give up his life for this nation.
Oh, young people, wake up and pay a little attention to Chota Nagpur.
Young people, wake up and pay a little attention to Chota Nagpur.

Composed by Mukund Nayak. 1976.

2) *Murga chap* ('Cockerel marker')

We sold our 'cockerel marker'.
We were slapped on both cheeks.
We became the labourers of others.
We gave what we cut from the forest to thieves.

When the cockerel was crowing,
sleeping people used to awaken and get up.
Lazy people changed their ways.
We gave what we cut from the forest to thieves.

We went and sold our mothers.
and today we bought *dains*.
Even then, we gave away our pride and sold our property for a profit.
We gave what we cut from the forest to thieves.

Oh Jharkhand, what have we done to you to get such punishment as this?
This clay [earth] is ours and this potters wheel is also ours.
But the potter is an outsider.
We gave what we cut from the forest to thieves.

Composed by Madhu Mansuri Hasmukh. n.d.

3) *Chala re Jharkhand raikhe mangay* (Let us go and demand a separate Jharkhand State)

Oh, Nagpuri people, get together and unite!
Let us go and demand a separate Jharkhand State.

Our weak bodies are getting tired.
Outsiders are profiting from many things in our area.
You must colour your views with your own blood.
Oh, Nagpuri people, get together and unite!
Let us go and demand a separate Jharkhand State.

Bribery and corruption are now taking away the life blood of this world.
It is the poor that are being made into a dough at the flour mill.
Oh, Nagpuri people, get together and unite!
Let us go and demand a separate Jharkhand State.

The government is cutting the forest.
The poor are worrying about how to obtain fuel wood.
Oh Jharkhandi people, where shall you go to beg?
Oh, Nagpuri people, get together and unite!
Let us go and demand a separate Jharkhand State.

Composed by A. Aansari, n.d.

Index

351

Printed in the United States
by Baker & Taylor Publisher Services